EMERALDS OF PAKISTAN
Geology, Gemology and Genesis

CONTRIBUTORS

Jawaid Anwar
Pakistan Mineral Development Corporation,
13-H/9,
Islamabad, Pakistan

Stephen J. Carter,
Oregon State University
Geology Department,
Corvallis, Oregon, 97331
U.S.A.

Hamid Dawood
Natural History Museum
Pakistan National Science Foundation,
Islamabad, Pakistan

Eugene E. Foord
U.S. Geological Survey
Box 25046, MS 905,
Denver Federal Center,
Denver, Colorado 80225,
U.S.A.

Edward J. Gübelin
Ratna Mahal
CH-6045 MEGGEN,
Lucern, Switzerland

Jane M. Hammarstrom
U.S. Geological Survey
959 National Center
Reston, VA 22092
U.S.A.

Brittain Hill
Oregon State University
Geology Department
Corvallis, Oregon 97331
U.S.A.

Shahid Hussain
House No. 693,
St. No. 47, G-9/2,
Islamabad, Pakistan

S. Sadaqat Ali Jafri
Geological Survey of Pakistan
42-R, Block-6, PECHS,
Karachi, Pakistan

Ali H. Kazmi
Geological Survey of Pakistan
Sariab Road,
Quetta, Pakistan

Tahseenullah Khan
Geological Survey of Pakistan
13-H/9,
Islamabad, Pakistan

Robert D. Lawrence
Oregon State University
Geology Department
Corvallis, Oregon 97331,
U.S.A.

Robert R. Seal, II
University of Michigan
Department of Geological Sciences
1006 C.C. Little Building
Ann Arbor,
Michigan 48109-1063
U.S.A.

Lawrence W. Snee
U.S. Geological Survey
Box 25046, MS 905
Denver Federal Center
Denver, Colorado 80225
U.S.A.

EMERALDS OF PAKISTAN
Geology, Gemology and Genesis

Edited by

ALI H. KAZMI
Geological Survey of Pakistan

LAWRENCE W. SNEE
U.S. Geological Survey

GSP Geological Survey of Pakistan

VNR Van Nostrand Reinhold Company

iv

This volume was made possible by grants from the
U.S. National Science Foundation, the Geological Survey
of Pakistan, and the U.S. Geological Survey. Any opinions,
findings, conclusions, or recommendations expressed in
this publication are those of the authors and do not
necessarily reflect the views of those organizations.

Printed in Pakistan by
⟨E⟩ Elite Publishers Limited, D-118, SITE, Karachi.

Van Nostrand Reinhold
115 Fifth Avenue
New York, New York 10003

Van Nostrand Reinhold International Company Limited
11 New Fetter Lane
London EC4P 4EE, England

Van Nostrand Reinhold
480 La Trobe Street
Melbourne, Victoria 3000, Australia

Macmillan of Canada
Division of Canada Publishing Corporation
164 Commander Boulevard
Agincourt, Ontario M1S 3C7, Canada

16 15 14 13 12 11 10 9 8 7 6 5 4 3 2 1

Library of Congress Cataloging-in-Publication Data
Main entry under title:

Kazmi, Ali H.
  Emeralds of Pakistan: geology, gemology & genesis/A.H. Kazmi
and L.W. Snee.
  Includes bibliographical references.
  ISBN 0-442-30328-9
  1. Emerald deposits—Pakistan. 2. Emeralds—Pakistan. I. Snee,
Lawrence W. II. Title.
TN997.E5K39   1990
553.8'6'095491—dc20   89-24827
                                                            CIP

CONTENTS

Foreword

Long before Pakistan emeralds became known to the world gem trade, I made my acquaintance with these gems in the year 1967 when a packet of faceted samples was sent to me for examination. I was indeed delighted to receive gems from an entirely new source. The discovery of a new gemstone deposit is always exciting news, capable of inciting intense curiosity in the mind of a gemologist. One is not only interested in the welcome increase of gem sources—causing expansion of the market and a resulting increase in demand—but equally so in discovering the means of identifying the new material in order to distinguish it from material originating from previously known deposits.

It was therefore with keen interest that I studied the fifty-seven Pakistani emeralds in the packet; determined their gemological constants; examined their internal features; and published the results in 1968 in "Der Aufschluss". This was the first published introduction of Pakistan emeralds to the outside world.

Most of the specimens examined by me were of a good to excellent quality and were outstanding for their vividness, their transparency and their saturated green color which I immediately found comparable with the prized Muzo emeralds from Colombia. As time went by we saw more of these emeralds on the world market, and in 1981 I was finally tempted to visit Pakistan and the Mingora mines which produced these exquisite gemstones.

It was during this and my subsequent visit to Pakistan (in 1983) that I met A. H. Kazmi, one of the founders of and the indefatigable Technical Director of the Gemstone Corporation of Pakistan. Together we toured the exciting and picturesque mountains of northern Pakistan and visited several gem deposits, accounts of which visits appeared in *Gems & Gemology*, Fall 1982 and Fall 1986.

In the course of these visits I had the pleasure of inspecting the geology of the various gem deposits extending from the emerald deposits of Mingora, Makhad and Gujarkili in Swat, along the Karakoram Highway and through the awe-inspiring Indus gorge up to the ruby deposits in Hunza. A. H. Kazmi showed me the spectacular suture zone along which the Indian and Asian continents had collided, and we examined the various geological structures and formations exposed in this region. Kazmi was at that time working on the geology of emerald deposits. He mentioned to me the shocking lack of published literature on the geology of gem deposits, and stated that he hoped to write a book on gem geology some day. I was therefore delighted to learn later that a group of distinguished scientists from the USA had joined him in research work on Pakistani emeralds, and that R. D. Lawrence led the studies on tectonic environment; L. W. Snee and Jane Hammarstrom contributed to the geochemistry of emeralds; and R. R. Seal worked on fluid inclusion studies. The results of this research work have been compiled in this book.

It has been known for a long time that the main chemical components of emerald, beryllium and the trace element chromium, without which beryl could not bear the name emerald, originate from entirely diverse and remotely separate geological regimes; beryllium is present abundantly in continental felsic rocks, while chromium only occurs in oceanic ultramafic rocks. Emerald can only form where these two rock types have met. It is, however, only with the present study by A. H. Kazmi, L.W. Snee and their associates that the true geological environment for the origin of emerald has been revealed and its association with suture zones highlighted.

In Pakistan the processes that led to emerald formation started during the great collision of the Indian and Asian continents in the early Cenozoic era, when the floor of the one-time Tethyan ocean started to be consumed and subducted and was ultimately squeezed and obducted on the Indian Plate to form the present Indus suture zone.

It was, however, many millions of years after the continental collision that the emerald crystallization began. The Indus suture rocks were overridden by the Kohistan island arc rocks to such an extent that they were metasomatized into a talc-dolomite melange. Burial of the frontal parts of the Indian Plate beneath these rocks—which have been estimated to be 12 km thick—generated anatectic granites, and the late pneumatolitic and hydrothermal fluids, which contained beryllium, seeped through the overlying rocks and marbles and along the shear zones, metasomatizing the chromium-rich ophiolites and marbles. In the pre-existing faults and rocks, the precious combination of beryllium and chromium, together with aluminium and silica from the continental source, was deposited and started to form the highly prized and valuable emerald crystals mined in Pakistan today.

The trace element chromium—responsible for the green color of the beryllium aluminium silicate—is accompanied by the trace elements iron, magnesium and sodium in relatively large amounts, and by manganese, cobalt, cesium and rubidium in variable small quantities. It is interesting to learn that the green beryls occurring at Gandao owe their color to a content of vanadium, which in my opinion renders them particularly interesting. The oceanic rocks are not only the source of chromium, but of magnesium, iron, manganese and cobalt. All the other chemical components of emerald originate from the continental felsic rocks.

Hammarstrom and Snee have done excellent work on the chemistry of Pakistan emeralds, which shows that the various emerald occurrences in Pakistan produce emeralds with variations in their chemical compositions. Thus the emeralds mined in Swat, Mohmand and Khaltaro can be distinguished from each other by chemical analysis. Even within the separate occurrences the emeralds are different. The Mingora mines alone contain at least three types of emeralds, differing in color and quality. As revealed in the study of the fluid inclusions, the differences among emeralds are also manifested in the characteristic inclusions to be observed under the microscope. But, equally fascinating, beautiful and revealing are the character and diversity of the internal features of the emeralds. These, too, have their unique signatures.

It is true that the geology of gemstone deposits has, until a few years ago, been unpardonably neglected. Only recently, within the last ten years or so, articles have been appearing in gemological magazines and journals considering the petrology and mineralogy of gem deposits, and either speculating upon or providing an explanation about the origin of the gemstones from these localities. This book, a detailed study of the emerald deposits of Pakistan, is an excellent example of this new trend in gemological literature, and truly goes into the depths of the matter. To my knowledge this book is the first attempt to conduct such a comprehensive and multi-disciplinary research on emerald of a definite region in the world. It corporates the results of extensive geological field research and detailed petrological, gemological, geochemical and physico-chemical researches on the emerald deposits of Pakistan. The book is well-planned and begins with a simple, easy to understand yet excellent account of the regional geological setting, followed by detailed accounts of the tectonic and geological environment of the emerald deposits. The physico-chemical conditions under which the Pakistani emeralds were formed are discussed, and a comparative study is made of the environment of Pakistan emeralds with those found in other parts of the world. All of these new data on Pakistani emerald deposits and the comparative data of world deposits are used to define a new classification of emerald deposits and to speculate on their origins. Finally, a list of selected references is a valuable resource for additional reading.

To conclude this foreword, I wish to express my conviction that this work on the emeralds of Pakistan will not only fill a vacuum which has existed for some time in gemological and mineralogical literature, but, on account of its wide scope of information, prove to be one of the most comprehensive reference books ever published about emeralds in general, and especially about emeralds from Pakistan. Few mineralogists and certainly no gemologist or gem geologist will forego the study of this book in future.

EDWARD GÜBELIN
Gemmologist, C.G., F.G.A.
Switzerland

July, 1989.

Preface

This book, a comprehensive study of the geology of major emerald deposits, is the result of a collaborative program among the Geological Survey of Pakistan, the Gemstone Corporation of Pakistan, Oregon State University and the U.S. Geological Survey. Careful geologic mapping of the known emerald deposits of northern Pakistan was begun by Gemstone Corporation under the direction of Ali H. Kazmi in 1979 in order to prove known reserves and to determine geological indicators for new deposits. At the beginning of 1982, Oregon State University, under the auspices of Robert D. Lawrence, collaborators, and students, was invited by Gemstone Corporation to become involved with study of the growing number of emerald deposits. At that time Lawrence was on Fulbright appointment at the University of Peshawar and, along with Robert Yeats of Oregon State University had a National Science Foundation grant to study the tectonics of northern Pakistan. In 1984, Ali H. Kazmi went to Oregon State on a Fulbright fellowship, at which time numerous analytical studies on the emeralds and their host rocks were begun at the U.S. Geological Survey and at Oregon State. The chapters of this book are the culmination of nearly ten years of work on the emerald deposits of Pakistan.

Our ten years of research has shown that the origin of Pakistani emeralds is directly linked to a Himalayan suture zone that marks the collisional belt between the Indo-Pakistan subcontinental plate and an island arc of the Asian plate. This geologic event brought chromium-bearing oceanic rocks in direct contact with beryllium-bearing continental rocks; the essential combination of elements needed to cause the formation of exquisite, green emerald. This fascinating setting led us to look into the geologic setting of other emerald deposits of the world. Our inquiry led to several interesting conclusions about world emerald deposits and about the origin of emeralds in general.

Accordingly, the present volume summarizes our work. Chapters 1, 2 and 3 describe the geologic setting of the emerald deposits and the geology and tectonics of northern Pakistan. Chapter 4 is a summary of the gemological and physical characteristics of Pakistani emeralds. Chapters 5, 6 and 7 investigate emerald chemistry and fluid inclusion characteristics. In chapter 8 we attempt to summarize existing literature on world emerald deposits in order to lay the foundation for a classification of emerald deposits in chapter 9 and a discussion of the origin of emeralds.

At the end of the book is a list of selected references that were assembled with the kind and painstaking assistance from the librarians of the U.S. Geological Survey in Reston, Virginia. We do not claim to have included all references to emeralds from the scientific literature, but we hope that the list is comprehensive enough for most readers.

Acknowledgments: This book, which is the result of many years of field and laboratory study, owes its existence to the hard work of many people and financial support of a number of organizations. Mapping by Geological Survey of Pakistan and Gemstone Corporation of Pakistan geologists provided the foundation for this work. National Science Foundation grants NSF INT 81-18403, INT 86-09914 and EAR 86-17543 to Robert Lawrence and Robert Yeats provided support for field and laboratory studies. The Geological Survey of Pakistan and NSF grant INT 80-13158 to Kees DeJong and 81-18403 covered costs of publication. Laboratory support by the U.S. Geological Survey and the Oregon State University Radiation Center is gratefully acknowledged. Encouragement and support by Waheeduddin Ahmed, formerly Director General, Geological Survey of Pakistan; Brigadier (retired) Kaleem ur Rahman Mirza, Managing Director, Gemstone Corporation of Pakistan, and R.A.K. Tahirkheli, formerly Director, National Center of Excellence in Geology, Peshawar University were very valuable. Lawrence and Kazmi both benefited from Fulbright fellowships during the course of this work. Snee and Kazmi were kindly given great latitude and encouragement by members of the U.S. Geological Survey and the Geological Survey of Pakistan.

Many other people and organizations helped us during the course of this work. Prof. William Kelly of the University of Michigan discussed fluid inclusion research with us and kindly provided use of his laboratory and personnel. Terry Ottaway of the Royal Ontario Museum engaged in valuable discussions. Dallas Peck, Director of the U.S. Geological Survey, made possible the invaluable "Georef" of emerald literature by Elizabeth Yeats of the U.S.G.S. library. Wally Griffiths, Joseph Taggart and James G. Crock of the U.S. Geological Survey provided chemical data. Scott Hughes and Roman Schmidt of Oregon State University assisted with INAA procedures.

Drafting of most of the illustrations was done by Abdul Subhan, Fahimuddin, Mohammad Siddique and Inayat Ali (GSP) and typing was done by Farzand Hussain, Kaleemullah Farooqi and Mehr Gul (GSP). Q. Z. Abid (GSP) helped with the proof reading, layout and printing of this book.

ALI H. KAZMI
Karachi, Pakistan

LAWRENCE W. SNEE
Denver, Colorado
USA

July, 1989

A Brief Overview of the Geology and Metallogenic Provinces of Pakistan

Ali H. Kazmi

INTRODUCTION

Located strategically at the head of the Arabian Sea and surrounded by Iran, Afghanistan, China, and India, the territories that compose Pakistan today (fig. 1.1) have lain through the ages at the main crossroads between the restless and effervescent Middle East and the relatively tranquil and quiescent Far East. Vast armies ranging from the ancient Aryans, Greeks, Huns, Arabs, and Mongols have criss-crossed its domain in quest of power, glory, and the fabulous riches of the Orient. Yet the country also has been the birthplace of ancient and glorious cultures such as the Indus Valley Civilization, which dates back to over 5000 BP, and the cradle of such venerable, widespread and lasting religions as Buddhism. The country is thus endowed with rich cultural heritage equally matched with a variety of climates and exotic geographic terrains (fig. 1.1).

Northern Pakistan is composed of three of the loftiest snow-capped mountain ranges of the world—the Himalayas, Karakorams (photo. 1.1), and Hindu Kush which coalesce to form part of the southern Pamirs, the roof of the world. Tucked within these great mountains of Pakistan is the largest concentration of the tallest (above 7200 m) mountain peaks in the world including K2 (8611 m). A part of this region is seismically one of the most unstable in the world and is almost constantly quivering.

These lofty mountain ranges of northern Pakistan make a series of southwest loops or syntaxial bends, and southwestward they loose their stature. Thus the western part of Pakistan comprises smaller, festooned hilly offshoots of the Himalayas in the form of arcs, oroclines, and syntaxes that bound the basin and range terrain of the Baluchistan Plateau.

The lofty snow-capped mountains of the north (photo. 1.2) form the headwaters of the mighty Indus River and its tributaries. South and east of the mountains, these rivers have formed a vast, flat, and fertile flood plain and a deltaic complex at the head of the Arabian Sea (Kazmi, 1977, 1984). Eastward, the Indus Plain is flanked by the formidable Thar Desert (Kazmi, 1985).

These geomorphic features reflect the geological setting which in many respects is unique in the world. Underneath the Thar Desert and the Indus Plain are concealed the platform areas of the Indo-Pakistan crustal plate. This plate was once part of the ancient supercontinent—Gondwanaland (Wegener, 1924; Athavale and others, 1970). Extensive rifting on a continental scale during the Permian (Klootwijk, 1979) dismembered Gondwanaland into several fragments ranging in size from small microcontinental blocks to large continental masses such as the African, Australian, and Indo-Pakistan blocks (fig. 1.2).

A vast ocean, the Tethys, initially separated Gondwanaland from Eurasia (also referred to as Laurasia). Paleomagnetic data indicate that some of the smaller fragmented blocks of Gondwanaland (Central Iran, Lut, Afghan, Pamir, and Lhasa) drifted away northward across the Paleotethys and collided with Eurasia in Early to Middle Jurassic time. This was followed about 130 million years ago by the northward migration of the Indo-Pakistan subcontinent initially at a pace of 3 to 5 centimeters per year. The movement increased to over 15 centimeters

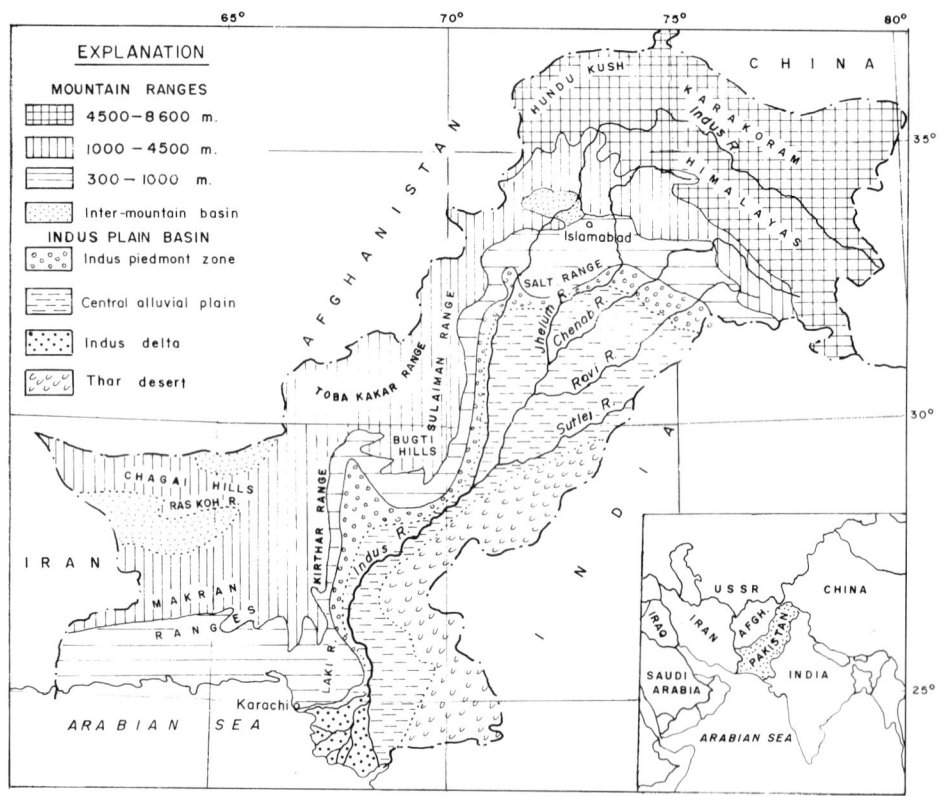

Figure 1.1. Map showing location and major physical features of Pakistan.

per year 80 to 53 million years ago (Powell, 1979). The northward drifting of the subcontinent gradually closed the Neotethys Sea between Eurasia and India. In its wake behind India, due to a complex process of seafloor spreading, the Indian Ocean opened. The subcontinent finally collided with Eurasia about 40 million years ago (Molnar and Tapponnier, 1975). This incredible event resulted in the formation of the loftiest mountains in the world. Geology of Pakistan enshrines all the telltale evidence of this spectacular cataclysmic scenario.

GEOLOGICAL FRAMEWORK

In Pakistan and adjacent regions of Iran and Afghanistan three distinct geotectonic domains, or terranes, may be recognized from south to north. These terranes are herein referred to as (1) southern or Gondwanic, (2) central or Tethyan and (3) northern or Eurasian (Stocklin, 1977; Powell, 1979; and Kazmi and Rana, 1982). The southern terrane comprises two fragmented elements of ancient Gondwanaland, namely the Arabian plate to the west and the Indo-Pakistan plate to the east (fig. 1.3). The northern terrane comprises the frontal elements of the Eurasian plate and is referred to as the Turon continental block. The central or the Tethyan terrane, which covers the greater part of Iran, western Baluchistan, and northern Pakistan, is a complex geologic belt that contains ophiolitic rocks, geosynclinal sediments, island arcs, and

Photo.1.1. The lofty Karakorams with K² Peak (8611m) in the background (photo *J. La Fortune*).

Photo.1.2. Nangaparbat—Haramosh massif with the Nangaparbat Peak (8125m) in the background (photo *P. Verplanck*).

microcontinents (fig. 1.3), which have been tectonically wedged in between the Eurasian and Gondwanic continental crustal plates. The microcontinents comprise four main mini-crustal blocks, namely Azerbaijan, Lut, Seistan-Helmand (or Afghan), and Pamir. These microcontinental blocks were formed as a result of the rifting of frontal promontories from the Arabian plate (Takin, 1972; Soffel and others, 1975) prior to the Mesozoic (Powell, 1979) and prior to the northward movement of India. These small blocks collided with the Turon plate along the southern edge of Eurasia in the Early Jurassic (Stocklin, 1974), forming the northern Alborz-Hindu Kush suture zone. This event marked the closure of the Paleotethys Sea, which had existed in front of the microcontinents, and the opening of the Neotethys in their wake. The Paleozoic to Early Mesozoic Tethyan sediments and shelf carbonates were deformed and wrapped around the microcontinental blocks forming relatively small and narrow fold and thrust belts (fig. 1.3).

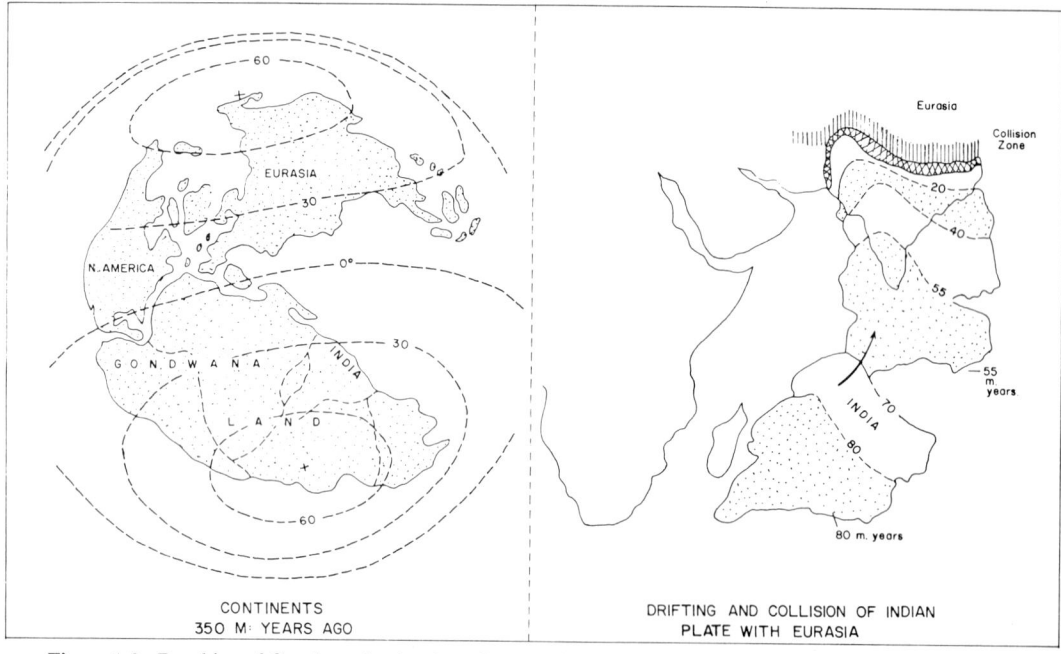

Figure 1.2. Breaking of Gondwanaland and northward drift of Indian subcontinent (Seyfert and Sirkin, 1973; Powell, 1979; M.=million years).

Pakistan largely comprises elements of the Gondwanic terrane and the structural zones that separate the Gondwanic terrane from the Tethyan and Eurasian terranes. Southeastern Pakistan, in the region of the Thar Desert and the Indus Plain, is part of the northern zone in the Indo-Pakistan crustal plate. In the eastern part of this zone a gently dipping relatively thin cover of Mesozoic to Cenozoic marine sedimentary rocks forms broad monoclinal zones or zones of upwarp and downwarp, and at places it contains monadnocks that are inliers of Precambrian metamorphic and igneous basement rocks (Balkrishna, 1977; Kazmi and Rana, 1982). The western part of the foreland zone is a foredeep filled in by Late Cenozoic molasse (fig. 1.4).

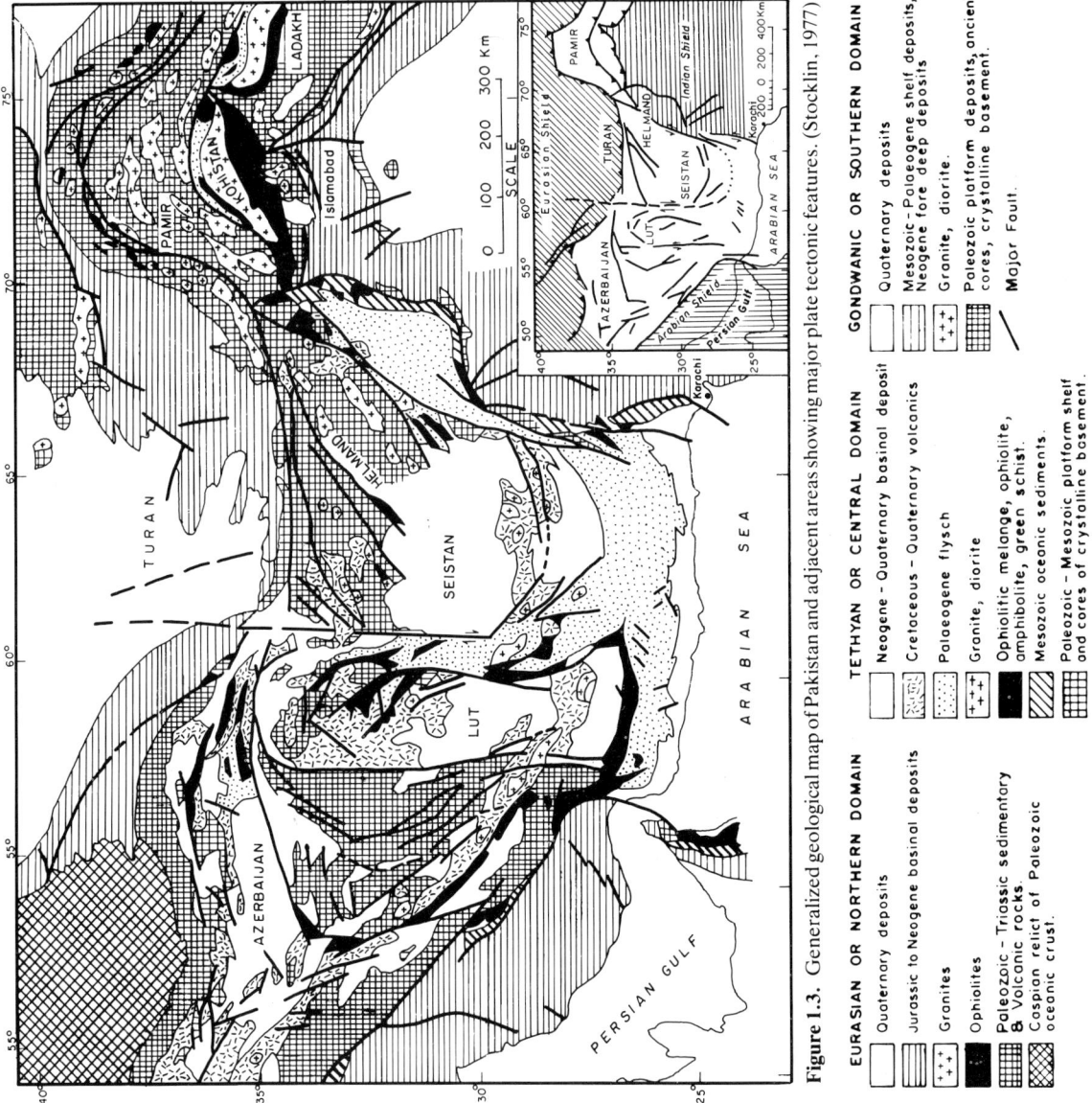

Figure 1.3. Generalized geological map of Pakistan and adjacent areas showing major plate tectonic features. (Stocklin, 1977).

EURASIAN OR NORTHERN DOMAIN

- Quaternary deposits
- Jurassic to Neogene basinal deposits
- Granites
- Ophiolites
- Paleozoic – Triassic sedimentary & Volcanic rocks.
- Caspian relict of Paleozoic oceanic crust.

TETHYAN OR CENTRAL DOMAIN

- Neogene – Quaternary basinal deposit
- Cretaceous – Quaternary volcanics
- Palaeogene flysch
- Granite, diorite
- Ophiolitic melange, ophiolite, amphibolite, green schist.
- Mesozoic oceanic sediments.
- Paleozoic – Mesozoic platform shelf and cores of crystalline basement.

GONDWANIC OR SOUTHERN DOMAIN

- Quaternary deposits
- Mesozoic – Palaeogene shelf deposits, Neogene fore deep deposits
- Granite, diorite.
- Paleozoic platform deposits, ancient cores, crystalline basement.
- Major Fault.

Beyond the foreland zone indelible imprints of continental collision are seen around the margins of the Indo-Pakistan plate, the foremost being the wide Himalayan orogenic belt. This belt formed along the collisional zone that borders the Tethyan terrane. Because the collisional zone is complex, it is subdivided into four major structural zones; these have been named the Himalayan imbricate crystalline zone, the Chaman transform fault zone, the Sub-Himalayan thrust belt, and the autochthonous shelf and foredeep zone (fig. 1.4). Before the Tethyan terrane is described, these structural zones will be reviewed.

The Himalayan imbricate crystalline zone (Kazmi and Rana, 1982), the first of these four structural zones, is located in northern Pakistan. This zone comprises Precambrian to Cretaceous metamorphosed shelf carbonates and pelitic rocks intruded by granites and granodiorites, which range in age from Precambrian to Cenozoic.

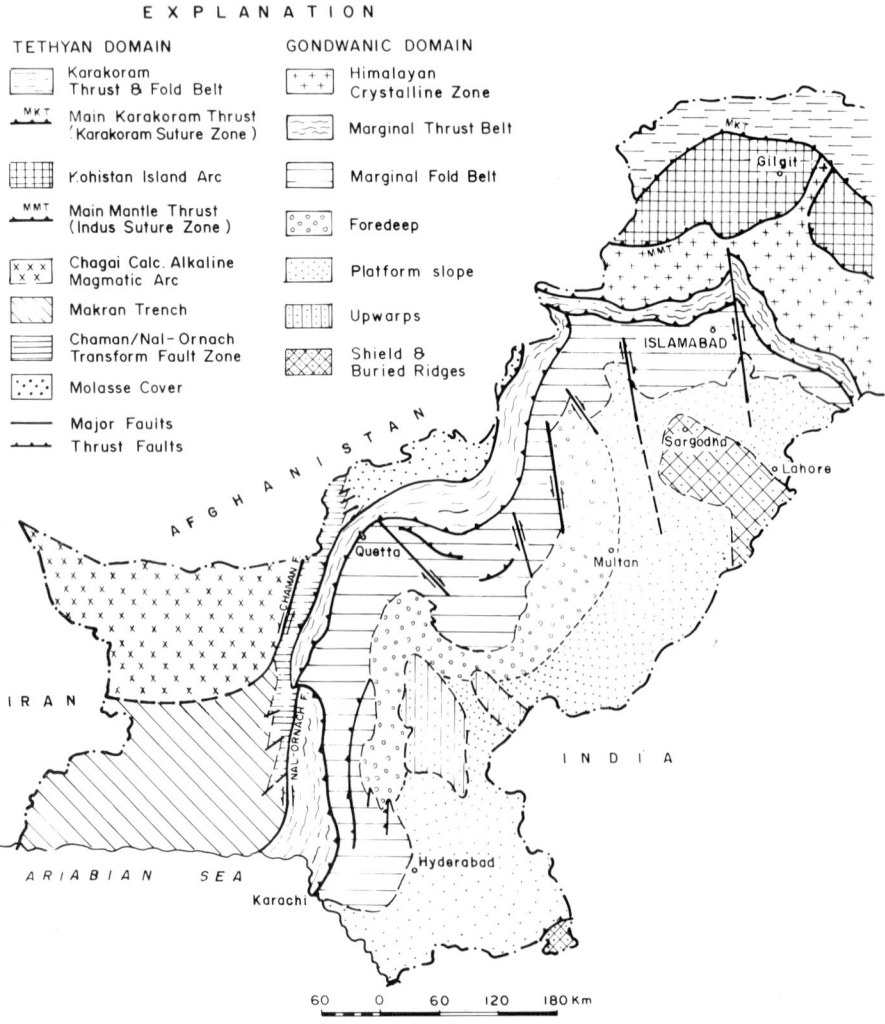

Figure 1.4. Map showing major tectonic zones of Pakistan.

Westward towards Afghanistan, the Himalayan imbricate crystalline zone is truncated by a zone of thrust and transform faults. One of the most significant tectonic features in this zone is the left-lateral Chaman transform fault (Griesbach, 1893; Jones, 1961). The Chaman fault trends from near the town of Chaman southwards into Kharan District where another north-trending left-lateral fault, the Nal-Ornach fault, begins en echelon to it and continues into the Arabian Sea. Apparently this complex zone of Chaman—Nal-Ornach faults forms the western transform margin of the Indo-Pakistan plate.

The Himalayan imbricate crystalline zone and the Chaman fault zone have been pushed southward and southeastward respectively over an allochthonous thrust and fold belt, the Sub-Himalayan thrust belt, which forms a double loop and extends several hundred kilometers along the outward margins of the Indo-Pakistan plate (fig. 1.4). Its western part is characterized by sheets and slivers of ophiolites, melanges, and volcanic rocks (Asrarullah and others, 1979; De Jong and Subhani, 1979; Kazmi, 1979) but the majority of the thrust belt is Late Paleozoic to Early Eocene marine pericratonic carbonates and shelf deposits.

Eastward and southward, the Sub-Himalayan thrust belt overlays autochthonous, broadly folded Jurassic to Pleistocene sedimentary rocks that were deposited on the shelf and foredeep zone of the Indo-Pakistan plate. This part of the Himalayan fold belt is characterized by oroclines, gravity faults, and strike-slip features (Sarwar and De Jong, 1979).

In Pakistan, elements of the Tethyan terrane may be seen north and west of the four structural zones described above. The Tethyan terrane includes three separate tectonized blocks, namely the Kohistan-Ladakh island arc lying to the north of the Himalayan imbricate crystalline zone, the Chagai calcalkaline magmatic and volcanic arc located west of the Chaman fault zone, and the Makran trench-arc subduction zone (Jacob and Quittmeyer, 1979) occurring west of the Nal-Ornach fault zone. Large megashears or suture zones separate these three extra-continental tectonic blocks from the elements of the Indian plate.

The Chagai calcalkaline magmatic belt occupies a convergence zone along the southern margin of the Helmand (Afghan) microcontinental block and contains Cretaceous to Neogene interlayered marine sedimentary and volcanic rocks, which were intruded by mafic and felsic igneous rocks. Eastward, the left-lateral Chaman transform fault separates this block from the Indo-Pakistan plate, whereas to the south, the Chagai magmatic belt has been thrust over the Eocene to Pliocene sediments of the accretionary prism in the Makran subduction zone (Jacob and Quittmeyer, 1979).

The Kohistan-Ladakh island arc consists of a relatively thick sequence of Late Cretaceous amphibolites, pyroxene granulites, and calcalkaline metavolcanics and metasediments, which were extensively intruded by diorites and granodiorites (Tahirkheli and Jan, 1979; Bard and others, 1979). Southward, the island arc has been thrust over the Himalayan imbricate crystalline zone along the Indus shear zone, which is probably the western extension of the Tsangpo suture zone in Nepal. The Indus suture is a complex, imbricate tectonic zone characterized by blueschist metamorphic rocks and a series of melanges and ophiolites which were obducted onto the Indo-Pakistan plate margin (Shams, 1980; Kazmer and others, 1983; Lawrence and others, 1983; Kazmi and others, 1984).

In Pakistan there is no trace of elements of the Eurasian terrane, all of which lies to the north of Pakistan. North of Kohistan, however, there is a small fragment of the Pamir microcontinental block, largely represented by the Karakoram fold and thrust belt (fig. 1.4). This belt comprises metamorphosed Paleozoic to Mesozoic intracratonic shelf carbonates and pelites and has been intruded by large and extensive syntectonic granitic batholiths (Stocklin, 1977; Kazmi and Rana, 1982; Coward and others, 1982). The Karakoram fold and thrust belt

Table 1.1. Mineral and Gemstone occurrences in Pakistan.

	GONDWANIC DOMAIN			TETHYAN DOMAIN					
	Platform & Marginal fold belt	Sub-Himalayan Thrust belt	Himalayan Crystalline Zone	Chaman Trans.fault Zone	Chagai Calc.alk. Magm. arc.	Indus Suture Zone	Kohistan Island arc.	Karakoram Suture Zone	Karakoram Thrust & Fold belt.
Oil & gas	●								
Coal	●								
Dolomite	●		●			●			●
Gypsum	●								●
Iron Ore	●		●		● ●		●	○	
Laterite/Bauxite	●								
Rock salt	●								
Sulphur	●				●				
Barite	●	●	●						●
Celestite	○	○							
Fluorite	●		●						
Galena	●		●				●		
Lead/Zinc	●								
Uranium	●						●		
Cu-Zn-Ag-Au			●		● ● ●				
Antimony			●	●					●
Poly. met. Sulphd.					○	●	●	●	●
Manganese		●							
Chromite		●			●		●		
Magnesite		●	●				●		
Asbestos		●					●		
Graphite			●				●	●	
RK. Phosphate	●		●						
GEMSTONES									
Aquamarine			●				●	●	●
Topaz			●				●	●	●
Tourmaline			●				●	●	●
Moonstone			●				●		
Quartz Crystal			●				●	●	●
Spinel								●	
Pargasite								●	
Ruby								●	
Emerald						● ●			
Epidote			●				●	●	
Garnet			● ●				●		●
Turquoise					●				
Agate	●								

φ OPHIOLITIC   ⊖ CONTACT METASOMATIC   ○ UNDIFFERENTIATED   φ MISSISSIPPI VALLEY TYPE

⊖ METAMORPHIC OR HYDROTHERMAL   φ PYROMETASOMATIC   φ SEDIMENTARY   ᛘ SANDSTONE TYPE URANIUM

⊖ PEGMATITIC   φ REPLACEMENT   φ MANTO TYPE   ℞ TRANSFORM FAULT RELATED

φ PORPHYRY TYPE   φ VOLCANOGENIC

has been thrust over the Kohistan-Ladakh island arc along a major megashear, the Main Karakoram Thrust (MKT), which is also a complex suture zone characterized by tectonic slices of ophiolites, melanges, and metavolcanic rocks.

METALLOGENY AND GEMSTONE OCCURRENCES

Deposits and showings of more than twenty-eight different kinds of minerals, including gemstones, have been reported in Pakistan (Ahmad, 1969; Ahmed and Abid, 1983). These deposits formed in geological environments unique to the various geotectonic zones in which they occur and which have been briefly described above. Table 1.1 summarizes the major mineral showings in Pakistan and the geological environment in which they occur. This list tends to give the impression that Pakistan should prove to be a gem and mineral rich country. However, on this question geologists have different views, which vary from cautious optimism to a disappointing prognosis. According to the present author, whose views are shared by Ahmed and Abid (1983) and Tahirkheli (1979a), the calcalkaline plutonic and volcanic rocks pre-date the major orogenic events. This calcalkaline igneous activity probably produced hydrothermal emanations that may be expected to have produced significant mineralization. The presently known showings should not be underestimated. Some showings in the Kohistan island arc are compatible with the possibility of finding Kuroko-type stratiform sulfide mineralization in the volcanic rocks, Solomon Island-type gold deposits in association with some of the diorites, and platinum, nickel, and chromium in the ultramafic complexes of the suture zones.

Sillitoe (1979) examined some of the significant mineral showings in northern Pakistan and briefly reviewed them. He placed these showings in two main groups (a) epigenetic, polymetallic deposits generated by metamorphic fluids in low-grade metamorphic environment and (b) minor mineralization affiliated with anatectic granites. He concluded that the deposits resulting from metamorphic mineralization are typically small and unlikely to support mining operations whereas the Neogene anatectic granites of crustal parentage are unfavorable for the generation of base- and precious-metal ores, including epithermal precious metal, contact-metasomatic, and porphyry copper deposits.

There is, however, no mistake that metamorphism and granitic rocks have generated concentrations of gemstones—notably ruby, emerald, topaz, and aquamarine (Okrusch and others, 1976; Tahirkheli, 1979a; Sillitoe, 1979; Gübelin, 1982; Gübelin and others, 1986; Kazmi and others, 1986). The prospects for the occurrence of larger gemstone deposits are indeed bright. The gemstone occurrences in Pakistan (Kazmi and O'Donoghue, in preparation) are almost entirely confined to its northern areas. The geodynamical setting of this region, as discussed earlier, has been particularly favorable for hosting gem mineralization.

Emerald is the principal gemstone of Pakistan. It is indeed the most unique gemstone in the world inasmuch as it comprises chemical constituents derived from two completely different geological domains, namely beryllium, which occurs in appreciable amounts only in continental felsic rocks, and chromium, which is found in significant quantities only in oceanic ultramafic rocks. Blending of these two such diversely and remotely occurring elements in one single crystal could have occurred only through exceedingly dramatic geological events and complicated geochemical processes. Indeed, it is therefore not surprising that in Pakistan all of its emerald deposits are located in the Indus suture zone, which largely comprises dismembered fragments of the former Tethyan oceanic crust caught up and crushed between the two colliding continents.

The geology and origin of these emerald deposits are discussed in the following papers.

REFERENCES

Ahmad, Z., 1969, Directory of mineral deposits of Pakistan: Geological Survey of Pakistan, Record, v. 15, 220 p.

Ahmed, W., and Abid, Q.Z., 1983, Mineral map of Pakistan: Geological Survey of Pakistan, Quetta.

Asrarullah, Ahmad, Z., and Abbas, S.G., 1979, Ophiolites in Pakistan: an introduction: *in* Farah, A., and DeJong, K.A., editors, *Geodynamics of Pakistan,* Geological Survey of Pakistan, p. 181-192.

Athavale, R.N., Verma, R.K., Bhalla, M.S., and Pullaiah, G., 1970, Drift of the Indian subcontinent since Precambrian times: *in* Runcorn, S.K., editor, *Palaeogeophysics,* London, Academic Press, p. 291-305.

Balkrishna, T.S., 1977, Role of geophysics in the study of geology and tectonics: AEG Progress, v. 1.

Bard, J.P., Maluski, H., Matte, P., and Proust, F., 1979, The Kohistan arc sequence: crust and mantle of an obducted island arc: Proceedings of the International Commission on Geodynamics, Group 6 Meeting, Peshawar, University of Peshawar, p. 87-94.

Coward, M.P., Jan, M.Q., Rex, D., Tarney, J., Thirlwall, M., and Windley, B.F., 1982, Geotectonic framework of the Himalayas of northern Pakistan: Journal of Geological Society of London, v. 139, p. 299-308.

DeJong, K.A., and Subhani, A.M., 1979, Note on the Bela ophiolites with special reference to the Kanar area: *in* Farah, A., and DeJong, K.A., editors, *Geodynamics of Pakistan,* Geological Survey of Pakistan, p. 263-270.

Griesbach, C.L., 1893, Notes on the earthquake in Baluchistan on the 20th December 1892: Geological Survey of India, Record, v. 26, p. 57-64.

Gübelin, E., 1982, Gemstones of Pakistan: emerald, ruby, and spinel: Gems and Gemology, v. 18, p. 123-139.

Gübelin, E., Graziani, G., and Kazmi, A.H., 1986, Pink topaz from Pakistan: Gems and Gemology, v. 22, p. 140-151.

Jacob, K.H., and Quittmeyer, R.L., 1979, The Makran region of Pakistan and Iran: trench-arc system with active plate subduction: *in* Farah, A., and DeJong, K.A., editors, *Geodynamics of Pakistan,* Geological Survey of Pakistan, p. 305-317.

Jones, A.G., 1961, Reconnaissance geology of part of west Pakistan, a Colombo Plan Cooperative Project: Hunting Survey Corporation Report, Oshawa, Government of Pakistan, 550 p.

Kazmer, C., Hussain, S.S., and Lawrence, R.D., 1983, The Kohistan-Indian plate suture zone at Javan Pass, Swat, Pakistan: Geological Society of America, Abstracts with Programs, v. 15, p. 609.

Kazmi, A.H., 1977, A review of the Quaternary geology of the Indus Plain: Presidential Address, Earth Sciences Section, Scientific Society of Pakistan, Multan.

Kazmi, A.H., 1979, The Bibai and Gogai nappes in the Ziarat area of northeastern Baluchistan: *in* Farah, A., and DeJong, K.A., editors, *Geodynamics of Pakistan,* Geological Survey of Pakistan, p. 334-339.

Kazmi, A.H., 1984, Geology of the Indus Delta: *in* Haque, B.U., and Milliman, J.D., editors, *Marine Geology and Oceanography of Arabian Sea and Coastal Pakistan,* New York, Van Nostrand Reinhold Company, p. 71-84.

Kazmi, A.H., 1985, Geology of the Thar Desert: Acta Mineralogica Baluchistan, University of Baluchistan, v. 1, p. 64-67.

Kazmi, A.H., and Rana, R.A., 1982, Tectonic map of Pakistan (1:2,000,000): Geological Survey of Pakistan, Quetta, Pakistan.

Kazmi, A.H., Lawrence, R.D., Dawood, H., Snee, L.W., and Hussain, S. S., 1984, Geology of the Indus suture zone in the Mingora-Shangla area of Swat, northern Pakistan: University of Peshawar, Geological Bulletin, v. 17, p. 127-144.

Kazmi, A.H., Lawrence, R.D., Anwar, J., Snee, L.W., and Hussain, S., 1986, Mingora emerald deposits (Pakistan): suture-associated gem mineralization: Economic Geology, v. 81, p. 2022-2028.

Kazmi, A.H., and O'Donoghue, M., *Gemstones of Pakistan.* (under publication).

Klootwijk, C.T., 1979, A review of paleomagnetic data from the Indo-Pakistan fragment of Gondwanaland; *in* Farah, A., and DeJong, K.A., editors, *Geodynamics of Pakistan*, Geological Survey of Pakistan, p. 41-80.

Lawrence, R.D., Kazmer, C., and Tahirkheli, R.A.K., 1983, The main mantle thrust: a complex zone: Geological Society of America, Abstracts with Programs, v. 15, p. 624.

Molnar, P., and Tapponnier, P., 1975, Cenozoic tectonics of Asia: effects of a continental collision: Science, v. 189, p. 419-426.

Okrusch, M., Bunch, T.E., and Bank, H., 1976, Paragenesis and petrogenesis of a corundum-bearing marble at Hunza: Mineralum Deposita, v. 11, p. 278-299.

Powell, C.M., 1979, A speculative tectonic history of Pakistan: *in* Farah, A., and DeJong, K.A., editors, *Geodynamics of Pakistan*, Geological Survey of Pakistan, p. 5-24.

Sarwar, G., and DeJong, K.A., 1979, Arcs, oroclines, syntaxes: the curvature of mountain belts in Pakistan: *in* Farah, A., and DeJong, K. A., editors, *Geodynamics of Pakistan*, Geological Survey of Pakistan, p. 341-350.

Seyfert, C.K. and Sirkin, L.A., 1973, Earth history and plate tectonics, Harper & Row, New York, 504 p.

Shams, F.A., 1980, Origin of the Shangla blueschist, Swat Himalaya, Pakistan: University of Peshawar, Geological Bulletin, v. 13, p. 67-70.

Sillitoe, R.H., 1979, Speculation on Himalayan metallogeny based on evidence from Pakistan: *in* Farah, A., and DeJong, K.A., editors, *Geodynamics of Pakistan*, p. 167-180.

Soffel, H., Forster, H., and Becker, H., 1975, Preliminary polar wander path of central Iran: Journal of Geophysics, v. 41, p. 841-843.

Stocklin, J., 1974, Northern Iran: Alborz Mountains: *in* Spencer, A.M., editor, *Mesozoic-Cenozoic Orogenic Belts, Data for Orogenic Studies*, Geological Society of London, Special Publication, v. 4, p. 213-234.

Stocklin, J., 1977, Structural correlation of the Alpine ranges between Iran and central Asia: Societe Geologique de France, Memoires, v. 8, p. 333-353.

Tahirkheli, R.A.K., 1979, The main mantle thrust: its scope in metallogeny of northern Pakistan: Proceedings of the International Commission on Geodynamics, Group 6 Meeting, Peshawar, University of Peshawar, p. 193-198.

Tahirkheli, R.A.K., and Jan, M.Q., 1979, Geology of Kohistan, Karakoram Himalaya, northern Pakistan: University of Peshawar, Geological Bulletin, v. 11, 187 p.

Takin, M., 1972, Iranian geology and continental drift in the Middle East; Nature, v. 235, p. 147-150.

Wegener, A., 1924, *The origin of continents and oceans:* Dutton and Company, New York, 212 p.

Geological Setting of the Emerald Deposits

Robert D. Lawrence, Ali H. Kazmi, and Lawrence W. Snee

INTRODUCTION

In Pakistan, emeralds are found in fascinating surroundings. The spectacularly rugged Himalayan ranges (photo. 2.1) capped by the lofty snow-capped peaks of Nanga Parbat (8128 m), Rakaposhi (7790 m) and Haramosh (7406 m) are dissected by the awe-inspiring canyon of the Indus River which reaches over 6100 meters of relief in places. Farther south the verdant, picturesque Swat valley between pine-clad mountains gives way to the barren, hilly and hostile terrain of the Malakand, Mohmand and Bajaur agencies. Though today the emerald occurrences are spread over a fairly extensive belt, their presence was not known until 1958 when the first emerald deposit was discovered at the northern edge of the city of Mingora in Swat (Kazmi, 1983). Since then, a number of deposits or showings of emeralds have been located in Mohmand Agency at Nawe Dand, Gandao (actually green beryl), Pranghar, Bucha, and Khanori; in the Bajaur Agency at Aman Kot and Maimola; in Swat District at Charbagh, Makhad and Gujarkili and in the Gilgit Agency at Khaltaro (fig. 2.1).

Detailed geological prospecting of the Mingora mines (Kazmi, 1983; Kazmi and others, 1984, 1986), has shown that they contain at least three distinct types of emerald deposits, which differ in quality. Thus, Pakistan emeralds offer a wide range of color and quality, with the best ones coming from Gujarkili and Mingora mines (Kazmi and others, this volume).

Exquisite as these emeralds are as gems; their geological setting is also wonderful: they are the products of collision between the Indian and Asian continental crustal plates. This history has been briefly described in the previous paper (Kazmi, this volume). The present study indicates that all the emerald deposits and showings in Pakistan except the Khaltaro occurrence are located exclusively in the metamorphosed ophiolitic melange of the suture zone. Although the Khaltaro emerald occurrence is within the Nanga Parbat gneiss, it also is found in close proximity to the suture zone. The geological setting of these deposits is briefly described below.

REGIONAL GEOLOGY

The primary emerald-bearing belt of rocks occurs in the northern part of Pakistan in the central one of three major tectonostratigraphic subdivisions. From north to south these are (1) the Kohistan island arc sequence which has been faulted against (2) the Indus suture melange group which had previously been obducted onto (3) the Indo-Pakistan subcontinent sequence (fig. 2.1).

The Kohistan sequence is composed of a southern ultramafic-amphibolitic zone structurally overlain by a layered mafic complex with abundant norite. Locally these rocks are overlain by marine metasedimentary rocks and calcalkaline volcanics. This sequence has been extensively intruded by diorites and granodiorites. It forms a vast thrust block which delimits the northern exposure of the Indus suture zone, containing the "emerald belt". For detailed descriptions of the Kohistan sequence see Jan (1977, 1980), Tahirkheli (1979a, 1983), Majid and Paracha (1980), Bard and others (1980), Bard (1983) and Coward and others (1987). Isotopic

Photo.2.1. View of the rugged Himalayan Ranges (photo *J. La Fortune*).

Photo.2.2. Photomicrograph of garnets in Mingora Schist (photo *A. H. Kazmi*).

Photo.2.3. Photomicrograph of Alpurai amphibolite (photo *L. W. Snee*).

Photo.2.4. Photomicrograph of Charbagh greenschist (photo *L. W. Snee*).

Photo. 2.5. Photomicrograph of Shangla blueschist (photo *A. H. Kazmi*).

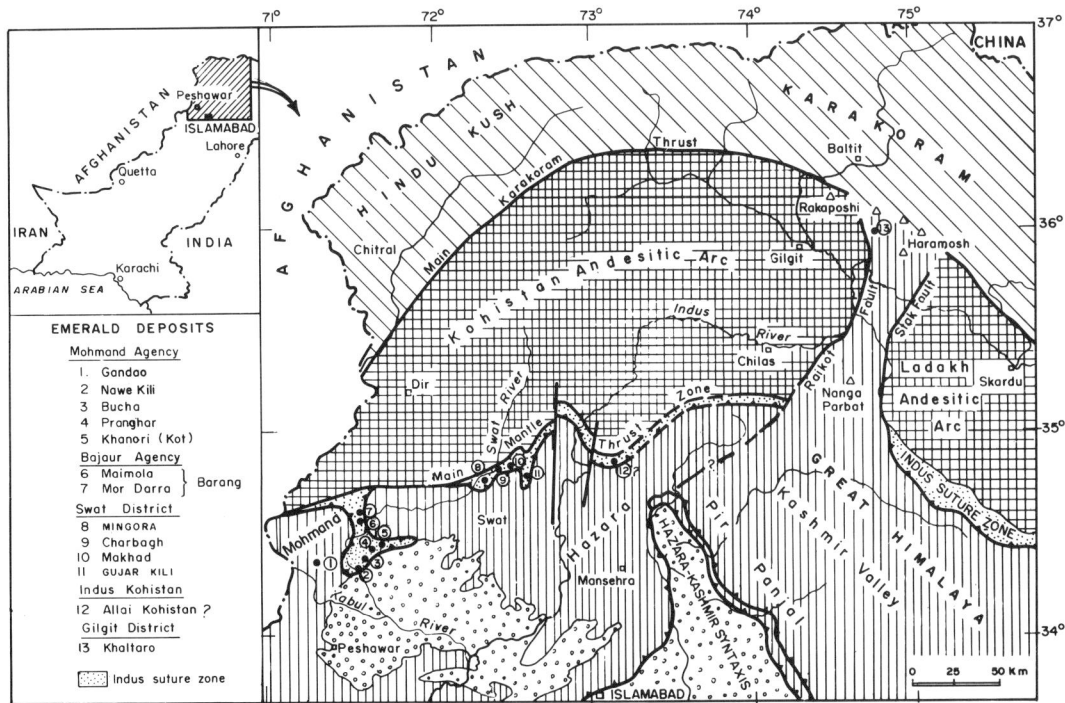

Figure 2.1. Emerald locations in northern Pakistan in relation to major tectonic subdivisions. Small circles indicate molasse of Murree and Siwalik Formations and sedimentary fill of the Peshawar Plio-Pleistocene basin. Vertical ruled lines indicate the Indo-Pakistan subcontinent sequence of Gondwana-related, folded and thrust faulted shelf sediments, basement, and their metamorphosed equivalents. Crossed lines indicate the Kohistan andesitic arc sequence of amphibolites, layered mafic complex, arc volcanosedimentary units, and arc-related, calc-alkaline, I-type intrusions. Diagonal ruled lines indicate Karakoram microplate sequence. Indus suture zone and Main Mantle thrust zone are principal Cenozoic suture between Indo-Pakistan subcontinent and Asian mass and are occupied by the Indus suture melange group indicated by fine dots.

dates of 80 to 150 m.y. including K/Ar and $^{40}Ar/^{39}Ar$ data constrain the time of metamorphism of the base of the Kohistan sequence (reviewed by Coward and others, 1986). In addition, zircon fission-track dates from north of the MMT in Swat show that this area was near its current crustal level by 40-50 m.y. (Zeitler and others, 1982b; Zeitler, 1985). These ages of metamorphism and uplift are prior to or simultaneous with amphibolite facies metamorphism of the rocks south of the MMT in Swat (Maluski and Matte, 1984; Lawrence and others, 1985) and indicate that the Kohistan rocks have little importance to the genesis of the emeralds.

The Kohistan arc is bounded on the north by the Main Karakoram thrust (MKT), also called the Northern Megashear. This is a less prominent suture, now generally interpreted as resulting from the collapse of a back-arc basin developed behind the Kohistan arc (Coward and others, 1986, 1987). The Hindu Kush and Karakoram Ranges north of the MKT are formed of a Gondwana microcontinent which separated as early as the Triassic and sutured to southern Asia in the Jurassic (see Kazmi, this volume).

The Indo-Pakistan subcontinent sequence was the northern edge of the Gondwana continent (Yeats and Lawrence, 1984). Metamorphosed and intruded shelf sediments of the edge of the continent were detached and thrust south over the Indian crustal basement. Locally fragments of the basement are present projecting up through the overlying sequence. In northern India and the Himalaya, the basement is overlain by a varied late Precambrian section into which some granites were intruded, and this in turn is unconformably overlain by Paleozoic and Mesozoic sediments (Gansser, 1964, 1981; LeFort, 1975; Valdiya, 1981). In the northwestern Himalaya two zones of metamorphism are present. In the Great Himalaya Range north of Kashmir Valley, the High Himalayan crystalline slab is metamorphosed and intruded by young granites. Farther north and to the east around Tso Morarai the Paleozoic section is also metamorphosed and intruded by young granites (Thakur, 1983; Valdiya, 1983; Searle, 1983). These metamorphic bands end against the line of the Nanga Parbat-Haramosh massif and Hazara-Kashmir syntaxis. Across the remainder of northern Pakistan, in Hazara, Swat, Malakand, and Mohmand (fig. 2.1), only one such metamorphic band is present and it is not clear which Himalayan metamorphic band it more nearly resembles.

The Indus suture melange group is a composite of remnants of the lithosphere of the Tethys Ocean that once intervened between the Indian subcontinent and the Kohistan arc. This through-going structure in northern Pakistan is known as the Main Mantle thrust (MMT), and is the equivalent in Pakistan of the Indus suture of Ladakh (Tahirkheli, 1979 b; Tahirkheli and others, 1979). The MMT is a complex zone with diverse rock assemblages separated by faults of different ages (Lawrence and others, 1983). We herein call the rock assemblages the Indus suture melange group and have been able to separately map different melanges within this group which developed (1) in association with the subduction zone along the northern Tethys boundary or (2) by obduction over the Indian continental margin south of the Tethys (Lawrence, 1982, 1984; Lawrence and others, 1983; Kazmi and others, 1984). These different melanges are separated by major faults which collectively form the MMT. In Swat where the melanges are most extensively exposed, we call the northern fault separating the melanges from the Kohistan rocks the Kohistan fault and the southern fault separating the melanges from the Indo-Pakistan subcontinent sequence the Kishora thrust (Kazmi and others, 1984). The Indus suture melange group is the host rock for all of the emerald deposits of Pakistan except Khaltaro and so is of great importance to understanding emerald genesis.

THE INDO-PAKISTAN SUBCONTINENT SEQUENCE

Along the Himalayan margin of the Indo-Pakistan plate in India there is a relatively thick pile of variably metamorphosed sedimentary rocks ranging in age from Precambrian to early Mesozoic. These rocks have been intruded by Early Paleozoic granitic plutons (500 to 600 m.y.; Valdiya, 1983). In the High Himalaya they are underlain by a crystalline slab of older metamorphic rocks containing intrusions that are both older (greater than 1700 m.y.; Valdiya, 1983) and younger (20 to 30 m.y.; Sharma, 1983, and Valdiya, 1983) than these. The High Himalaya are separated from the Lesser Himalaya by the Main Central thrust. In the Lesser Himalaya the late Precambrian to early Mesozoic section is presumed to be detached from the older rocks by the basal decollement of the foreland thrust system. No extension of the two metamorphic belts nor the Main Central thrust westward beyond the Nanga Parbat-Haramosh massif and the Hazara-Kashmir syntaxis has been documented. Instead in northern Pakistan, the Indo-Pakistan plate sequence is divided into four geological provinces or zones, that reflect this geologic change along structural trend from northern Kashmir to northern Pakistan. From east to west these are the Nanga Parbat, Hazara-Mansehra, Besham, and Swat-Malakand-Mohmand zones. Emerald deposits are known in the Nanga Parbat and Swat-Malakand regions (fig. 2.1).

The Nanga Parbat Region

The Nanga Parbat massif is a major re-entrant into the Kohistan arc terrain (fig. 2.1). It is entirely composed of Precambrian metasedimentary and plutonic rocks (Wadia 1933; Misch, 1936, 1949; Calkins and others, 1975; Verplanck and others, 1985; Madin, 1986; Verplanck, 1987; Madin and others, in press) and is the site of the Khaltaro emerald deposit. In the southern part of the region (near Babusar Pass in the southwest and near Burzil Pass in the southeast) the metasediments were considered part of the Salkhala Formation by Wadia and Misch and are largely made up of quartzites, slates, phyllites, marble beds, graphite-, quartz-mica-, chlorite-, and talc-schists. However, the type area of the Salkhala is in the Hazara-Kashmir syntaxis area to the south and extension of the term to the north is based only on preliminary reconnaissance work. For many years most of the metamorphic rocks south of the suture have been routinely correlated with the Salkhala Formation. Our recent work shows that the metamorphic stratigraphy of this region is complex and that correlation of distant sections is not straightforward. Northeastward, a steady increase in the metamorphic grade occurs and the rocks change to garnetiferous calc-gneisses and feldspathic, banded, garnet-biotite paragneisses locally containing kyanite and/or sillimanite. These paragneisses are intruded by a foliated biotite orthogneiss. The two gneisses make up the Nanga Parbat granite gneisses. They are similar in composition and the two rocks commonly have a strong external resemblance. The intrusive rocks have an 1800 m.y. U/Pb zircon age (Zartmann, personal communication, 1985). As the paragneisses are older than this, they must be Proterozoic or Archean rocks of the Indian basement and are certainly not equivalent to the Salkhala Formation of Kashmir.

The Nanga Parbat massif is multiply deformed (Madin, 1986; Verplanck, 1987; Madin and others, in press); three sets of major folds can be recognized (fig. 2.2). The earliest folds are very tight isoclinal structures with sizes ranging up to several kilometers in amplitude and largely intrafolial in style. Axial plane and fold axis orientations are quite variable due to younger superimposed structures. Highest metamorphic temperatures occurred after this folding event, as mineral textures are annealed. These folds were probably tightened during this metamorphism. The second fold event to affect these rocks involves large kink folds with approximately east-west axes. The deep gorge at the Indus valley west of Skardu partially follows the trough of one of these folds (Indus kink on fig. 2.2). The last set of folds is a strong set of very large anticlines on north-south axes that are a major control in developing the massif as a syntaxial feature. This last set of folds is genetically related to the Raikot fault which forms the west margin of the massif.

The rocks of the Nanga Parbat massif are cut by both mafic and silicic dikes (fig. 2.2). Amphibolites derived from cross-cutting mafic dikes cut the Nanga Parbat granite gneisses and are folded and metamorphosed with them. In addition, there are abundant young, undeformed injections of true granites into the paragneisses; coarse-grained biotite-muscovite granite pegmatites, locally mined for gem-quality green tourmaline, aquamarine, topaz and garnet (Kazmi and others, 1985; Verplanck and others, 1985; Verplanck, 1987) cross-cut and parallel the foliation of all of the older rocks. In addition to pegmatite medium- to fine-grained tourmaline granite is present, either as selvages to the pegmatites or as independent dikes (Madin, 1986).

The Raikot fault on the western margin of the massif was previously considered to be part of the MMT. However, this structure is active (Lawrence and Ghauri, 1983a; Lawrence and Shroder, 1984), has the opposite sense of motion to the MMT, and contains no suture melange group rocks. Thus it appears that the MMT has been displaced by the younger Raikot fault (fig. 2.2).

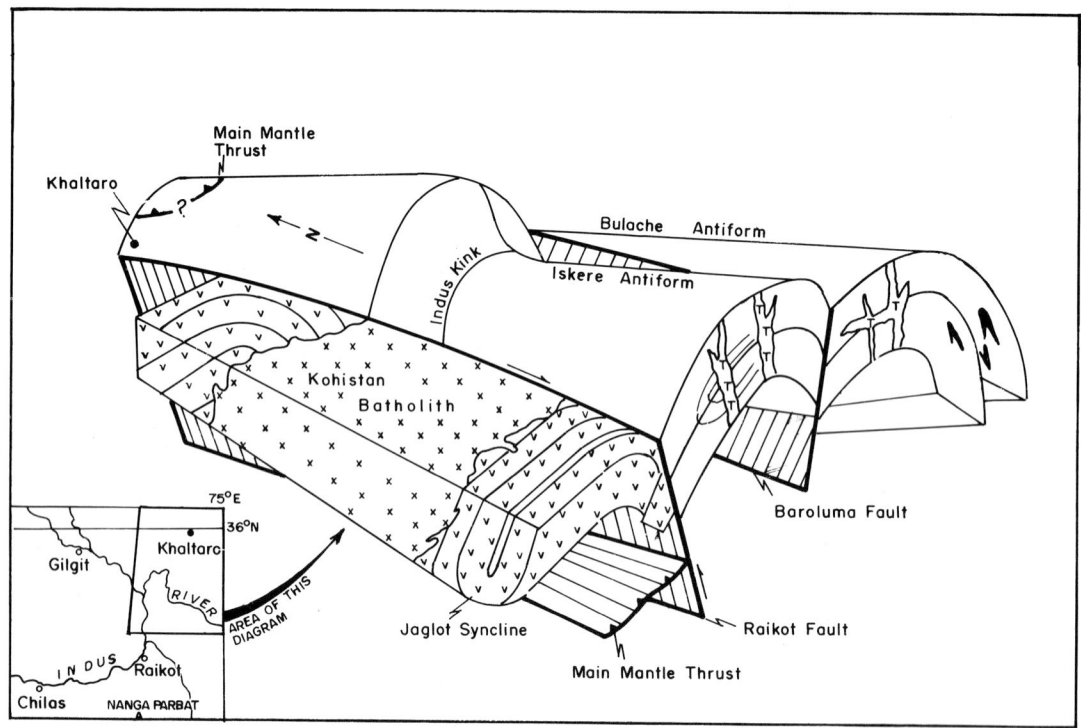

Figure 2.2. Structural diagram of the northwestern portion of the Nanga Parbat-Haramosh massif modified after Madin (1986). In the Kohistan terrain west of the Raikot fault "v" symbol indicates metavolcanic and metasedimentary rocks of the andesitic arc, "x" symbol indicates granodiorite and quartz diorite plutons of the Kohistan batholith. The eastern end of the Jaglot syncline shows drag by oblique right-lateral reverse motion of the Raikot fault. In the Nanga Parbat-Haramosh terrain east of the Raikot fault the Nanga Parbat granite gneisses are shown without pattern, isoclinally folded and transposed amphibolite dikes are shown in solid black, and post-tectonic, cross-cutting tourmaline granite dikes and sills are indicated by "T" symbols. The Khaltaro emerald deposit is located within the Nanga Parbat massif adjacent to the Raikot fault.

The Khaltaro emerald deposit occurs at high altitude among ice fields near the northern end of the Nanga Parbat massif. It is close to the Raikot fault and the MKT (fig. 2.1). Field workers report mafic and ultramafic rocks in the vicinity of this poorly studied deposit and emeralds formed in amphibolites derived from their metamorphic equivalents. The field setting is unknown at present, however, three possible sources for these ultramafic rocks may be noted. (1) The Indus suture melanges of the MMT should be above the Nanga Parbat massif after displacement on the Raikot fault. The structure of the massif plunges to the north, and on top of its northern end it is possible that these ophiolitic melanges crop out in a small area (fig. 2.2). (2) Small fragments of ultramafic rock are known to occur in the MKT. The MKT passes somewhere near the Khaltaro deposit. However, its actual location is not at all well known in this area of ice field. Thus while Tahirkheli and Jan (1979) show the MMT in a loop that terminates south of the MKT, Gansser (1981) shows the MMT terminating against the MKT. Thus it is possible that ultramafic material related to the MKT occurs at Khaltaro. (3) Lastly it is possible that some small slivers of ultramafic rock occur in the Raikot fault. Occurrence of ultramafic rocks along cross faults that cut the MMT is not uncommon because slivers of ultramafic rocks are known to occur along the Stak fault(the eastern border fault of the Nanga Parbat massif; Verplanck and others, 1985; Verplanck, 1987) and it is now known that the Stak fault is also a younger cross fault that cuts the MMT.

Hazara-Mansehra and Besham Region

In the northern part of the Hazara District near Mansehra, the Indo-Pakistan plate metamorphic sequence includes a psammitic sequence correlated with the Precambrian Tanawal Formation (Calkins and others, 1975) which has been intruded by the Mansehra granite (516 m.y. Rb/Sr isochron age; Le Fort and others, 1980), and a unit of poorly known calcareous metasediments. Medium-to high-grade metamorphosed shelf sediments have been intruded by the Mansehra granite (Shams, 1969, 1983) which forms extensive outcrops in the area. This granite is largely foliated, light colored, medium- to coarse-grained, and porphyritic. Its main constituents are microcline, oligoclase and quartz with small amounts of biotite, muscovite, accessory magnetite, apatite and sphene. Within this granitic complex a tourmaline granite also has been described by Shams (the Hakale granite). These rocks have close similarities to the rocks of the Swat-Malakand-Mohmand area discussed below.

Along the Indus River near Besham, a quite different sequence of rocks is exposed. A strongly migmatized terrain of high-grade metasediments is unconformably overlain by graphitic phyllites (Ashraf and others, 1980; Butt, 1983; Fletcher and others, 1986; LaFortune, 1988). These are faulted up along large newly discovered faults which cut the entire MMT zone (M.S. Baig, personal communication, 1987). The migmatitic gneisses appear to represent true Proterozoic Indian crustal basement. In contrast the rocks of both Hazara-Mansehra and Swat-Malakand-Mohmand appear to be part of the supracrustal shelf sequence of the northern margin of Gondwana.

Swat-Malakand-Mohmand Region

We have studied the metamorphic rocks of the Indo-Pakistan subcontinental sequence in the area near Mingora in Swat (Kazmi and others, 1984; Lawrence and others, 1985) and established new insight into the stratigraphy of these units (fig. 2.3). The most important previous work was that of Martin and others (1962). In this region the Indo-Pakistan plate sequence is composed of Manglaur crystalline schist of probable Precambrian age, intruded by Swat granite gneiss which may be early Paleozoic. These two units are unconformably overlain by schists and marbles of the Alpurai group of probable Paleozoic age and the Saidu calcareous graphitic schist (Kazmi and others, 1984). The Alpurai group is intruded by syntectonic tourmaline granites and post-tectonic Malakand and related granites.

Manglaur Schist

East of Mingora and south of Manglaur village, this schist crops out as tectonized and at places mylonitized granoblastic quartz-mica-garnet schist (fig. 2.3). At places it has been feldspathized and tourmalinized. It is mainly a heterogeneous mix of quartz-feldspar schist and quartz-mica-garnet schist (table 2.1) with thin layers of graphitic schist, amphibolite and calcsilicate (S. Carter, in prep.). Rare kyanite occurs in the mica schists (J. Di Pietro, personal commun., 1987). These schists contain relics of garnet porphyroblasts, largely as crushed pseudomorphs replaced by penninite and mica, stretched parallel to foliation or as scattered grains in the matrix. Fresh, undeformed garnet porphyroblasts have grown across these and form a second generation (photo. 2.2). Thus the Manglaur schists have experienced at least two periods of metamorphism separated by a retrograde episode.

The Manglaur schist is exposed in the core of a large anticline (fig. 2.3, cross-section). The schist has been extensively tectonized by differential shear strains during folding and by several intersecting gravity faults of small displacement. The schist has been intruded by the Swat granite gneiss. Xenolithic inclusions of Manglaur schist (from a few centimeters to several hundred meters across) are commonly present in the gneiss (Shams, 1963; S. Carter, in prep.) and migmatitic veins and injected lenses of granite are common in the schist.

The Manglaur schist is not known to be present in other localities, but the description of the area northwest of Malakand Pass by Chaudhry and others (1974, 1976) suggests that it may be present there (fig. 2.3). They report extensive xenolithic inclusions in the Malakand granite gneiss and intrusive contacts between the gneiss and other units.

Swat Granite Gneiss

The Swat granite gneiss intrudes the Manglaur schist and unconformably underlies the Alpurai group. In the area of interest a number of separate bodies are present which may be referred to by local names (fig. 2.3). All of these are orthogneisses of granite composition. The easternmost body, the Choga granite (CH on fig. 2.3), occurs east of lower Alpurai village east of Shangla Pass (Martin and others, 1962). The Loe Sar granite (LS) is south of Manglaur and the Ilum granite (IL) is south of Mingora (Palmer-Rosenberg, 1985; Ahmad, 1986; Carter, in prep.; DiPietro, personal commun., 1987). West of Mingora the Dopialo Sar granite (DS) is south of the village of Kotah (Ahmad, personal commun., 1987). North of the Swat River, the Chakdara granite (CK) is widely exposed (Chaudhry and others, 1974, 1976; Hussain and others, 1984). The Kot granite (KT), which is located northwest and west of Malakand Pass, was previously called the Malakand granite gneiss but this name is easily confused with the younger Malakand granite which intrudes it, so we abandon the usage. The Kalangai granite (KL) is the westernmost body and occupies the beginning of the narrow gorge of the Swat River which separates the Swat Valley from the Peshawar Plain.

The Ilum granite near Mingora is largely a coarse-grained muscovite augen gneiss with 5 to 30% megacrysts (relict phenocrysts up to 5 cm across) of potassium feldspar and less commonly of quartz (photo.2.3). Near the margins of the body a strong foliation developed with alternating discontinuous quartzofeldspathic and micaceous layers in which the megacrysts have been destroyed. Throughout the body the groundmass shows some degree of foliation development and locally, cataclastic zones in the body obliterate the megacrysts (Palmer-Rosenberg, 1985; Ahmad, 1986; S. Carter, in prep.). The Ilum granite is chemically a peraluminous granite indistinguishable from undeformed peraluminous porphyritic granites in other settings. Major element, trace element and rare earth element data all indicate that this granite is highly chemically evolved (Carter, in prep.).

Less information is available on the other granite bodies. The Dopialo Sar granite is very similar in appearance to the Ilum granite. The Loe Sar and Choga granites have a substantial biotite content, but are otherwise similar to the Ilum granite. The Chakdara granite is also biotite rich, but is otherwise similar to the Ilum granite (Chaudhry and others, 1974, 1976) except near its contact with the Alpurai where it is a quartz-feldspar-magnetite leucocratic gneiss (J.Di Pietro, personal commun., 1988). All of these bodies appear to be closely related and all are similar to the Mansehra granite of Hazara (Jan and others, 1981a; Shams, 1983). We

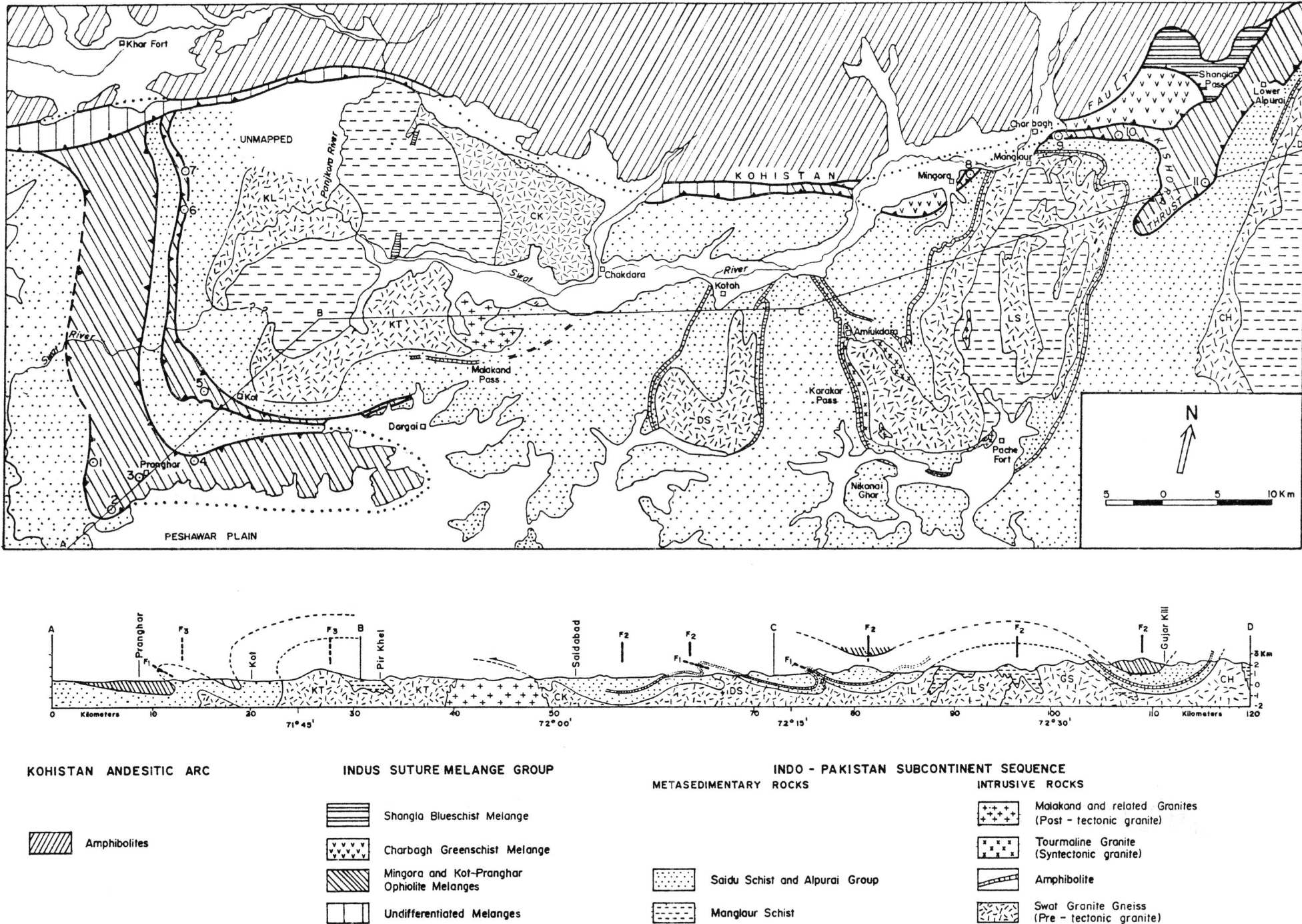

KOHISTAN ANDESITIC ARC

▨	Amphibolites

INDUS SUTURE MELANGE GROUP

☰	Shangla Blueschist Melange
⌄⌄⌄	Charbagh Greenschist Melange
⧄	Mingora and Kot-Pranghar Ophiolite Melanges
☐	Undifferentiated Melanges

INDO - PAKISTAN SUBCONTINENT SEQUENCE

METASEDIMENTARY ROCKS

⣿	Saidu Schist and Alpurai Group
⤶	Manglaur Schist

INTRUSIVE ROCKS

✛✛	Malakand and related Granites (Post - tectonic granite)
∷∷	Tourmaline Granite (Syntectonic granite)
▱	Amphibolite
⟩⟨	Swat Granite Gneiss (Pre - tectonic granite)

Figure 2.3 Geologic map of Swat-Malakand-Mohmand area. Numbered emerald deposits named on Fig. 2.1. Individual plutons of the Swat granite gneiss are KL=Kalangai, KT=Kot, DS=Dopialo Sar, IL=Ilum, LS=Loe Sar, GS=Guli Sar, and CH=Choga. Presence of Manglaur schist northwest of Malakand is an interpretation of published mapping that did not recognize the unconformity between the Manglaur and Alpurai units. Data sources: Field mapping by authors; compilation of work in progress by J. DiPietro, S. Carter, I. Ahmad, A. Hussain; published maps by Ahmad, 1986, Chaudhry and others, 1976, Hussain and others, 1984., Kazmi and others, 1984, Martin and others, 1962, Palmer-Rosenberg, 1985, Rafiq, 1984, Tahirkheli and Jan, 1979.

Structural cross section of the Swat-Malakand-Mohmand area shown along line A-B-C-D. Structural interpretation east of Saidabad based on extensive field work by authors and colleagues, but that west of Saidabad re-interprets published mapping. F₁ folds in Kot-Pranghar melange can alternatively be interpreted as imbricate thrust slices using available field data. Relative proportions of granite gneiss versus Manglaur schist in subsurface highly speculative.

Table 2.1 Metamorphic mineralogy of the rocks of the Swat area (after Kazmi and others, 1984).

Main mineral assemblages	Manglaur Crystalline Schist	Swat Granite Gneiss	Alpurai Garnetiferous Schist	Saidu Graphitic Schist	Mingora Ophiolitic Melange – Serpentinite	Mingora Ophiolitic Melange – Talc-chlorite-dolomite Schist	Mingora Ophiolitic Melange – Dolomite	Charbagh Greenschist Melange – Greenschist – Meta-tuff	Charbagh Greenschist Melange – Greenschist – Meta-basalt	Charbagh Greenschist Melange – Greenstone – Meta-mafic ign rks	Shangla Blueschist Melange – Meta-basalt	Shangla Blueschist Melange – Meta-tuff
Quartz	≡◇	≡◇	≡	≡		—	—	=	—	—	—	=
Plagioclase	—	—						=	=	=	=	=
K-Feldspar	—	=◇										
Biotite	=	=	=	=								
Muscovite	—	—	—	—				—				
Fuchsite						—	—				=	—
Hornblende												
Pyroxene					....							
Olivine					....							
Tourmaline	—	—										
Beryl		—				—e	—e					
Fe Ore	—	—	—	—	=	=		—	—	—	=	—
Chromite					—							
Sphene								—	=	=	=	=
Actinolite								--		=	=	=
Tremolite								—				=
Epidote	—	—						=		—		—
Piedmontite												—
Zoisite											=	—
Clinozoisite										=	=	
Glaucophane											≡	=
Serpentine					≡	—	—					
Chlorite	—				—	≡	=	=	=	≡	≡	≡
Talc					—	≡	=					
Dolomite					—◇	=◈	=◈				—	—
Siderite					—	—	=					
Calcite	—		—	—	—	—	—	—		—	—	
Graphite	—		—	≡								
Garnet	=◈	—◈	=◇									
Kyanite	—											
Sillimanite	≡											

EXPLANATION

≡   Abundant
=   Common
—   Present
◇   Porphyritic or porphyroblastic
◈   Crushed porphyroblast
∞   Two generations of porphyroblast
....   Relicts
—e   Emerald

suggest that they all may be part of the same magma series, which for purpose of this paper is called the Swat granite gneiss.

The ages of the Manglaur schist and Swat granite gneiss are uncertain; no useful isotopic dates of crystallization are yet available from Swat. Le Fort and others (1980) have published a 516 ± 16 m.y. Rb/Sr whole rock isochron date on the very similar Mansehra granite; Robert Zartman (personal commun., 1985) has determined a U-Pb zircon date of about 580 m.y. This granite intrudes siliceous schists which resemble the Manglaur schist; these schists display staurolite-to-sillimanite-zone metamorphism. If this correlation is correct, the Manglaur schist is Precambrian and the Swat gneiss intruded it probably in the early Paleozoic. An early metamorphism may have been synchronous with this intrusive event.

Alpurai Group

The Alpurai group is continuously exposed from east of Alpurai village to west of Malakand Pass with a similar stratigraphic section throughout this 80 km distance. Previously the inferred intrusive relationship with the Swat gneiss was taken to indicate a Precambrian age for these formations and these units were equated to the Salkhala Formation of Kashmir (Tahirkheli, 1979 b). However, on a regional basis the unconformity between the Alpurai group and the Manglaur schist-Swat granite gneiss is shown by (1) the complete lack of any intrusive apophyses of the granite into the Alpurai group even though intrusive features in the Manglaur are abundant and (2) by the same basal Alpurai unit overlying a diverse section of older units throughout the region. This contact is everywhere overprinted by metamorphism and commonly has been sheared, probably because of high strain between the competent granite gneisses and the incompetent stratified sediments. In many places, tourmaline granites intruded along the contact. Thus normal sedimentary features of an unconformity are revealed only by regional mapping. In a few locations truncation of Manglaur beds by basal Alpurai beds can be seen after careful mapping (J. Di Pietro, personal commun., 1987) and the unconformity can be demonstrated in a single outcrop, for example, in the hill about 1 km west of Pacha Fort (fig. 2.3).

The Alpurai group can be divided into two major units, a lower section dominated by quartz-mica schists, quartzofeldspathic gneisses, metapelites, and amphibolites (photo. 2.3) and an upper unit of calcschists and marble (table 2.1). It varies from a few hundred to a few thousand meters in thickness. The lower section locally has a schistose marble at the base. The contact between the lower and upper sections is marked by a thick amphibolite layer. In places, as near Karakar Pass and on the south side of Malakand Pass, amphibolite shows clear intrusive relations with other units. On the basis of geochemisty, it was derived from intrusive tholeiitic basalt magma (Ahmad, 1986). However, the high quartz content of some of the amphibolite and hornblende schists and their consistent stratigraphic position suggest that some of this material may be derived from mafic tuffaceous sediments, contaminated basalt, or quartz diorite of calcalkaline composition. The amphibolites form one of the key marker horizons in field mapping.

South of Karakar Pass the top of the unit is a massive marble locally called the Nikanai Ghar marble (fig. 2.3) which contains unidentifiable fossils (Palmer-Rosenberg, 1985). Units below the Nikanai Ghar marble have been traced southward into outcrops in which Lower Carboniferous conodonts have been identified (K. Pogue and J. Di Pietro, personal commun., 1987). Thus the Alpurai group appears to correlate with at least a portion of the Paleozoic sedimentary section exposed along the eastern margin of the Peshawar Basin which ranges in

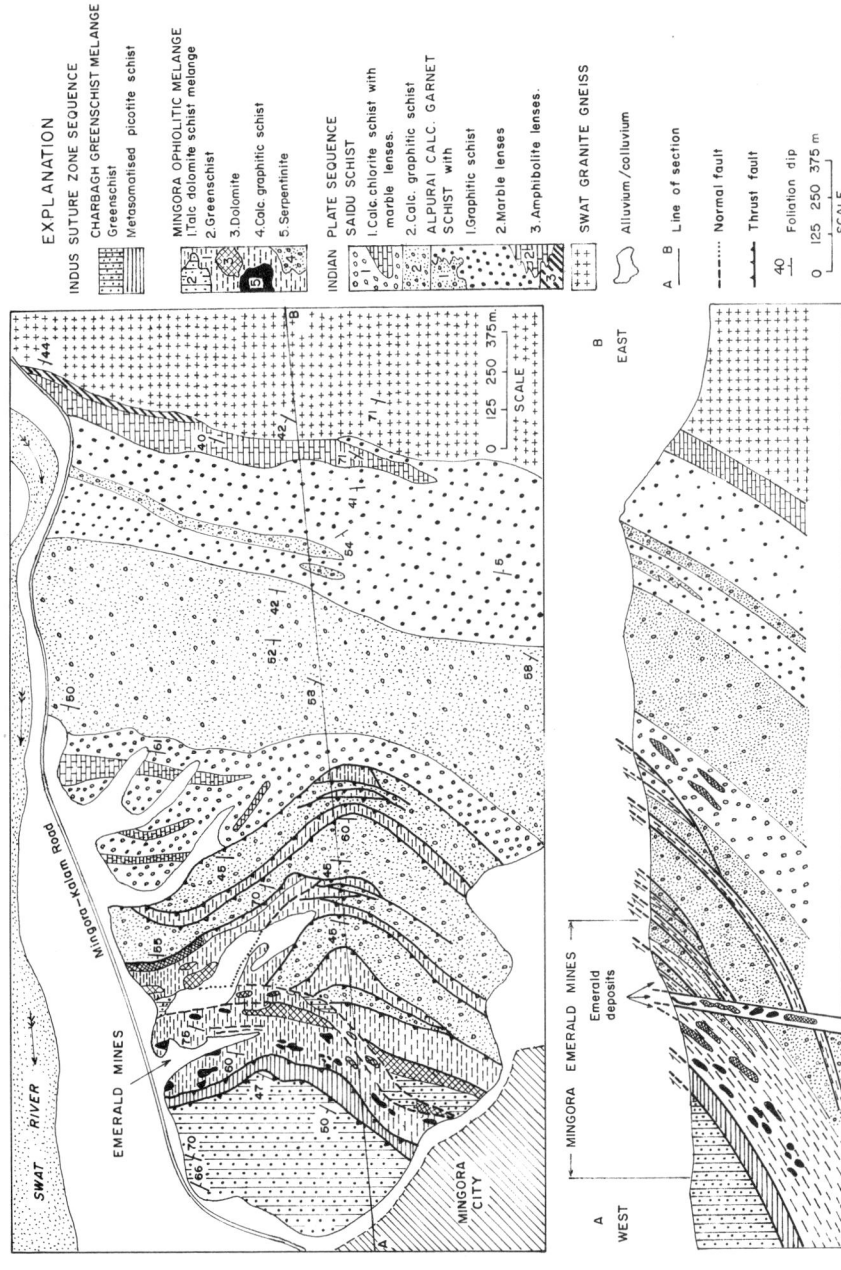

Figure 2.4. Geologic map and cross-section of the Mingora emerald mine area (after Kazmi and others, 1984). Geologic map shows details of melange in Mingora emerald mine and imbricate interleaving of talc-carbonate melange slices with Saidu schist.
Cross-section shows melange structure and younger cross-cutting normal fault with which emerald mineralization is associated.

age from Cambrian to Carboniferous (Pogue and Hussain, 1986). This is compatible with the ages of the Manglaur schist and Swat granite gneiss suggested above.

Saidu Schist

In the Mingora area and elsewhere this schist overlies the Alpurai schist. The contact appears to be gradational, but an unconformity is possible. The Saidu schist is gray to dark gray, calcareous and pelitic. Most of the unit is graphitic (fig. 2.4), but chlorite schists are locally present (see table 2.1). The chlorite schists are easily confused with parts of the Mingora ophiolite melange and have caused mapping errors in the past. The metamorphic facies is upper greenschist to lower amphibolite.

In the Mingora-Alpurai region, the upper part of the Saidu schist has been overthrust by the Indus suture melanges. An intensely tectonized imbricate zone has formed with several overlapping slices of Saidu calc-graphite schist alternating with slices of talc-dolomite schist and greenschist of the Mingora ophiolite melange (fig. 2.4a and b). The thrust emplacement of thin slices of the overlying unit into the overridden substrate requires that successive faults cut higher into the section as the fault system developed. Thus the faults at the western end of the cross section (fig. 2.4b) are younger than those at the eastern end. Note that the movement direction of the thrust sheet was probably southward, that is, nearly perpendicular to the section line shown. It is possible that the structure seen here is a metamorphosed and flattened melange. If this is the case, one would have to consider the possibility that the Saidu schist is a much younger unit than the Alpurai group, possibly equivalent to one of the Mesozoic units of the Indus suture zone in Ladakh such as the Lamayuru unit or the Indus flysch (Frank and others, 1977; Thakur, 1981; Thakur and Bagati, 1983).

Younger Granites

In addition to the Swat granite gneiss, the Swat-Malakand-Mohmand region includes at least two sets of younger intrusions. These are all small bodies, rarely more than a few kilometers in longest direction. One unit consists of syntectonic tourmaline granite gneisses and the other of post-tectonic biotite granite-granodiorite.

The tourmaline granites are found mainly near the contact between the Alpurai group and the underlying Swat granite gneiss. It has previously been considered to be a late phase of the Swat granite gneiss which it consistently intrudes. However, in the areas near Karakar Pass and Amlukdara (fig. 2.3) detailed field relations demonstrate that the tourmaline granite intrudes the quartz-mica schist, quartzofeldspathic gneiss, and amphibolites of the Alpurai group. Thus it is late- to post-Paleozoic in age and it is possible that these tourmaline granite gneisses are a syntectonic, synmetamorphic, early Himalayan intrusive unit.

The tourmaline granite gneisses are generally medium-grained and composed of potassium feldspar (mostly microcline, 30-35%), plagioclase (An_{5-15}, 12-25%), quartz (20-40%), muscovite (5-12%) and black tourmaline (3-15%). Numerous tourmaline aplite and quartz-feldspar-tourmaline ($\pm$ muscovite-biotite) pegmatite dikes are present (Ahmad, 1986; Palmer-Rosenberg, 1985). The tourmaline granites are moderately to strongly foliated and pegmatite dikes are transposed and preserved only as fold noses. Major element geochemistry shows that the tourmaline granites are peraluminous. Rare earth element data confirm that the tourmaline granites are magmatically distinct from the Swat granite gneiss (Carter, in prep.).

The biotite granodiorite to granite is best known as the Malakand granite, a circular body (fig. 2.3) about 3 km across exposed on the north side of Malakand Pass (Chaudhry and others, 1974, 1976). Modally the rock is a granodiorite to granite. It is medium-grained with a hypidiomorphic granular texture. It shows no internal deformation effects. Contacts with the surrounding rocks are sharp and excellent contact metamorphic effects are present. Xenoliths and screens are common in the Malakand granite. An Ar^{40}-Ar^{39} muscovite date of 23 m.y. on the Malakand granite is reported by Maluski and Matte (1984). The undeformed character of the rock suggests that this cooling age is close to the magmatic age of the unit.

Pegmatites are common in the Malakand granite and range in size from a few centimeters to 5 meters (Chaudhry and others, 1974). The pegmatites may be divided into three types: feldspar-quartz pegmatites, mica-feldspar-quartz pegmatites, and tourmaline-fluorite-bearing pegmatites.

Recently a small body (less than 1 km across) of undeformed biotite granite was mapped by S. Carter (in prep.) east near Manglaur (fig. 2.3). This body is probably genetically related to the Malakand granite which it resembles in geochemistry.

Blue beryl has been reported in one place about 1 km west of Pacha Fort (fig. 2.3) in hydrothermal veins near the margin of the Ilum granite (Butt and Shah, 1985). This locality is mainly an outcrop of tourmaline granite (J. Di Pietro, personal commun.,1987) which indicates that these veins are more likely related genetically to tourmaline or Malakand-age granite and not to the Ilum granite. Blue beryl has also been reported east of Kot (fig. 2.3) in small pegmatites, usually in association with fluorite (Hussain and others, 1984). These occurrences indicate that beryllium-bearing fluids were associated with at least some of the granitic rocks emplaced in the northern margin of the Indo-Pakistan subcontinent sequence.

Structure and Metamorphism

Deformation within the rocks of the Indo-Pakistan subcontinental sequence has largely been by ductile processes during Himalayan metamorphism. Three different sets of major folds with different styles and geometry are recognized. (Rosenberg and others, 1984; Lawrence and others, 1985; Palmer-Rosenberg, 1985; Ahmad, 1986). Minor folds are developed in association with all three of these major fold sets. Structural mapping of these features has been based on the recognition of the internal stratigraphy of the Alpurai group with its basal unconformity and particularly on the mapping of the amphibolite horizon within the Alpurai as a stratigraphic marker (fig. 2.3). The structural importance of this marker has not been previously recognized, and it has not been separately mapped in the western half of the map area (fig. 2.3). In this area our interpretation is tentative. Near Malakand Pass, Chaudhry and others (1976) have suggested a thrust fault where we place our unconformity. Structural complications may well exist between here and Chakdara.

Based on more recent mapping (J. Di Pietro, personal commun., 1988), the earliest folds, F_1 are recumbent, isoclinal nappes on northwest-southeast axes. These are large folds with fold-widths of more than a kilometer and fold-heights of at least 5 kilometers. They developed at the height of the metamorphism for which we have hornblende Ar^{40}-Ar^{39} age-spectrum dates of 39 to 40 m.y. (Lawrence and others, 1985). Metamorphism reached epidote-amphibolite facies in the northern part of the area indicating approximate conditions of 550°C and 5.5 Kbar (Palmer-Rosenberg, 1985). These folds have been mapped in the eastern portion of figure 2.3 by Di Pietro (personal commun., 1988). Detailed structural studies are not available in the west, but the map pattern of the Kot-Pranghar melange, Saidu schist and Alpurai group strongly suggests that the melange is preserved in the troughs of large F_1 synclines. We have shown this interpretation on figure 2.3. Alternatively this structural pattern may be explicable in terms of large-scale imbricate faulting in the Saidu schist and melange similar to what is seen on a small scale east of the Mingora mine (fig. 2.4).

The second set of folds F_2 are upright, open folds developed on approximately north-south axes. These have fold widths of about 12 to 18 kilometers and fold heights of 3 to 5 kilometers in the Mingora area. They tighten and become more intense towards the east, and in the Besham area are the only set of folds clearly discerned. They are also synmetamorphic, and may not represent a long interval of time separation from the earlier folds. The Mingora melange is clearly preserved in the troughs of these synclines and so has been folded by them (fig. 2.3).

The last set of folds, F_3, is developed on east-west axes. In the southern part of the metamorphic area of Swat-Malakand-Mohmand, a single, very large anticline overturned towards the south is developed with a fold width of up to 25 kilometers and fold height of over 5 kilometers. A portion of this fold west of Malakand Pass is shown on figure 2.3, A-B portion of the cross section.

It is important to note that the axes of both F_1 and F_2 are oblique to the MMT. These are very large structures that show no evidence of having been rotated into parallelism with a tectonic transport direction. They strongly suggest oblique motion with a strike-slip component in the early history of collision between the Indo-Pakistan subcontinent and the Kohistan arc. In the area north of Kotah, bending of F_2 folds suggests drag that was related to a right-lateral component of motion.

Minor normal faults have developed in the Indo-Pakistan subcontinental sequence subsequent to the metamorphism and folding discussed above. These are most evident in the Manglaur area where displacement of Manglaur schist and Swat granite contacts locate these structures which fall in two groups, one trending NW or NE, the other, EW or NS. The latter group is more prominent. Displacements on these features appears to be small, on the order of a few hundred meters, but they are important because emerald mineralization is localized along such a fault in the Mingora mine area (fig. 2.4b) (Kazmi and others, 1984, 1986). Others, *e.g.* in the Malakand and Mohmand areas, were hydrothermal fluid pathways along which fluorite and topaz were precipitated.

The Indus Suture Melange Group

The Indus suture melange group contains a complex sequence of imbricated melanges, largely composed of tectonic blocks of ophiolites, blueschists, greenschists, metavolcanics and metasediments in a matrix of sheared and variously metamorphosed fine-grained sediments and/or serpentinite. The rocks occupy the Main Mantle thrust zone (MMT) (Tahirkheli and others, 1979; Lawrence, 1982; Lawrence and others, 1983; Kazmi and others, 1984) between the Kohistan andesitic arc and the northern margin of the Indo-Pakistan plate and are separated from the Indus-Tsangpo suture zone of Ladakh and Tibet by the Nanga Parbat massif. The Himalayan suture system is offset along the western side of the Nanga Parbat-Haramosh massif by the active Raikot fault, and the MMT is the name used for the suture west of this fault. It extends west from near Babusar Pass through Allai Kohistan, Besham, Swat, Dir and Mohmand Agency into western Afghanistan.

From east to west between Haramosh Mountain, which forms the northeastern part of the Nanga Parbat massif, and the Mohmand Agency (on the Afghan border) we distinguish five structural segments. The first is the Haramosh-Raikot segment in which the Raikot fault is present and which is associated with the Khaltaro emerald deposit. The remaining three segments are parts of MMT zone. Their emerald associations are as follows:

Raikot-Alpurai	— Unconfirmed reports
Alpurai-Chakdara	— Gujarkili, Makhad, Charbagh, Mingora
Chakdara-to Khar Fort	— Barang, Khanori, Bucha, Nawe Dand, Tora Tigga, Pranghar, Kot, Maimola, Mor Darra (including Gandao green beryl).

The Haramosh-Raikot Segment

This segment starts from the western part of the Haramosh Range and trends southwards to the bend in the Indus River near Raikot. It is characterized by an active dextral reverse fault, the Raikot fault (Lawrence and Ghauri, 1983a; Lawrence and Shroder, 1984; Madin, 1986; Madin and others, in press; Shroder and others, in press). East of this fault, gneisses and schists of the Nanga Parbat group of rocks are exposed, while to the west, amphibolites, gabbros, diorites and metasediments of the Kohistan sequence of rocks are exposed (fig. 2.1 and 2.2).

The fault zone itself is 0.5 to 3 km wide and is characterized by hydrothermal alteration, a discontinuous band of mylonite (up to 300 m thick) and numerous fault strands. Despite local, extreme mylonitization, protoliths are generally identifiable and a sharp transition from Nanga Parbat group protoliths to Kohistan sequence protoliths is recognizable (Madin, 1986). Amphibolites, which were likely ultramafic blocks, are found within the mylonite zone near Khaltaro. Along the Raikot fault, sediments of the early Pleistocene Jalipur till are folded into overturned synclines (Shroder and others, in press). Neotectonic activity is indicatd by offset Holocene talus cones, knick points in the Indus and Astor Rivers and tributary streams, and major earthquake-induced landslides such as that in 1841. Based on fission-track evidence (Zeitler and others, 1982a; Zeitler 1985), metamorphic facies, and movement direction indicators, movement on the Raikot fault and associated folding may have uplifted the Nanga Parbat massif over 20 km in the late Cenozoic (Madin and others, in press). The Raikot fault, therefore, appears to be a major crustal structure. It may be a tear fault which terminates the Main Central thrust of the High Himalaya (Lawrence and Ghauri, 1984; Madin and others; in press, 1984).

From the emerald standpoint this segment is significant because emeralds have been discovered within Nanga Parbat gneisses in the Khaltaro area close to the northern part of this segment. It seems likely that the emeralds at Khaltaro were found where pegmatites cut metamorphosed ultramafic rocks (see Kazmi and others, this volume for detailed geologic descriptions). However, even though no suture melange group rocks are exposed along the Raikot fault, suture melange rocks may occur on top of the Nanga Parbat massif in the north (fig. 2.2). Thus, emeralds of Khaltaro alternatively may have been found where pegmatites cut suture melange rocks near the Raikot fault.

The Raikot—Alpurai Segment

The greater part of this segment runs along the inaccessible terrain that forms the upper ridges of the Nanga Parbat Range and Allai Kohistan to the Indus canyon near Jijal (fig. 2.1). Due to the extremely rugged terrain this segment of the suture zone is the least known. In this sector the Indus suture melange group is largely composed of relatively thin tectonized blocks and slivers of blueschist, greenschist, serpentinite, gabbro and peridotite (Majid and Tahir Shah, 1985) except in the Jijal area where a large mass of peridotite and garnet granulite is present (Jan, 1979, 1980; Jan and Howie,1981). These rocks have been thrust over the Indian plate sequence of the south and, at places along the thrust, breccias with serpentinite matrix and blastomylonites derived from paragneisses formed (Lawrence and Ghauri, 1983b).

There are unconfirmed reports of the occurrence and mining of emeralds from this segment of the Indus suture melange group in the Allai Kohistan area of the Indus Canyon.

Alpurai to Chakdara Segment

In this sector the rocks of the Indus melange group attain their maximum thickness and are well exposed (fig. 2.3). They are easily accessible and contain the best and the largest known emerald deposits occurring in Pakistan. The Mingora area may be taken as the type locality for this group of rocks.

Near Mingora, the Indus suture melanges occur as a chaotic assemblage of rocks derived from oceanic crust, volcanic arcs, trenches, and other Tethyan oceanic settings. They are tectonically milled and strung together as fragmented blocks, sheets and lenses, from a few meters to a few kilometers in extent, and set in a ductile matrix of volcaniclastic to pelitic schist, serpentinite, and talc-dolomite schist. The role of gravity processes in their creation is difficult to discern through the overprint of continental collision. They were created in processes at the margins of the Tethys Ocean and combined during collision. Major thrust faults, the Kohistan, Shangla, Makhad, and Kishora thrusts, slice this melange complex into three melange units (fig. 2.3). From north to south these are (a) the Shangla blueschist melange, (b) the Charbagh greenschist melange, and (c) the Mingora ophiolitic melange (Kazmi and others, 1984). The MMT zone of Tahirkheli (1979b) encompasses all of these structures and units.

The Mingora ophiolitic melange. Near Mingora this unit occurs as a narrow north-trending wedge which contains the Mingora emerald deposits (fig. 2.3a). Northeastward it may be traced beyond Alpurai. Westward it intermittently crops out near Kabal (Kazmer and others, 1983) and extends to the Dir District. Its thickness varies from a few hundred meters between Mingora and Makhad to several thousand meters between Kishora and Alpurai. In decreasing order of abundance, it is composed of talc-dolomite schist, greenstone metabasalt, greenschist metapyroclastics, metagabbro, metasediments, and metachert. All of this is set in a matrix of talc-chlorite-dolomite schist (abundant near Mingora). At some places the matrix is calcareous quartz-mica-chlorite schist. The latter occurs as slivers and thrust sheets squeezed in between the other tectonized blocks. Three main features distinguish this melange from the others: (1) abundance of the ophiolite suite of rocks, (2) presence of talc-dolomite schist, and (3) emerald mineralization.

The Mingora ophiolite melange has been metamorphosed with the underlying metasediments of the Indo-Pakistan subcontinent sequence. Mineral assemblages are representative of greenschist facies conditions and melange is preserved primarily in the troughs of north-plunging synmetamorphic F_2 synclines indicating that it was already in place by the time that this fold event took place. The sinuous outcrop pattern of the Kishora thrust at the base of the melange reflects its folding by F_2 (fig. 2.3).

Charbagh greenschist melange. This melange crops out between Charbagh and Shangla Pass (fig. 2.3) and forms a thick tectonic wedge between the overlying Shangla blueschist melange to the north and the Mingora ophiolitic melange to the south. It is characterized by greenstone metabasalts and greenschist metapyroclastics (photo. 2.4), with minor tectonized layers and wedges of metasediments. The latter have been extensively sheared and mylonitized, particularly along the lower thrust contact with the ophiolitic melange, where about 100-meter-long, en enchelon lenses of talc-dolomite schist have been strung out parallel to the Makhad thrust. No emeralds are found in this melange. It is tentatively interpreted as volcanic arc material although no geochemical data are available.

Shangla blueschist melange. This melange crops out in a relatively small area around Shangla Pass (fig. 2.3). It forms a tectonic wedge between the Kohistan arc sequence to the north and the underlying Charbagh greenschist melange to the south. It is characterized by blocks of glaucophane-crossite-bearing blueschists (photo. 2.5) (Shams, 1972, 1980; Jan, 1985). It is composed of large dismembered masses (2—4 kilometers in extent) of metavolcanics and phyllitic schists with smaller lensoidal masses of serpentinite, metagabbro, metadiabase, metagraywacke, metachert and marble.

The metavolcanics are composed of metatuffs and metabasalts. The tuffs retain relict bedding, commonly contain lapillae and volcanic bombs, and largely occur as soft schistose rocks of variegated color. Near Shangla Pass the tuffs are gray or grayish pink in color and contain piedmontite, albite, quartz and mica (Jan and Symes, 1977). The metabasalts are brown to brownish-green in color and, rarely, contain relict pillow structure. They are resistant to erosion and commonly occur as large blocks or lensoidal phacoids in the softer, sheared metatuffs (table 2.1). Jan and others (1981b) indicate pressure-temperature conditions during the blueschist metamorphism of approximately 7 kb and 380°C on the basis of its mineralogy; they also indicate the possibility of a later higher temperature overprinting. Isotopic dates on the blueschist metamorphism are 84 m.y. (K/Ar, Desio and Shams, 1980) and 83.6 m.y. (Ar^{40}/Ar^{39}, Maluski and Matte, 1984). Kazmer and others (1983) report fossils of Jurassic to middle Cretaceous age from a limestone block in the Shangla blueschist melange. These Cretaceous ages indicate that the Shangla melange was largely developed in a subduction zone adjacent to the Kohistan andesitic arc while the Tethys Ocean was rapidly closing and the Indo-Pakistan sub-continent was thousands of kilometers to the south. This subduction complex developed approximately concurrent with the end of metamorphism of the Kohistan arc norites between 80 and 150 m.y. ago (Coward and others, 1986).

The northern boundary of the Shangla melange against the Kohistan arc rocks is the Kohistan fault. This fault is not deformed by the F_1 or F_2 folds of the Indo-Pakistan subcontinent sequence. This suggests that the Kohistan sequence was not in its current position during the metamorphism of the Mingora ophiolite melange and underlying metasediments and that older blue schists were juxtaposed against younger amphibolites by strike-slip movement along a portion of the MMT.

Chakdara to Khar Fort Segment

West of Mingora, the Indus suture melange group wedges out, and southwest of Jawan Pass near Chakdara the MMT zone is narrow and largely covered by Quaternary alluvium. Between Chakdara and Dir, the Indus suture melange group nearly pinches out and the exposed section shows the MMT as a narrow fault zone along which amphibolites and hornblende gneisses of the Kohistan arc sequence have been thrust southward over the Saidu schists and marbles (fig. 2.3). Here the Saidu schist and Alpurai group overlie the Kalangai granite gneisses and Manglaur schist (?).

The above sequence may be traced westward to Khar Fort and beyond into Afghanistan. However, about 20 kilometers southeast of Khar Fort along the Swat River canyon, a north-

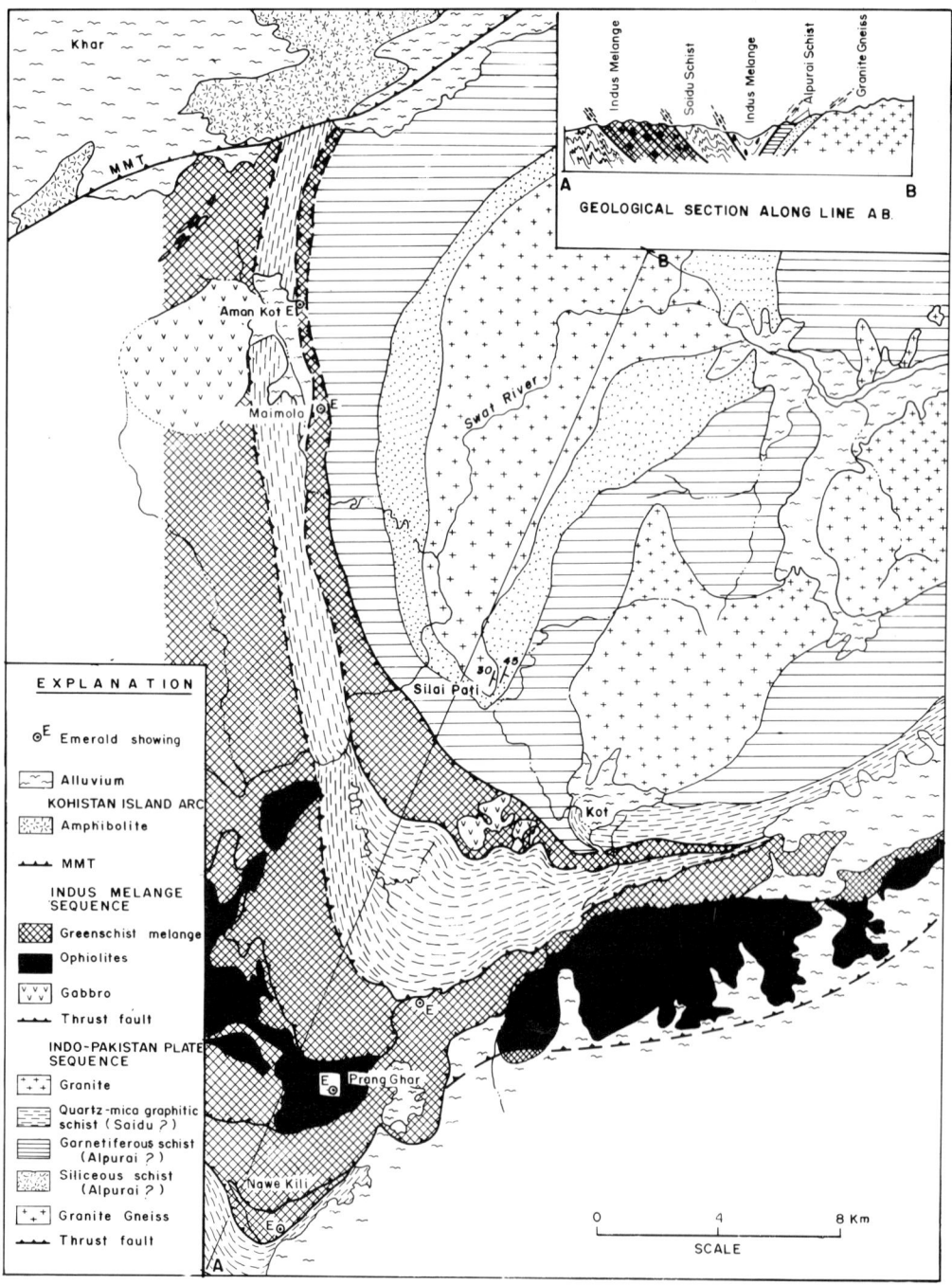

Figure 2.5. Geologic map of the Kot-Pranghar emerald mine area (after Hussain and others, 1984).

northeast-trending antiformal body of Kalangai granite gneiss and Manglaur schist(?) is exposed. It is overlain and surrounded by the north-trending Alpurai and Saidu schists over which a vast pile of melange complex has been thrusted (fig. 2.5) (Hussain and others, 1984; Rafiq, 1984).

The Kot-Pranghar melange complex consists of randomly oriented blocks of greenstone, greenschist, metavolcanics, serpentinite, pyroxenite, peridotite and talc-chlorite-dolomite schist. The complex is interspersed with a thick pile of graphite schist, garnet-mica schist and siliceous schists which are thrust slices of Saidu and Alpurai schists. It may be traced southward to beyond Pranghar and from there eastward to Dargai. Between Pranghar and Dargai it contains a large thrust slice of ophiolites previously referred to as the Dargai klippe (fig. 2.5).

South of Khar Fort the east-trending MMT truncates the north-trending Kot-Pranghar melange complex and the Saidu and Alpurai schists. This melange complex is in fact equivalent to the Mingora melange of the Alpurai-Chakdara segment of the MMT zone. The cross-cutting relationship of MMT with the Kot-Pranghar melange clearly confirms that the northern fault of the MMT (the Kohistan thrust) is a younger structural feature than the southern fault of the MMT (Kishora thrust). The distribution of the Kot-Pranghar melange suggests that this unit is involved in all of the folds affecting the Indo-Pakistan subcontinental sequence (fig. 2.3) although detailed structural mapping is not available in this area. The north-trending portion of the melange is probably caught in the core of an F_1 fold which is preserved on the western flank of an F_2 anticline. The southern outcrop area is preserved in the trough of an F_3 syncline. This structural interpretation implies a minimum of 40 kilometers of displacement (the distance moved prior to folding of these units).

The Kot-Pranghar melange complex contains emerald showings at several localities: notably at Aman Kot (Barang), Maimola, Khanori, Pranghar, Bucha, Nawe Dand and Gandao (Tora Tigga) (fig. 2.1).

CONCLUSIONS

The available geologic data are most informative about the origin of the emeralds in Swat, Malakand, and Mohmand. The difficult to reach, less studied Khaltaro deposit remains an anomaly in its possible non-ophiolite setting. However, it is close to locations where suture zone rocks may be slivers in the Raikot fault or MKT or are eroded ophiolite thrust sheets above the Nanga Parbat massif. Certainly the Nanga Parbat area has abundant beryllium-rich pegmatites from which the mineralizing fluids probably emanated.

The remaining emerald deposits in Pakistan are intrinsically associated with Mingora ophiolitic melange of the Indus suture zone. The ophiolitic rocks of this suture are the source of the chromium which colors the beryl to make emerald (Snee and others; Hammarstrom; this volume). The Indus suture zone in Pakistan is composed of a complex set of distinguishable melange units separated by major thrusts. Separate melange units initially developed in quite distinct areas of the Neotethys Ocean. The Mingora ophiolitic melange is oceanic ridge material that was obducted onto the southern margin of the Neotethys Ocean in the early stages of collision. On this northern margin of Gondwana, Paleozoic and Mesozoic(?) shelf sediments are unconformably underlain by Precambrian crystalline schists and Paleozoic granitic intrusions. The Mingora melange was involved in metamorphism and folding of these rocks during the early Tertiary and was present during intrusion of the tourmaline granites and younger granitic rocks. Part of the early collision during which the early folds developed was probably oblique producing strike-slip motion along the Kohistan fault on the northern margin of the MMT.

Emeralds formed late in the history of this area after major deformation had been completed as they are nowhere crushed, stretched or otherwise deformed. Emerald mineralization apparently relates to the final stages of the metasomatic alteration of the ophiolite melange to talc-dolomite melange. Beryl mineralization is typically associated with the late stages of cooling of granitic intrusive bodies and is a typical pegmatitic and hydrothermal mineral. The Swat granite gneiss is too old to be the source of mineralizing fluids, and the tourmaline granite is syntectonic with the metamorphism and hence also probably too old. Muscovite $^{40}Ar/^{39}Ar$ ages from metamorphic micas in the Alpurai schist of 30 m.y. (Maluski and Matte, 1984; Lawrence and others, 1985) reflect the waning of this metamorphism, and show temperatures (300°C) at the lower range of homogenization of the emerald fluid inclusions (Seal; this volume). The absence of granitic or pegmatitic veins in the melange, the presence of beryl pegmatites and tourmaline in the granitic rock near Kot and Pacha Fort and the occurrence of early Miocene anatectic granites in the Malakand Pass region and perhaps near Manglaur suggest that the mineralizing fluids came from these youngest granitic rocks. Emerald developed only in the Mingora melange which provided the source of the chromium.

Late metamorphic beryllium-bearing pneumatolitic and hydrothermal fluids passing through the older granitic rocks and overlying schists and marbles metasomatized the chromium-bearing ophiolite. These exhalations found easy access through the porous and permeable talc-dolomite melange matrix, especially near the normal faults where they deposited exquisite, verdant green emerald crystals through an intricate sequence of events.

ACKNOWLEDGMENTS

We acknowledge financial support from NSF INT 81-18403 and 86-09914 and NSF EAR 86-17543 for field and laboratory studies. Laboratory support by the U.S. Geological Survey and the Oregon State University is appreciated. Encouragement and support by Brigadier Kaleem Ur Rahman Mirza, Chairman, Gemstone Corporation of Pakistan, Waheeduddin Ahmed, Director-General, Geological Survey of Pakistan, and R.A.K. Tahirkheli, Director, National Center of Excellence in Geology were very valuable. Lawrence and Kazmi both benefited from Fulbright scholarships during the course of this work. R.A.K. Tahirkheli, A.A.K. Ghauri, M.Q. Jan, M. Majid, K.A. DeJong, J. Anwar, S.S. Hussain, A. Hussain, S. Rehman, all were generous in discussions of the geology and in field excursions. Oregon State University and University of Peshawar students whose thesis projects have contributed to this work include S. Carter, J. Di Pietro, Imtiaz Ahmad, Shafiq Ur-Rehman, C. Kazmer, P. Verplanck, J. LaFortune, I. Madin, and P. Palmer-Rosenberg. Field assistance to these students was provided by Gulzar Aziz. Mohammed Riaz. and Munir Hamayun. Field work by Gemstone Corporation geologists Javaid Anwar, Shahid Hussain, Tahseenullah Khan, and H. Dawood contributed to our understanding of these gemstone deposits. Critical reviews and comments by J. Di Pietro and K. Pogue were helpful.

REFERENCE

Ahmad, I., 1986, Geology of Jowar area, Karakar Pass, Swat District, N.W.F.P., Pakistan: Unpublished M Phil thesis, University of Peshawar, 144p.

Ashraf, M., Chaudhry, N.M., and Hussain, S.S., 1980, General geology and economic significance of the Lahor granite and rocks of southern ophiolite belt in Allai Kohistan area: University of Peshawar, Geological Bulletin, v. 13, p. 207-213.

Bard, M.A., Maluski, H., and Proust, F., 1980, The Kohistan sequence: crust and mantle of an obducted island arc: University of Peshawar, Geological Bulletin, v. 13. p. 87-94.

Bard, J.P., 1983, Metamorphic evolution of an obducted island arc: Example of the Kohistan sequence (Pakistan) in the Himalayan collided range: University of Peshawar, Geological Bulletin, v. 16, p. 105-184.

Butt, K. A., 1983, Petrology and geochemical evolution of Lahor pegmatoid-granite complex, northern Pakistan, and genesis of associated Pb-Zn-Mo and U mineralization: *in* F.A. Shams, editor, *Granites of the Himalayas, Karakoram and Hindu Kush,* Institute of Geology, Punjab University, Lahore, Pakistan, p. 309-326.

Butt, K.A., and Shah, Z., 1985, Discovery of blue beryl from Ilum granite and its implications of the genesis of emerald mineralization in Swat District: University of Peshawar, Geological Bulletin, v. 18, p. 75-81.

Calkins, J.A., Offield, T.W., Abdullah, S.K.M., and Ali, S.T., 1975, Geology of the southern Himalayas in Hazara: U.S. Geological Survey, Professional Paper 716-C, 29p.

Carter, S., in prep., Petrology and structure of the Swat granite gneisses near Manglaur, Swat, Pakistan: Unpublished M.S. thesis, Oregon State University, Corvallis, Oregon.

Chaudhry, M.N., Jafferi, S.A., and Saleemi, B.A., 1974, Geology and petrology of the Malakand granite and its environs: Punjab University, Geological Bulletin, v. 10, p. 43-58.

Chaudhry, M.N., Ashraf, M., Hussain, S.S., and Iqbal M., 1976, Geology and petrology of Malakand and a part of Dir (Toposheet 38 N/14): Punjab University, Geological Bulletin, v. 12, p. 17-40.

Coward, M.P., Windley, B.F., Broughton, R.D., Luff, I.W., Peterson, M.G., Pudsey, C.J., Rex, D.C., and Asif Khan, M., 1986, Collision tectonics in the NW Himalaya: *in* M.P. Coward and A.C. Ries, editors, *Collision Tectonics,* Geological Society Special Publication 19, Blackwell Scientific Publications, London, England, p. 203-219.

Coward, M.P., Butler, R.W.H., Asif Khan, M., and Khipe, R.H., 1987, The tectonic history of Kohistan and its implications for Himalayan structure: Geological Society of London, Journal, v. 144, p. 377-391.

Desio, A., and Shams, F.A., 1980, The age of the blueschists and the Indus-Kohistan suture line, NW Pakistan: Accademia Nazionale dei Lincei, s. 8, v. 68, p. 74-79.

Fletcher, C.J. N., Leake, R.C., and Haslam, H.W., 1986, Tectonic setting, mineralogy, and chemistry of a metamorphosed stratiform base metal deposit within the Himalayas of Pakistan: Geological Society of London, Journal, v. 143, p. 521-536.

Frank, W.A., Gansser, A., and Trommsdorff, V., 1977, Geological observations in the Ladakh area (Himalayas): Schweizerische Mineralogische und Petrographische Mitteilungen, v. 57, p. 89-113.

Gansser, A., 1964, *Geology of the Himalaya:* Wiley Interscience, New York, 289 p.

Gansser, A., 1981, The geodynamic history of the Himalaya: *in* H.K. Gupta and F. Delany, editors, *Zagros-Hindu Kush-Himalaya Geodynamic Evolution,* American Geophysical Union Geodynamics Series 3, Washington D.C., p. 111-121.

Hammarstrom, J.M., 1989, Mineral chemistry of emeralds and some associated minerals from Pakistan and Afghanistan: An electron microprobe study: *in* A.H. Kazmi and L.W. Snee, editors, *Emeralds of Pakistan: Geology, Gemology, and Genesis,* Van Nostrand Reinhold, New York, p. 125-150.

Hussain, S. Khan, T., Dawood, H., and Khan, I., 1984, A note on Kot-Prang Ghar melange and associated mineral occurrences: University of Peshawar, Geological Bulletin, v. 17, p. 61-68.

Jan, M.Q., 1977, The Kohistan basic complex, a summary based on recent petrological research: University of Peshawar, Geological Bulletin, v., 9-10, p. 36-42.

Jan, M.Q., 1979, Petrography of the Jijal complex, Kohistan: University of Peshawar, Geological Bulletin, v. 11, p. 31-50.

Jan M.Q., 1980, Petrology of the obducted mafic and ultramafic metamorphites from the southern part of the Kohistan island arc sequence: University of Peshawar, Geological Bulletin, v. 13, p. 95-107.

Jan, M.Q., 1985, High-P rocks along the suture zones around Indo-Pakistan plate and phase chemistry of blue schists from eastern Ladakh: University of Peshawar, Geological Bulletin, v. 18, p. 1-40.

Jan M.Q., and Symes, R.F., 1977, Piedmontite schist from upper Swat, N.W. Pakistan: Mineralogical Magazine, v. 41, p. 537-540.

Jan M.Q., and Howie, R.A., 1981, The mineralogy and geochemistry of the metamorphosed basic and ultrabasic rocks of the Jijal complex, Kohistan, northwest Pakistan: Journal of Petrology v. 22, p. 85-126.

Jan, M.Q., Asif, M., Tahirkheli, T., and Kamal, M., 1981a, Tectonic subdivision of granite rocks of northern Pakistan University of Peshawar, Geological Bulletin, v. 14, P. 159-182.

Jan, M.Q., Kamal, M., and Khan, M.I., 1981b, Tectonic control over emerald mineralization in Swat: University of Peshawar, Geological Bulletin, v. 14, p. 101-109.

Kazmer, C., Hussain, S.S., and Lawrence, R.D., 1983, The Kohistan-Indian plate suture zone at Jawan Pass, Swat, Pakistan: Geological Society of America, Abstracts with programs, v. 15, p. 609.

Kazmi, A.H., 1983, Report on the economic geology and development prospects of Swat emerald mines, Mingora: Pakistan Gemstone Corporation, Unpublished report, 105p.

Kazmi, A.H., 1989, A brief overview of the geology and metallogenic provinces of Pakistan: *in* A.H. Kazmi and L.W. Snee, editors, *Emeralds of Pakistan: Geology, Gemology, and Genesis*, Van Nostrand Reinhold, New York, p. 1-12.

Kazmi, A.H., Lawrence, R.D., Dawood, H., Snee, L.W., and Hussain S.S., 1984, Geology of the Indus suture zone in the Mingora-Shangla area of Swat: University of Peshawar, Geological Bulletin, v. 17, p. 127-144.

Kazmi, A.H., Peter, J.J., and Obodda, H.P., 1985, Gem pegmatites of the Shengus Dusso area, Gilgit, Pakistan: The Mineralogical Record, v. 16, p. 393-411.

Kazmi, A.H., Lawrence, R.D., Anwar, J., Snee, L.W., and Hussain, S., 1986, Mingora emerald deposits (Pakistan): Suture-associated gem mineralization: Economic Geology, v. 81, p. 2022-2028.

Kazmi, A.H., Anwar, J., Hussain S., Khan, T., and Dawood, H., 1989, Emerald deposits of Pakistan: *in* A.H. Kazmi and L. W. Snee, editors, *Emeralds of Pakistan: Geology, Gemology, and Genesis*, Van Nostrand Reinhold, New York, p. 39-74.

LaFortune, J., 1988, Geology and geochemistry of rocks beneath the Indus suture, Besham area, N. Pakistan: Unpublished M.S. thesis, Oregon State University, Corvallis, Orgeon.

Lawrence, R.D., 1982, West end Tibetan collision zone in Pakistan, American Geological Union, EOS, v. 63, p. 1112.

Lawrence R.D., 1984, Suture tectonics of Pakistan: First Pakistan Geological Congress, 28-31 October 1984, Volume of Abstracts, p. 49.

Lawrence, R.D., and Ghauri, A.A.K., 1983a, Evidence of active faulting in Chilas district, northern Pakistan: University of Peshawar, Geological Bulletin, v. 16, p. 185-186.

Lawrence, R.D., and Ghauri, A.A.K., 1983b, Observations on the structure of the Main Mantle thrust at Jijal, Kohistan, Pakistan: University of Peshawar, Geological Bulletin v. 16, p. 1-10.

Lawrence, R.D., and Ghauri, A.A.K., 1984, Tectonics of the western Indus suture in Pakistan; EOS, v. 65, p. 1094.

Lawrence, R.D., Kazmer, C., and Tahirkheli, R.A.K., 1983, The Main Mantle Thrust: a complex zone, northern Pakistan: Geological Society of America, Abstracts with Programs, v. 15, p. 624.

Lawrence, R.D., and Shroder, J.S., 1984, Active fault northwest of Nanga Parbat: First Pakistan Geological Congress, 28-31 October 1984, Volume of Abstracts, p. 50.

Lawrence, R.D., Snee, L.W., and Rosenberg, P.S., 1985, Nappe structure in a crustal scale duplex in Swat, Pakistan: Geological Society of America, Abstracts with Programs, v. 17. p. 640.

LeFort, P., 1975, Himalayas: the collided range: American Journal of Science, v. 275A, p. 1-44.

LeFort, P., Debon, F., Sonet, J., 1980, The "Lesser Himalayan" cordierite granite belt, typology and age of the pluton of Mansehra, Pakistan: University of Peshawar, Geological Bulletin, v. 13, p. 51-62.

Madin, I., 1986, Structure and neotectonics of the northwestern Nanga Parbat-Haramosh Massif: Unpublished M.S. thesis, Oregon State University, Corvallis, Oregon, 160 p.

Madin, I., Lawrence, R.D., and Rehman, S., 1984, Neotectonics of the western Haramosh Range: American Geophysical Union, EOS, v. 65, p. 1094.

Madin, I., Lawrence, R.D., and Rehman, S., (in press), Geology and structure of the northwestern Nanga Parbat-Haramosh massif; crustal uplift along a terminal tear fault on the MCT?: *in* L.L. Malinconico and R.J. Lillie, editors, *Tectonics and Geophysics of the Western Himalaya*, Geological Society of America, Special Paper, Boulder, Colorado.

Majid, M., and Paracha, F.A., 1980, Calc-alkaline magmatism at destructive plate margin in Kohistan, northern Pakistan: University of Peshawar, Geological Bulletin, v. 13, p. 109-120.

Majid, M. and Tahir Shah, M., 1985, Mineralogy of the blueschist facies metagraywacke from the Shergarh

Sar area, Allai Kohistan, N. Pakistan: University of Peshawar, Geological Bulletin v. 18, p. 41-52.

Maluski, H., and Matte, P., 1984, Ages of alpine tectonometamorphic events in the northwestern Himalaya (northern Pakistan) by ^{39}Ar-^{40}Ar method: Tectonics, v. 3, p. 1-18.

Martin, N.R., Siddiqui, S.F.A., and King, B.H., 1962, A geological reconnaissance of the region between the lower Swat and Indus Rivers of Pakistan: University of Punjab, Geological Bulletin, v. 2, p. 1-14.

Misch, P., 1949, Metasomatic granitization of batholithic dimensions: American Journal of Science, v. 247, p. 209-245.

Misch, P., 1936, Einiges zur metamorphose des Nanga-Parbat: Geologische Rundschau, v. 27, p. 79-81.

Palmer-Rosenberg, P.S., 1985, Himalayan deformation and metamorphism of rocks south of the Main Mantle thrust, Karakar Pass area, southern Swat, Pakistan: unpublished M.S. thesis, Oregon State University, Corvallis, Oregon, 68 p.

Pogue, K., and Hussain, A., 1986, New light on the stratigraphy of Nowshera area and the discovery of Early to Middle Ordovician trace fossils in N.W.F.P., Pakistan: Geological Survey of Pakistan, Information Release No. 135, 15 p.

Rafiq, M., 1984, Extension of Sakhakot-Qila ultramafic complex in Utman Khel, Mohmand Agency, N.W.F.P., Pakistan: University of Peshawar, Geological Bulletin, v. 17, p. 53-59.

Rosenberg, P.S., Lawrence, R.D., and Khan, I.A., 1984, Deformation and metamorphism of rocks south of the Main Mantle thrust in southern Swat, Pakistan: American Geophysical Union, EOS, Pakistan: v. 65, p. 1094.

Seal, R. R. II., 1989, A reconnaissance study of the fluid inclusion geochemistry of the emerald deposits of Pakistan and Afghanistan: *in* A.H. Kazmi and L.W. Snee, editors, *Emeralds of Pakistan: Geology, Gemology and Genesis,* Van Nostrand Reinhold, New York, p. 151-164.

Searle, M.P., 1983, Stratigraphy, structure and evolution of the Tibetan-Tethys zone in Zanskar and the Indus suture zone in the Ladakh Himalaya: Royal Society Edinburg Transactions, Earth Sciences, v. 73, p. 205-219.

Sharma, K.K., 1983, Granitoid belts of the Himalaya: *in* F.A. Shams, editor, *Granites of Himalayas, Karakorum and Hindu Kush,* Institute of Geology, Punjab University, Lahore, Pakistan, p. 11-37.

Shams, F.A., 1963, Reactions in and around a calcareous xenolith lying within the granite-gneiss of Manglaur, Swat State, West Pakistan: Punjab University, Geological Bulletin, v. 3, p. 7-18.

Shams, F.A., 1969, The geology of the Mansehra-Amb State area, northern West Pakistan: Punjab University, Geological Bulletin, v. 8, p. 1-31.

Shams, F.A., 1972, Glaucophane-bearing rocks from near Topsin, Swat, first record from Pakistan: Pakistan Journal of Science, v. 24, p. 343-345.

Shams, F.A., 1980, Origin of the Shangla blueschists, Swat Himalaya, Pakistan: University of Peshawar, Geological Bulletin, v. 13, p. 67-70.

Shams, F.A. 1983, Granites of the NW Himalayas in Pakistan: *in* F.A. Shams, editor, *Granites of Himalayas, Karakorum and Hindu Kush,* Institute of Geology, Punjab University, Lahore, Pakistan, p. 341-354.

Shroder, J.F., Saqib Khan, M., Lawrence, R.D., Madin, I.P., and Higgins, S.M., (in press), Quaternary glacial chronology and neotectonics in the Himalaya of northern Pakistan: *in* L.L. Malinconico and R.J. Lillie, editors, *Tectonics and Geophysics of the Western Himalaya,* Geological Society of American, Special paper, Boulder, Colorado.

Snee, L.W., Foord, E.E., Hill, B., and Carter, S.J., 1989, Regional chemical differences among emeralds and host rocks of Pakistan and Afghanistan: implications for the origin of emerald: *in* A.H. Kazmi and L.W. Snee, editors, *Emeralds of Pakistan: Geology, Gemology, and Genesis,* Van Nostrand Reinhold, New York, p. 93-124.

Tahirkheli, R.A.K., 1979a, Geotectonic evolution of Kohistan: University of Peshawar, Geological Bulletin, v. 11, p. 113-130.

Tahirkheli, R.A.K., 1979b, Geology of Kohistan and adjoining Eurasian and Indo-Pakistan continents, Pakistan: University of Peshawar, Geological Bulletin, v. 11, p. 1-30.

Tahirkheli, R.A.K., 1983, Geological evolution of Kohistan island arc on the southern flank of the Karakoram-Hindu Kush in Pakistan: Bollettino Geofisica Teorica Applicata, v. 25, p. 351-364.

Tahirkheli, R.A.K., and Jan, M.Q., 1979, A preliminary geological map of Kohistan and adjoining areas, N. Pakistan, 1:1,000,000 University of Peshawar, Geological Bulletin, v. 11, in pocket.

Tahirkheli, R.A.K., Mattauer, M., Proust, F., and Tapponier, P.,1979, The India-Eurasia suture zone in northern Pakistan: synthesis and interpretation of recent data at plate scale: *in* A. Farah and K.A. DeJong, editors, *Geodynamics of Pakistan,* Geological Survey of Pakistan, Quetta, Pakistan, p. 125-130.

Thakur, V.C., 1981, Regional framework and geodynamical evolution of the Indus-Tangpo suture zone in the Ladakh Himalayas, Royal Society of Edinburgh Transactions, v. 72, p. 89-97.

Thakur, V.C., 1983, Deformation and metamorphism of the Tso Morari crystalline complex: *in* V.C. Thakur and K.K. Sharma, editors, *Geology of Indus Suture Zone of Ladakh,* Wadia Institute of Geology, Dehra Dun, India, p. 1-8.

Thakur, V.C., and Bagati, T.N., 1983, Indus Formation: an arc trench gap sediments: *in* V.C. Thakur and K.K. Sharma, editors, *Geology of the Indus Suture Zone of Ladakh,* Wadia Institute of Himalayan Geology, Dehra Dun, India. p. 9-19.

Valdiya, K.S., 1981, Tectonics of the central sector of the Himalaya: *in* Gupta and F. Delany, editors, *Zagros-Hindu Kush-Himalaya Geodynamic Evolution,* American Geophysical Union, Geodynamics ser. 3, Washington D.C., p. 87-110.

Valdiya, K.S., 1983, Tectonic setting of Himalayan granites: *in* F.A. Shams, editor, *Granite of Himalayas, Karakorum and Hindu Kush,* Institute of Geology, Punjab University, Lahore, Pakistan p. 39-53.

Verplanck, P.L., Snee, L.W., and Lund, K., 1985, The boundary between the Nanga Parbat massif and the Ladakh island arc terrain, northern Pakistan; a cross fault on the Main Mantle Thrust: American Geophysical Union, EOS, v. 66, p. 1074.

Verplanck, P., 1987, Petrology and structure of the eastern portion of the Nanga Parbat-Haramosh massif along the Indus River, northern Pakistan, with special attention to the gem-bearing pegmatites: Unpublished M.S. thesis, Oregon State University, Corvallis, Oregon, 138 p.

Wadia, D.N., 1933, Note on the geology of Nanga Parbat (Mt. Diamir) and adjoining portions of Chilas, Gilgit District, Kashmir: Geological Survey of India, Records v. 66, p. 212-234.

Yeats, R.S., and Lawrence, R.D., 1984, Tectonics of the Himalayan thrust belt in northern Pakistan: *in* B.U. Haq and J.D. Milliman, editors, *Marine Geology and Oceanography of Arabian Sea and Coastal Pakistan,* Van Nostrand Reinhold, New York, p. 177-198.

Zeitler, P.K., 1985, Cooling history of the NW Himalaya, Pakistan: Tectonics, v. 4, p. 127-151.

Zeitler, P.K., Johnson, N.M., Waeser, G.W., and Tahirkheli, R.A.K., 1982a, Fission-track evidence for Quaternary uplift of the Nanga Parbat region, Pakistan: Nature, v. 298, p. 255-257.

Zeitler, P.K., Tahirkheli, R.A.K., Waeser, G.W., and Johnson, N.M., 1982b, Unroofing history of a suture zone in the Himalaya of Pakistan by means of fission-track annealing ages: Earth and Planetary Science Letters, v. 57, p. 227-240.

Emerald Deposits of Pakistan

Ali H. Kazmi, Javed Anwar, Shahid Hussain,
Tahseenullah Khan, and Hamid Dawood

INTRODUCTION

Occurrence of emerald in Pakistan initially became known in 1958 when the Mingora deposit was discovered near the city of Mingora in Swat District. The occurrence was first reported by Davies (1962). Jan (1968) and Jan and others (1981b) discussed the petrography of the deposit. W. Ahmed examined the reported occurrence of emerald at Nawe Dand, in Mohmand Agency in 1966. A comprehensive gemological description of the Mingora emeralds was given by Gübelin (1968, 1982) who reported that "the best specimens were of a good to excellent quality and were outstanding for their vividness, transparency, and saturated color which immediately called to mind a comparison with the prized Muzo emeralds from Colombia". Nevertheless, for more than 20 years Mingora emeralds were clandestinely sold in the world markets by names other than Mingora (or Pakistan). They were, however, officially introduced in the world market when the Gemstone Corporation of Pakistan (GEMCP) exhibited a large consignment of cut and polished stones in the USA at the Tuscon Gem and Mineral Show and the Dallas Gem and Jewellery Show in 1981 (Herzberg, 1981). They have been since commonly known as the Pakistan or Swat emeralds.

Soon after the discovery of emeralds at Mingora, more deposits were discovered at Charbagh and in the tribal areas of the Mohmand Agency at Gandao (Tora Tigga). With the establishment of the government-owned Gemstone Corporation of Pakistan in 1979, an extensive geological exploration program was carried out* which resulted in the discovery of new emerald showings at Pranghar, Bucha, and Khanori in Malakand and Mohmand Agencies (Hussain and others, 1984) and at Aman Kot and Maimola (Barang) in Bajaur Agency and brought to light new emerald deposits at Charbagh, Makhad, and Gujarkili in the Swat District and at Khaltaro in the Gilgit Agency (fig. 2.1).

The best and the largest emerald deposit is at Mingora, followed by the Gujarkili deposit, both of which are being regularly mined by the Gemstone Corporation of Pakistan. Gem-quality green beryl has been produced from Gandao and emerald from Barang area (Aman Kot and Maimola), but these deposits have been sporadically worked by the local tribal chieftains and the supply has been irregular.

The Khaltaro deposit has not yet been opened up, though it promises to be a good sizeable deposit. Other emerald localities mentioned above are largely in the form of small showings with unproven reserves and uncertain prospects of large-scale development. A detailed geological description of the major emerald deposits is given in the following pages; the deposits are discussed in order of their importance and significance.

---

*Under the guidance and supervision of the senior author as Technical Director of the Corporation, assisted by Javed Anwar, Chief Geologist, Shahid Hussain, Geologist, Tahseenullah Khan, Hamid Dawood, Khalid Aziz, Abid Murtaza, and Bakhtiar, Assistant Geologists.

THE MINGORA EMERALD DEPOSIT

The Mingora emerald deposit is located in the picturesque valley of the Swat River amongst the lower hill ranges southeast of the Hindu Kush Mountains (photos. 3.1 and 3.2). The deposit occurs on the northern edge of Mingora (Swat District), 200 km northeast of Peshawar (fig. 2.1). These are the only urban gem deposits in the world, not only located in close proximity to a city but gradually being surrounded by it. The emerald deposit spreads over an area of about 180 acres. The city is now spreading rapidly along the perimeter of the mine area (photo. 3.3). Presently emerald is being mined from five points which, from south to north, are known as Islamia Trench, Farooq Mine or Mine 1, Carrel's Trench, Mine 2, and Mine 3 (fig. 3.1).

Geology

The Mingora emerald deposit occurs in the Mingora ophiolitic melange (Kazmi and others, 1986; Lawrence and others, this volume) which is a highly tectonized fault block. It has a north-south strike and dips westward. To the west the Charbagh greenschist melange has been thrust over this block along the Makhad thrust, whereas eastward, the Kishora thrust fault has pushed the greenschist melange over the Saidu schist (fig. 2.4). The Charbagh greenschist melange is largely composed of fine-grained, unfoliated to poorly foliated metabasalts and well-foliated metapyroclastic rocks. The Makhad thrust, beneath the Charbagh greenschist melange, is marked by phacoids of metagabbroic and metadioritic rocks. Metasomatic exchange with the Mingora ophiolitic melange has occurred since the motion on the thrust fault. The zone is marked by black-rock metamorphism with the development of chlorite-serpentinite-biotite schists with local large undeformed porphyroblasts of actinolite, biotite, and spinel. The fault is much obscured by this later metamorphism.

The Saidu calc-graphite schist occupies the area east of the mine and has been described earlier (Lawrence and others, this volume). In the eastern part of the mine, the main slab of the Mingora ophiolitic melange overlies the zone of imbrication in which wedges of ophiolitic melange and calc-graphite schist are repeated many times (fig. 2.4).

Emerald mineralization occurs only in the Mingora ophiolitic melange, making this a particularly crucial unit from an economic point of view. In the mine area it occurs in the form of a talc-dolomite melange. This rock unit is described in some detail in the following paragraphs.

Talc-Dolomite-Schist Melange

This unit is composed of a thick, heterogeneous mass of talc-dolomite schist containing tectonized clasts and phacoids of serpentinite, dolomite, greenschist, and graphitic schist, ranging from less than 2 centimeters to hundreds of meters in size. These clasts are set in a fine-to coarse grained, ductilely deformed matrix.

The talc-dolomite schist melange is medium to fine grained, grayish green, and massive to foliated. Clasts of serpentinite occur abundantly in the upper and western part of the melange zone and are practically absent in the lower and eastern part (fig. 3.1). At several points they form a cluster of closely spaced, relatively fresh blocks 5 to 10 meters across and similar in color and texture suggesting fragmentation from larger pieces subsequent to their incorporation in the melange. Some of the clasts are relatively fresh, while others show varying degrees of alteration. As seen under the microscope, some of the clasts entirely comprise a felted mass of serpentinite (photo. 3.4) showing minor alteration to chlorite and talc along microscopic shear planes. These

Photo.3.1. A view of the Mingora emerald mine 2, looking north (1982, photo *A. H. Kazmi*).

Photo.3.2. A view of the Mingora emerald mine 2, looking southeast along the mineralized zone (1980, photo *A. H. Kazmi*).

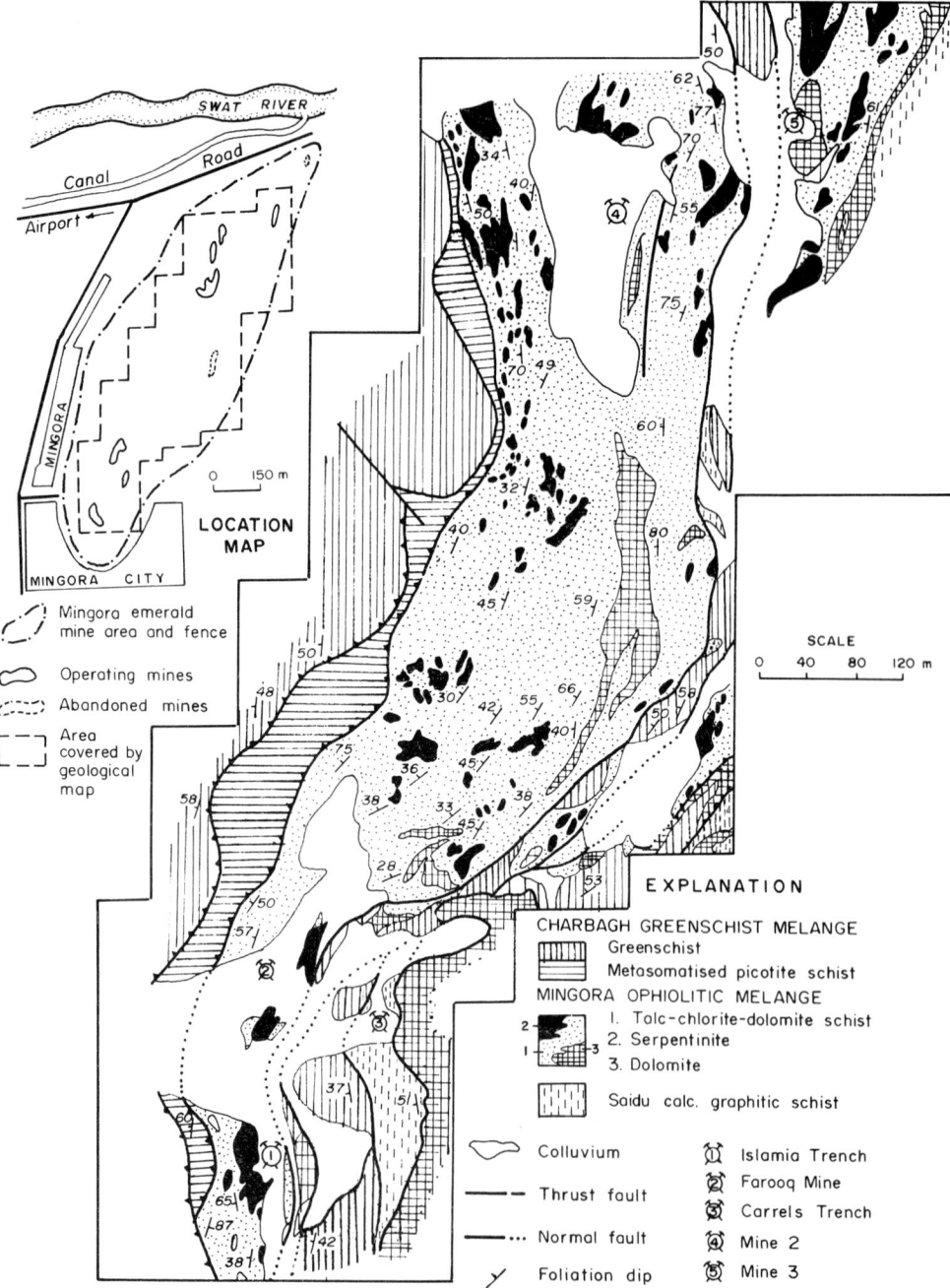

Figure 3.1. Geological map of Mingora emerald mines.

Photo.3.3. View of a part of Mingora emerald mine (foreground) and Mingora, city (1980, photo *A. H. Kazmi*).

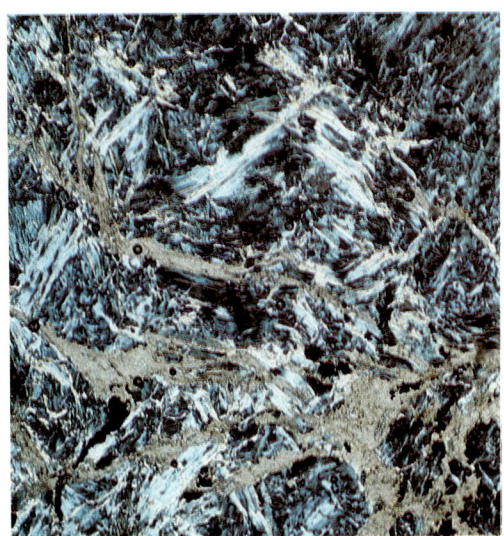

Photo.3.4. Photomicrograph of Mingora ophiolitic melange showing felted mass of serpentinite and talc cut by carbonate veins, field 2.3 x 3.3 mm. (photo *A. H. Kazmi*).

Photo.3.5. Photomicrograph of Mingora Ophiolitic melange showing tremolite-chlorite schist from metasomatised zone, field 4.6 x 6.6 mm. (photo *L. W. Snee*).

shear planes have been consistently calcitized. Commonly in the field and in a few thin sections, relicts and pseudomorphs of olivine and pyroxene may be seen. In others, extensive alteration to chlorite and talc is noted and at some points tremolite developed (photo. 3.5). Such rocks display an incipient foliation and may be referred to as semischists. Altered serpentinite contains ideoblastic to subideoblastic porphyroblasts of dolomite (photo. 3.6). Opaque ores (chromite, ilmenite, and/or magnetite) are common.

Greenschist

The greenschist of the Mingora ophiolitic melange is green to dark green, medium to coarse grained, and occurs as tectonized lenses and slivers in the talc-dolomite-schist melange. Greenschist clasts bear a granoblastic (photo. 3.7) to schistose texture and range in composition from chlorite-antigorite schist (metadunite) to actinolite-tremolite-chlorite schist (metavolcanic) to albite-biotite-clinozoisite-sphene semischist (metadiabase) to albite-chlorite-quartz granofels (metagabbro—photo. 3.8).

Dolomite

In the melange, the dolomitic clasts are cream gray to light grayish brown, medium to coarse grained, and vary from a few centimeters to hundreds of meters in length. They are largely composed of dolomite grains with lesser amounts of quartz, fuchsite, and talc. Commonly they are sheared and contain stockworks of milky white quartz veins (photo. 3.9). At places, for example in Mine 3, the stockwork in the dolomite contains thin veins of chlorite and fuchsite in which chrome tourmaline and emerald are found (photo. 3.10). In thin sections of such rocks, fine ideoblastic crystals of emeralds are easily seen. There is pervasive alteration of the dolomite to ferruginous dolomite and siderite.

Talc-Chlorite-Dolomite Schist

This is the most prominent rock in the Mingora emerald mines. It is a grayish to greenish white, greenish gray or greenish brown, fine- to medium-grained, soft, well foliated and sheared rock that forms the matrix of the talc-dolomite melange. Talc, chlorite, and dolomite are the main constituents (photo. 3.11). Fuchsite, quartz, pyrite, antigorite, actinolite, chromite, and magnetite occur in lesser amounts. The rock color is usually a good indicator of the dominant mineral constituent. Thus light whitish gray rocks contain abundant talc, the light brown variety is rich in dolomite, the light green schist is full of chlorite, the darker green type contains serpentine, and the darker brown schist is ferruginous. More commonly the rock has a variegated, mottled appearance and contains lenticular specks of various shades of gray, green, and brown up to 3 by 10 mm in size. This microscopic cataclastic texture mimics the melange fabric of the talc-dolomite melange as a whole.

In this schist the talc-to-dolomite ratio varies from place to place. Rarely quartz is present in the groundmass. It is more common as veins and minor stockworks where it is milk white in color. Dolomite occurs pervasively as single crystals in the groundmass and as polycrystalline phacoids a few millimeters to 2-3 meters in length arranged along the foliation (photo. 3.12). Cream-colored calcite nodules are commonly present but cut across the foliation. Several small limonitized faults and shear planes pervade the outcrops of this schist.

Thin sections of the talc-chlorite-dolomite schist show that the rock contains 40-60% talc

Photo.3.6. Photomicrograph of talc-dolomite schist, note dolomite porphyroblast in serpentinite, field 4.6 x 6.6 mm. (photo *A. H. Kazmi*).

Photo.3.7. Photomicrograph of Charbagh green schist melange (metabasalt) with porphyroblast of epidote in a matrix of plagioclase and epidote, field 2.3 x 3.3 mm. (photo *L. W. Snee*).

Photo.3.8. Photomicrograph of a block of metagabbro in green schist melange, field 4.6 x 6.6 mm. (photo *A. H. Kazmi*).

Photo.3.9. Quartz vein with ankerite and emeralds, from stockwork of Mingora mine 3 (photo *A. H. Kazmi*).

Photo.3.10. Photomicrograph showing green chrome tourmaline (hexagonal), in fuchsite vein (green) in a dolomitic matrix, field 2.3 x 3.3 mm. (photo *A. H. Kazmi*).

Photo.3.11. A close up view of the talc-chlorite-dolomite schist; talc light grey, chlorite and serpentine darker grey, dolomite yellowish brown and siderite brown (photo *A. H. Kazmi*).

with chlorite arranged parallel to the foliation. Relict grains of serpentine (in various stages of alteration) are common. Some of these are large enough to be considered relict clasts of serpentinite or peridotite. Dolomite crystals comprise up to 30% of the rock. They are ideoblastic to subideoblastic, with sharp straight edges and commonly cut across the talc and chlorite fibers giving a porphyroblastic texture to the rock (photo. 3.6). Opaque ores (Fe and Cr) are common (5%).

In some thin sections dolomite is seen as microscopic lenticular clasts 2 to 3 mm across and composed of aggregates of dolomite crystals, crushed and arranged parallel to the rock foliation (photo. 3.12). The crystals in the aggregate show fracturing and slippage along one of the cleavages (oblique to the foliation) resulting in a fibrous structure, showing gradual alteration from one end to the other and formation of talc in optical continuity with these fibers. These microscopic clasts are surrounded by iron stains, clusters of minute specks of iron oxide, and a reaction rim which has formed talc fibers growing normal to the outline of the dolomite clasts. Amidst the felted mass of talc fibers, small patches of antigorite also occur. At places the dolomite in these clasts is altered to siderite. Part of the dolomite is therefore early and deformed; it was incorporated as microscopic clasts in the talc-dolomite melange. Part of the dolomite is late, metasomatic, and ideoblastic to subideoblastic.

Structure

The schistose rocks in the mine area have a general northerly strike with the foliation dipping 25° to 75° westward. Several westerly dipping thrust fault have cut these rocks into several tectonic slices (fig. 2.4). A set of two parallel, north-trending normal faults runs through the mine area, straddling Mine 3, Carrels trench, Farooq mine, and Islamia trench (fig. 3.1). A set of conjugate normal faults extends from these larger faults through the Islamia trench. Another set of two normal faults, parallel to the main faults runs through the Mine 2 area. These normal faults are believed to have been the principal avenues for movement of mineralizing fluids.

Geologic Controls to Emerald Mineralization

In the Mingora mines, emerald mineralization is confined to the talc-dolomite melange (fig. 3.2). It is neither a regular nor a stratabound deposit, but largely occurs in disseminations. The emeralds are commonly associated with one or more of the following features:

1. Faults and fractures.
2. Limonite zones.
3. Calcite nodules and veinlets.
4. Quartz veins and stockworks.

At the Mingora mines the following four distinct modes of emerald occurrence are present.

1. Emeralds Along Shear Planes

This type of emerald occurrence is characteristic of the central quarry of Mine 2 (fig. 3.3). Here host rock is composed of gray to greenish gray sheared talc-chlorite-dolomite schist which is interspersed with talc-rich and dolomite-rich layers and lenses. Small shear joints, with no

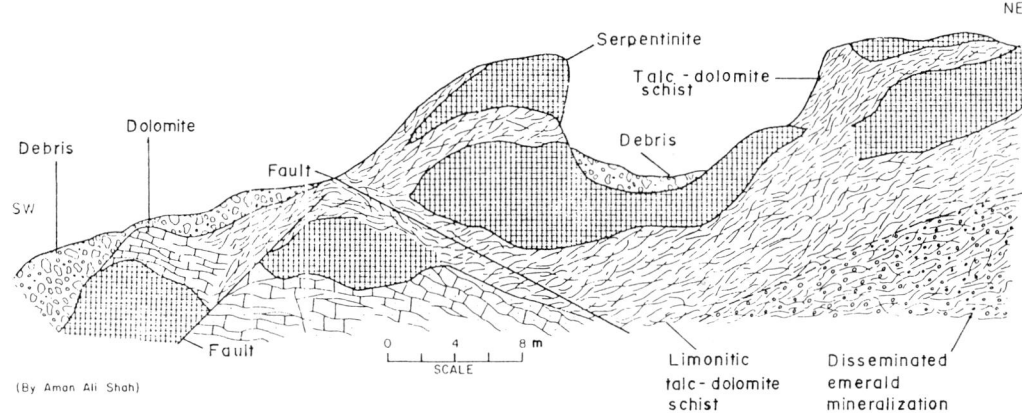

Figure 3.2. Sketch showing geological section at Farooq mine, Mingora.

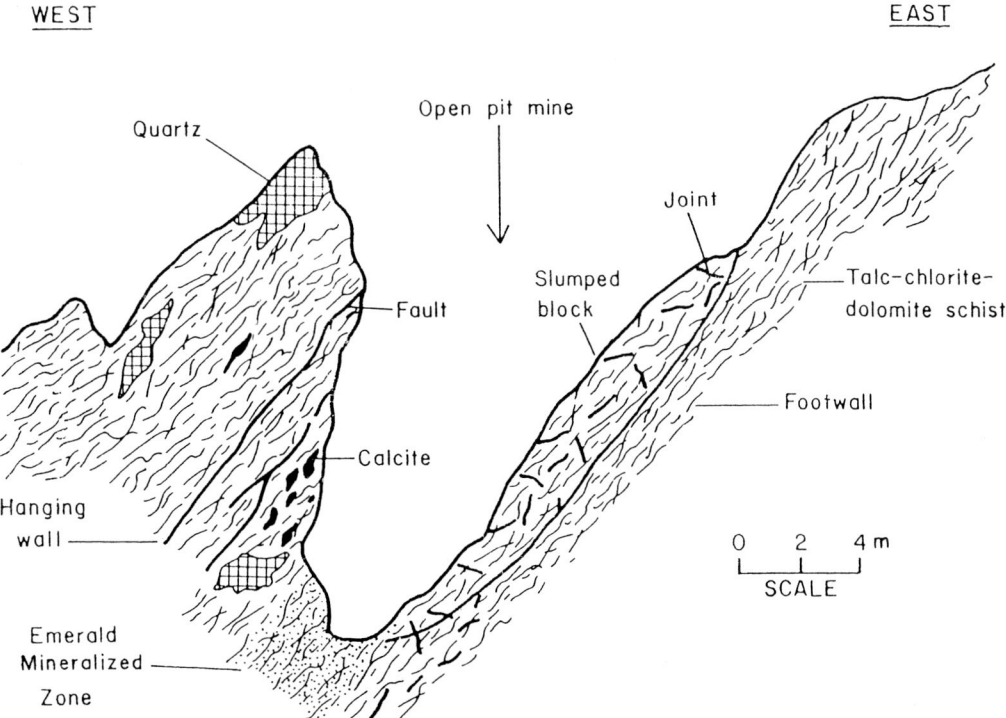

Figure 3.3. Sketch showing geological section across central quarry of Mine 2, Mingora.

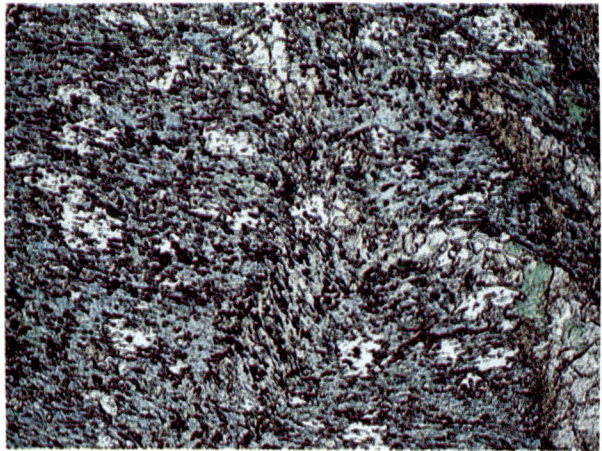

Photo.3.12. Photomicrograph of talc-chlorite-dolomite schist showing porphyroblasts of dolomite strung along the foliation, field 4.6 x 6.6 mm. (photo *A. H. Kazmi*).

Photo.3.13. Mingora emeralds 2 to 4.5 ct. size range (photo *A. H. Kazmi*).

Photo.3.14. Emerald crystals in quartz, Mingora mines (photo *A. H. Kazmi*).

significant displacement, crisscross the mineralized zone with numerous exposed slickensided planes. Emeralds occur along a well-defined shear zone. This shearing has resulted from two parallel normal faults straddling the deposit of Mine 2. The shear zone has been limonitized and contains a few scattered quartz veins. In this zone, emerald occurs largely in ideoblastic crystals which are found in sporadically scattered pockets or nests 5 to 15 cm across (fig. 3.4). Smaller ideoblastic crystals are disseminated pervasively throughout the matrix of the mineralized zone. The emerald is commonly associated with fuchsite and tourmaline. This type of deposit is fairly consistent in its mineralization and enrichment and has yielded the best production quantitatively and qualitatively (photo. 3.13).

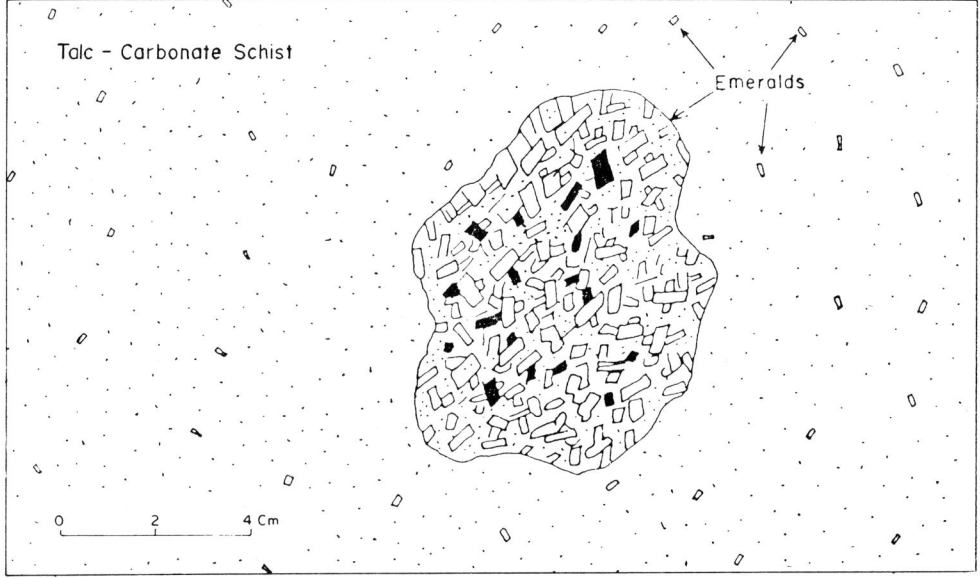

Figure 3.4. Emerald occurrence in talc-carbonate schist as dissemination and as a pocket of crystals (Mingora).

2. Fracture Fillings

This kind of emerald occurrence is typical of the northern and southern quarries of Mine 2. There talc-chlorite-dolomite schist, the host rock, shows crenulate folds, shearing and dismemberment by numerous small faults filled in by limonite and veinlets or stringers of quartz and calcite. Emeralds are largely confined to limonitized fault and joint planes and better production is commonly obtained from points where two mineralized fractures intersect (fig. 3.5). Emerald crystals also occur in quartz and calcite (photo. 3.14 and 3.15) along these fractures. This type of deposit has yielded excellent ideoblastic stones of good quality and size (up to 200 carats), but the production tends to be relatively sporadic.

3. Emeralds in Stockworks

Mine 3 of the Mingora deposit exemplifies this type of emerald occurrence where intensely fractured dolomite has been filled with stockworks of quartz (fig. 3.6). The emeralds are

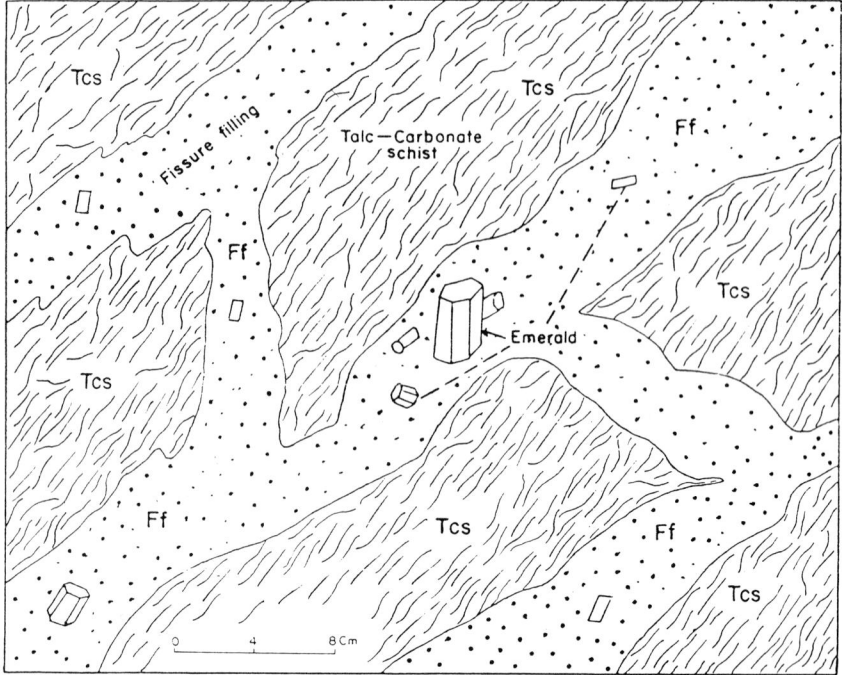

Figure 3.5. Emerald occurrence in fissure fillings (Ff) in talc-carbonate schist (Tcs) Gujarkili, Swat.

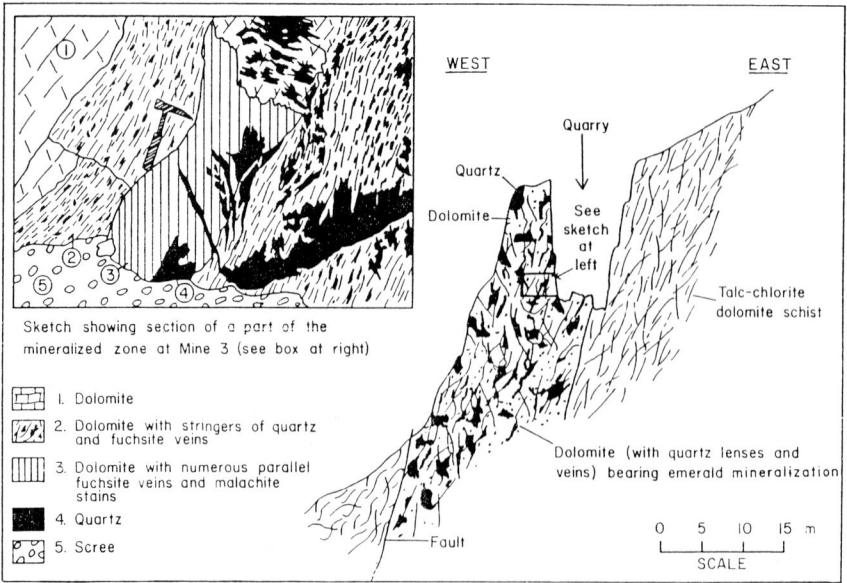

Figure 3.6. Sketch showing section of the mineralized zone of Mine 3, Mingora.

associated with fuchsite and tourmaline. Emeralds from such deposits are commonly of lighter color than elsewhere.

4. Emeralds Along Tension Gashes

The Islamia trench of the Mingora deposits is a good example of this type of emerald occurrence. There talc-chlorite-dolomite schist was faulted against a block of greenschist and dolomite. Large tension gashes in the talc-chlorite-dolomite schist were formed and were filled in by quartz lenses up to 2 meters in diameter and 1 meter thick. Emerald mineralization has occurred in a 15- to 30-cm-thick layer of talcose rock surrounding the quartz lenses (fig. 3.7). Large (commonly up to 30 carat size) ideoblastic emerald crystals of good deep green color and good clarity have been obtained from such deposits. Each of these mineralized lenses has yielded 1000 to 5000 carats of good stones.

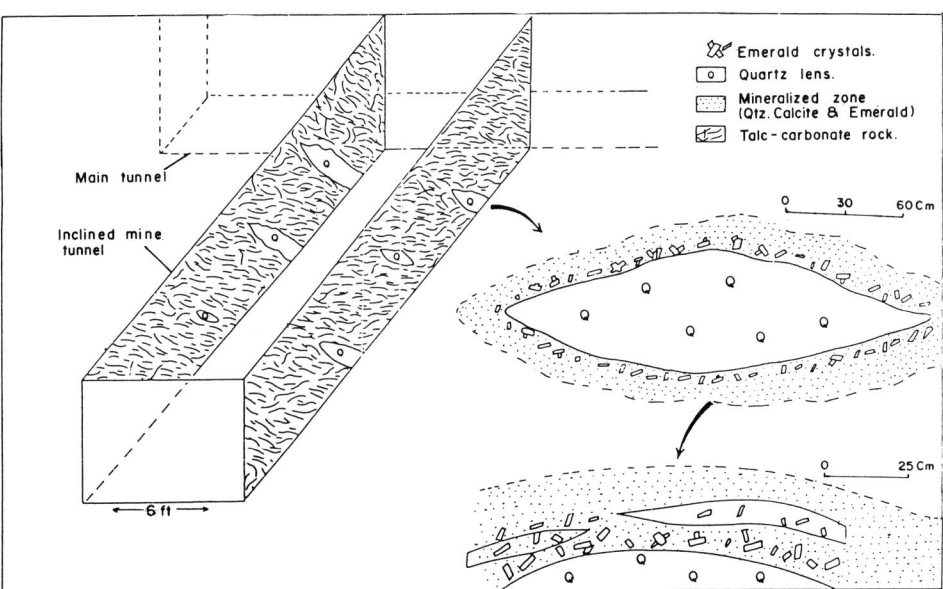

Figure 3.7. Sketch showing mode of emerald mineralization at Islamia mine, Mingora.

A Brief Description of the Mingora Mines

Islamia Mine

This is probably the earliest mine to be worked after the discovery of emeralds at Mingora. Here private parties began mining emeralds through an open trench which ultimately became 140 meters long, 12 meters wide and more than 15 meters deep. It is a northwest-trending trench at the southern end of Mine 1, and it has yielded good quality emeralds. Work in this trench was abandoned for some time as a result of debris accumulation and rock falls. Recently a north-trending adit has been started by GEMCP followed by cross cuts and inclines and the mine is in production again.

Fine-quality emeralds are found around white milky quartz veins and lenses which form infillings of secondary fractures or tension gashes caused due to faulting. The quartz is surrounded by soft talcose rock which is white to greenish gray in color. Emerald crystals have developed in talc surrounding the quartz lenses and veins. There is an apparent zonation in the arrangement of quartz crystals as the green emerald crystals are commonly present in whitish talc, whereas greenish talc contains light to colorless crystals.

Emerald is also found in talc which is present along fractures in quartz veins and lenses. The mineralized quartz veins are both parallel as well as oblique to the trend of foliation of the mother rock. Mineralization has been observed more commonly along the oblique quartz veins where the latter intersect the foliation of the host rock at an angle.

Emerald also occurs along the limonitic zone and in reddish brown clayey material along the fault planes where it is found surrounding quartz veins in narrow zones, which are 5 to 15 cm thick. In such instances emeralds have not been found embedded in the quartz itself.

Emeralds are found in the form of pockets as well as disseminated grains. Euhedral crystals of emeralds are generally surrounded by talc. The emeralds from this mine are comparatively less fractured, well developed and larger in size. Disseminated crystals are commonly better in quality as compared to those found in bunches and pockets.

Farooq Mine (Mine 1)

In this mine the host rock is a white, greenish white to grayish white talc-carbonate schist. It is highly sheared and lacks phacoids of carbonates. Quartz lenses are present. The rock has been subjected to severe tectonic movements, as a result of which many faults and crenulations have formed. Limonitization is very prominent along faults and shear zones.

Fair-to good-quality emeralds occur in the form of pockets in host rock of the Farooq mine as well as disseminated crystals in association with talc, quartz, and calcite. Emerald is found in limonitic zones associated with talc (fig. 3.2). The characteristic feature of mineralization in this part of the mine is that the emeralds are largely found in a relatively more disseminated form in the rock than elsewhere. Extension of the mineralized zone is in a north-south direction and the width of the mineralized zone is approximately 4 meters.

The emerald-bearing talc-carbonate schist has a faulted contact with the older siliceous schist. The emeralds are mainly along this fault. There are numerous other smaller local faults which are concentrated along the major fault zone. It has been observed that along some of these smaller faults the emeralds have been crushed and found in the form of a coarse powder of green color which suggests that these faults are of a later origin than, or were reactivated after, the emerald mineralization.

Carrel's Trench

The rocks in the area near Carrel's trench are mostly dolomite and dolomitic schist with abundant hydrothermal quartz and fuchsite. The host rock is massive and jointed. Quartz and talc-carbonate veins are found along these joints. Emeralds are mostly found in quartz veins and talc-carbonate veins present along the fractures within the carbonates but the emeralds in the area are poor, probably due to its distance from the major fault and also due to the nature of the host rock. Emeralds recovered from this area are mostly fractured, flawed and smaller in size. Due to the relatively poor quality of the emeralds, the work has been abandoned in this area.

3.15

3.16

3.17

3.18

Photo.3.15. Emerald crystal in calcite, Mingora mines, (photo *A. H. Kazmi*).

Photo.3.16. Emeralds from Mingora mine 1, larger crystals in the 8-12 ct size range (photo *A. H. Kazmi*).

Photo.3.17. Cut and polished emeralds from Mingora mine 1, 3-8 ct. size range (photo *A. H. Kazmi*).

Photo.3.18. Mellee emeralds from Mingora mines (photo *A. H. Kazmi*).

Mine 2 Area

Mine 2 covers the northwestern tip of Mingora Emerald Mines (fig. 3.1). The schist is mainly composed of talc, dolomite, siderite, mica, and quartz. The foliation of the rock is cut by numerous intersecting faults. The rock is thus much sheared and drag folded and contains prominent limonitic zones and quartz veinlets and stringers. Some of the quartz veins bear emeralds. Brownish gray and gray lenses of ankerite are present along the foliation and the size of these lenses ranges from a few centimeters to a few meters.

Good quality emeralds have been recovered from western and southern parts of the northern quarry. Emeralds are mainly confined to limonitic zones. The emeralds found are in the form of pockets and associated minerals are calcite, quartz, fuchsite and tourmaline. Crystals are usually found embedded in reddish brown limonitic material and in white to green talc-rich rock and yellowish clayey gouge, in close proximity to the faults. Crystals recovered from this mine are mostly well developed, up to 2 cm in size and have good color. Close to western part of the northern quarry there is a quarry face referred to as N3. In this face, mineralization is confined to the faults and limonitic zones and is more abundant at points where faults intersect each other. Previously good quality emeralds, though smaller in size, have been recovered from the crenulated parts of the rock.

It has been observed that emeralds recovered from carbonate-rich parts are mostly embedded in calcite or quartz and are fractured, flawed and poor in quality, while those obtained from talc-rich parts are well developed and less fractured, and variable in size. Emeralds occur either along the faults or close to the faults. In the mineralized zone the associated minerals are quartz, calcite, mica, pyrite and tourmaline.

Central Quarry

Rocks from the central quarry of Mine 2 are mainly comprised of greenish white, white, and gray colored talc schist which is rich in carbonates. Milky white quartz is present throughout the rock body in the form of stockwork, as well as lenses and veins and also in the groundmass (fig. 3.3). Calcite is abundantly present in the form of nodules or large crystals surrounded by talc-rich rocks. Replacement of carbonate by silica in calcite nodules is observed.

One of the characteristic features which distinguishes rocks of this area from the rest of the rocks of Mine 2 is the absence or lesser frequency of limonitization. Faults in this mine are common, but little or no limonitization or gouge material has been encountered. Small carbonate lenses and patches are also present along the general trend of the rock whereas others are oblique to the foliation. Joints are prominent in the carbonate-rich parts; some of the carbonate lenses are jointed and secondary calcite and/or quartz are found along the joints. Slickensides are common and conspicuous along the faults. The emeralds in this zone occur mainly in the form of pockets in carbonate and talc-rich rocks and are mainly associated with quartz, calcite, fuchsite and tourmaline. In carbonate-rich rocks, mineralization is largely confined to quartz veins. However, emeralds found in quartz are commonly fractured, and of poorer quality.

Southern Quarry

The host rock in this quarry is a greenish white talc-carbonate schist. The major portion of the quarry is occupied by talc-rich schist. Quartz, though not very common, occurs in the form of broken veinlets and nodules. Calcite is abundant both in the form of veinlets and nodules. At places these nodules contain greenish white flaky talc along fractures.

This part of Mine 2, like the southern part of the north face, is highly limonitized and criss-crossed by numerous faults. Rocks in the limonitic zones are reddish brown, fine grained, clayey and deformed. Water content is high and shearing is prominent, especially in the eastern part of the southern quarry. Along the faults, white and yellowish clayey gouge is present in which broken pieces of calcite and quartz are observed. Calcite nodules are usually arranged in rows and are surrounded by talc-rich talc-carbonates. Besides calcite nodules, talc-carbonate and magnesite nodules are also found. In some parts of the area the rocks are drag-folded and crenulated. Similarly quartz and calcite veins are also folded and at places these have been boudinaged.

This quarry contains good quality emeralds, which occur mainly in the form of pockets along fault planes and within the limonitic zones. It has been observed that emeralds occur on either side of the fault planes, either on the hanging wall or the footwall. In the talc-rich rock, crystals are mostly euhedral, larger in size and have unique clarity and transparency. Color is characteristically bright green.

At places emeralds also occur in veinlets and nodules of cream-colored calcite. Some of these emeralds are euhedral and flawless.

The rocks of the western flank of Mine 2 are largely comprised of phacoidal talc-carbonates. Phacoids of carbonate (magnesite ?) range from about one centimeter to a few centimeters. They are arranged along the foliation of the talc-carbonate schist. This part of the mine contains sheared and foliated serpentinite bodies which are arranged in parallel to sub-parallel rows along the rock foliation. Alteration of serpentinite to talc is common.

Mine 3 Area

Mine 3 is the northernmost quarry of the Mingora Emerald Mines (fig. 3.1). Talc-carbonates and serpentinites crop out in the mine area. A north-trending fault zone separates Mine 3 from Mine 2. In this zone blocks of siliceous schist, greenstone, greenschist, graphitic schist and dolomite are present. The carbonates are largely brownish gray, fine- to medium-grained, massive and jointed dolomites. In some parts, weak schistosity developed. Presence of abundant fuchsite and quartz are the characteristic features of Mine 3 area (fig. 3.6). The dolomite rock is randomly cut by quartz veins and stockwork. Malachite and chlorite are also present in the quartz veins and carbonates.

Emeralds are found in quartz stockwork in dolomitic rocks. Emeralds recovered from this mine are poorer in quality. Crystals are small in size, more fractured, and lighter in color.

Quality of Mingora Emeralds

The emeralds from the Mingora Emerald Mines have now become world famous for their exquisite deep green color, exceptional clarity and excellent quality (photos. 3.16, 3.17, and 3.18). However, the best material comes from Mine 2 and the Farooq Mine and the name "Mingora emerald" should be reserved only for the stones from these two mines or for stones of similar quality. A brief description of the stones from various mines is as follows:—

Mine 2 and Farooq Mine: These mines produce deep to medium deep green emeralds of good clarity that are mostly well-formed euhedral crystals in the size range of two carats and less.

Clear stones above 2 carat weight are rare. Larger stones tend to be aggregates or of lower quality and occluded with opaque inclusions.

Carrel's Trench: The medium deep green stones which have been produced from this mine are of good clarity and commonly are well-formed euhedral crystals. In general, however, the stones are small in size (mellee to about one carat).

Islamia Mine: Medium deep green to medium green, well-formed euhedral crystals, mostly in the size range of 0.2 to 2.0 ct. occur in this mine. Stones larger than 5 ct are common. Their clarity is good but inclusions are common. Inclusions of talc and other amorphous mineral impurities adversely affect the quality of these stones.

Mine 3: The emeralds in this mine are mostly light green though well formed. Euhedral crystals of good clarity in the size range of 0.2 to 4.0 ct are common.

The average size distribution of emeralds in the total mine production is approximately of the following order:

	Size:	*Percent*
(a)	Mellee	45.8
(b)	0.5 to 1 ct.	32.7
(c)	1 to 2 ct.	6.8
(d)	above 2 ct.	14.7

Emeralds from Swat are being graded by GEMCP in the following six main quality grades:

Excellent: Deep green, transparent, exceptionally clear and bright stones with sparkle and fire, unoccluded, unfractured, and unblemished, euhedral to subhedral crystals.

Very Good: Deep to medium green, transparent stones of good clarity and with considerable brilliance, with few inclusions, largely euhedral to subhedral crystals.

Good: Deep to medium bright green stones of fair clarity, with few inclusions and fractures, comprising euhedral to anhedral crystals.

Fair: Light medium to medium green or deep green, translucent to semitransparent stones with inclusions, largely euhedral pieces.

Mellee: Rough emeralds below 0.2 ct commonly of exceptionally good deep green color and clarity, and largely comprised of euhedral crystals with good luster and brilliance.

Industrial grade: Non-gem grade, light to medium green, opaque, dull stones, ranging from euhedral crystals to broken chips.

Average gradewise distribution of the rough gem-grade emeralds produced from the

Mingora mines (excluding the industrial grade) is approximately as follows:

Grade	Percent
Excellent	5
Very good	12
Good	25
Fair	13
Mellee	45

Mining and Production

It is fortunate that Mingora emeralds occur in very soft talc-carbonate schist which can be easily broken and excavated. Thus, for mining and quarrying a minimum use of explosives is warranted. Excavation of the mineralized zone is almost entirely done with pneumatic picks.

Most of the mining is done through open pits. Mine 2 has been recently redesigned, mechanized, and developed on modern lines with neatly excavated terraces and benches. The unwanted mine debris is mechanically dumped outside the mine. The mineralized zone is carefully excavated with pneumatic picks. At the quarry face the excavated mineralized rock fragments are gently broken down with wooden mallets and searched for enclosed emeralds. The grains thus obtained are immediately dropped into a locked red box through a slot. The mineralized debris discarded at this stage is then transferred to a mechanized washing plant where the debris is gently crushed, run through trommels, washed and passed through a centrifugal classifier which removes most of the talc and other fine material. The tailings are then spread on a white plastic sheet and emeralds are removed by hand picking. A limited amount of underground mining is being carried out at Islamia Mine and at Farooq Mine.

When the Mingora emerald mines were transferred to the Gemstone Corporation of Pakistan (1978-79), the average annual production of gem-grade emeralds was about 5000 carats. Under the management of the GEMCP, the production has steadily increased significantly as may be seen from the production data given below:

1978-79	4,371.72 cts	1983-84	23,434.73 cts
1979-80	22,882.55 cts	1984-85	29,280.25 cts
1980-81	20,736.92 cts	1985-86	53,295.43 cts
1981-82	20,669.40 cts	1986-87	38,112 cts
1982-83	28,242.96 cts	1987-88	39,646 cts

A peak production of 53,295 cts of gem-grade emeralds was reached in the Fiscal Year 1985-86. This production is almost entirely from the Northern and Central quarries of Mine 2. The other mines are expected to be revamped, mechanized and developed shortly on the same pattern as Mine 2. It may be therefore expected that after complete development of all the mines the production would increase manifold.

GUJARKILI EMERALD DEPOSIT

The Gujarkili emerald deposit is located near Gujarkili village 24 km east-northeast of Mingora, 12 km south-southwest of Alpurai and about 3.2 km south of Bazarkot village

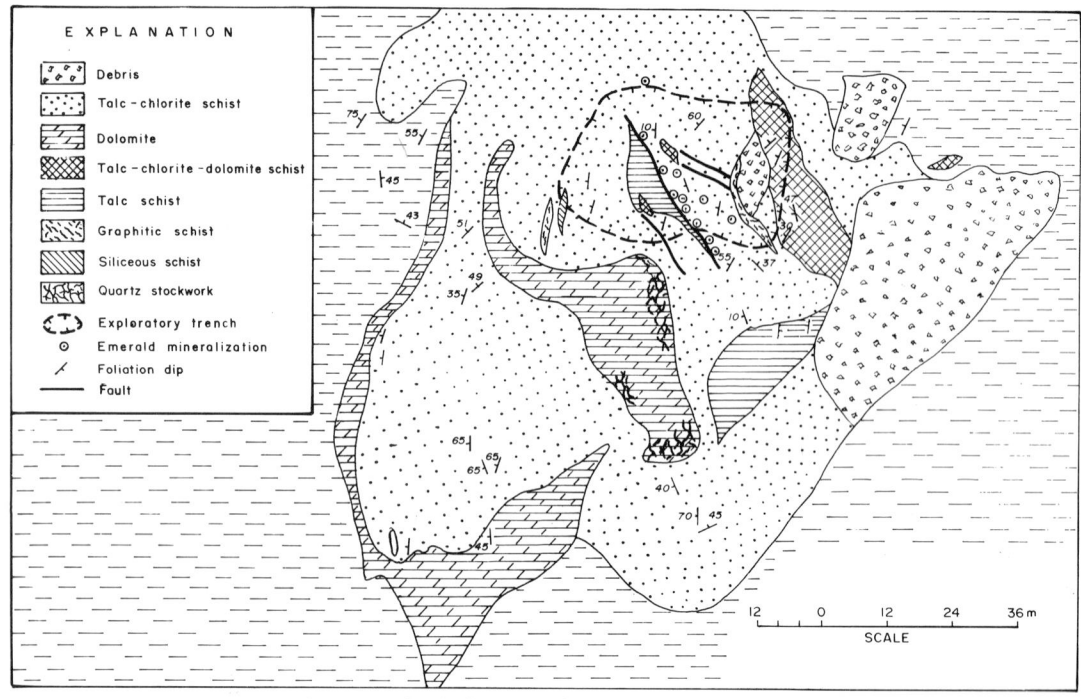

Figure 3.8. Geological map of Gujarkili emerald deposit.

(fig. 2.4). A pony track links the deposit with Bazarkot, whereas the latter is accessible by a metalled road from Mingora. The emerald deposit occurs in a hilly terrain at an altitude of approximately 1925 m above sea level, amidst a picturesque pine forest (photo. 3.19) through which flows a clear sparkling brook.

The deposit was discovered in 1981 as a result of extensive geological exploration by the Gemstone Corporation of Pakistan and it has been mined ever since. Bowersox and Anwar (1989) have recently described this deposit.

Geology

The emerald deposit occurs in a small triangular outcrop of ophiolitic melange, which is surrounded by Saidu graphitic schist. This outcrop is barely 3 acres in extent and forms the western slope of the north-trending Gujarkili valley (fig. 3.8).

The ophiolitic melange largely comprises talc-carbonate rocks which, depending upon the relative abundance of talc, range from almost pure talc schist to massive nonfoliated dolomitic carbonate rocks. These rocks are briefly described below in relative order of their abundance.

Talc-Chlorite Schist

This schist covers most of the outcrop area and has a steep westward dip. Emeralds are largely confined to this brown to yellowish green, medium-grained unit which is mainly composed of talc and chlorite with small amounts of muscovite, fuchsite, siderite, magnesite,

Photo.3.19. A view of Gujarkili emerald mines, 1984 (photo *A. H. Kazmi*).

Photo. 3.20. Cut and polished emeralds from Gujarkili emerald mines, 3-6 ct size range (photo *A.H. Kazmi*).

Photo. 3.21. Cut and polished mellee size emeralds from Charbagh mines (photo *A. H. Kazmi*).

and calcite. The schist is well jointed and traversed by several small faults and fractures. Quartz veins are common. The talc-chlorite schist contains large tectonized clasts of the following rocks.

Carbonate Rock

Massive, hard, non-foliated exotic blocks of carbonate rock which are 3 to 60 meters in extent, are found in the talc-chlorite schist. These are grayish brown and medium grained and are riddled with quartz veins. Limonitization is widespread. The common mineral constituents in these rocks are dolomite, magnesite, siderite, and calcite, with small amounts of quartz, talc, chlorite, muscovite, and fuchsite.

Carbonate-Talc Schist

A large tectonized block of this rock (10 meters wide and 40 meters long) occurs within the talc-chlorite schist, in the northeastern part of the mine area. Small blocks are also found scattered within the talc-chlorite schist. This rock is yellowish to greenish brown and yellowish green. It largely comprises magnesite, siderite, and calcite with smaller quantities of talc, muscovite, and quartz. Quartz veins are common. The rock has been extensively limonitized.

Talc Schist

The talc schist is white, greenish white to light greenish gray, in color. It is highly sheared, crumpled, and well jointed, with drag folds and crenulations. Two tectonized blocks (20 x 4 meters and 30 x 10 meters across) of this rock are found in the central part of the ophiolitic melange outcrop, within the mass of talc-chlorite schist. It is almost entirely composed of flaky talc grains with very small amounts of quartz, magnesite, calcite, chlorite, muscovite, and fuchsite. Milky white quartz veins are common and cut across the foliation. Small fragments and clasts of greenschist and graphitic schist are common.

Emerald Occurrences

The emeralds are confined to a relatively small, narrow block about 40 meters long and 10 meters wide, with a northwest-southeast orientation. This block is comprised of talc-chlorite schist, which also contains small clasts of graphitic schist, carbonate rock, and talc schist. A number of small northwest-trending faults with cross-cutting fractures and joints cut through the mineralized block (fig. 3.8). In fact, emeralds largely occur along these faults and fractures. A large tectonized block of talc schist with a westward dip of over 70° forms the hanging wall of the deposit, while a narrow faulted block of graphitic schist followed by a larger block of carbonate rock forms the foot wall.

Fault breccia occurs commonly along the larger faults. There is extensive limonitization along fault and joint planes. Emeralds are largely confined to these criss-crossing limonitized fault and joint planes. At the Gujarkili deposit the emeralds occur mainly as scattered isolated crystals often associated with fuchsite, euhedral quartz, and calcite crystals. Their occurrence in the form of pockets, bunches, and aggregates is extremely rare.

Quality of Emeralds

The rough gemstones mined from the Gujarkili deposit are largely in the form of well-formed euhedral crystals ranging in size from a few millimeters to large crystals weighing 100 to 200 carats. The smaller crystals are deep bluish green, transparent to translucent and contain few inclusions. Exceptionally clear crystals are not uncommon and several excellent large stones have been found in this mine. The Gujarkili emeralds are of an even deeper green color than the Mingora emeralds.

Some of the larger stones, in the 100 to 200 carat size range, appear as relatively clear excellent euhedral crystals, but unfortunately they are traversed by fine cracks which do not permit cutting of the crystals into large gems. Furthermore, the larger crystals commonly contain talc inclusions which give a pitted appearance to the crown of the facetted gems. Nevertheless, several good gems above 5 carat size have been fashioned out of these gemstones by the Gemstone Corporation of Pakistan (photo. 3.20).

Mining and Production

An open cast mine has been developed at Gujarkili by the Gemstone Corporation of Pakistan. The area is covered with snow during winter. Mining is therefore possible only for seven to eight months each year. Soft talcose rock is easily excavated with the help of pneumatic picks without resorting to the use of explosives. This enables extraction of gemstones without fracturing or damage.

The Gujarkili deposit has been in production since 1982. Initially during the first two years the production was limited and was mainly the by-product of exploration and prospection. During the next 5 years the mine produced about 12,000 carats of emeralds. However, its production has now increased significantly. During 1987-88 about 28,645 cts of gem quality emeralds were mined. In the last six months of 1988 about 17,000 cts were produced.

CHARBAGH EMERALD DEPOSIT

The Charbagh emerald deposit occurs about 2 km south of Charbagh village, about 14.5 km northeast of Mingora and about 180 meters east of the metalled Mingora-Charbagh road (fig. 3.9). It is a very small deposit that was discovered soon after the Mingora emerald deposit. It has been intermittently worked by local people and more recently by the Gemstone Corporation of Pakistan.

Geology

South of Charbagh there are east-trending outcrops of the Charbagh greenschist melange (see Lawrence and others, this volume). The exposed tectonostratigraphic sequence is as follows:

Charbagh greenschist melange
————————Makhad Thrust————————
Mingora ophiolitic melange
————————Kishora Thrust————————
Saidu calc-graphitic schist
Alpurai calc-mica-garnet schist
————————Unconformity————————
Swat granite gneiss

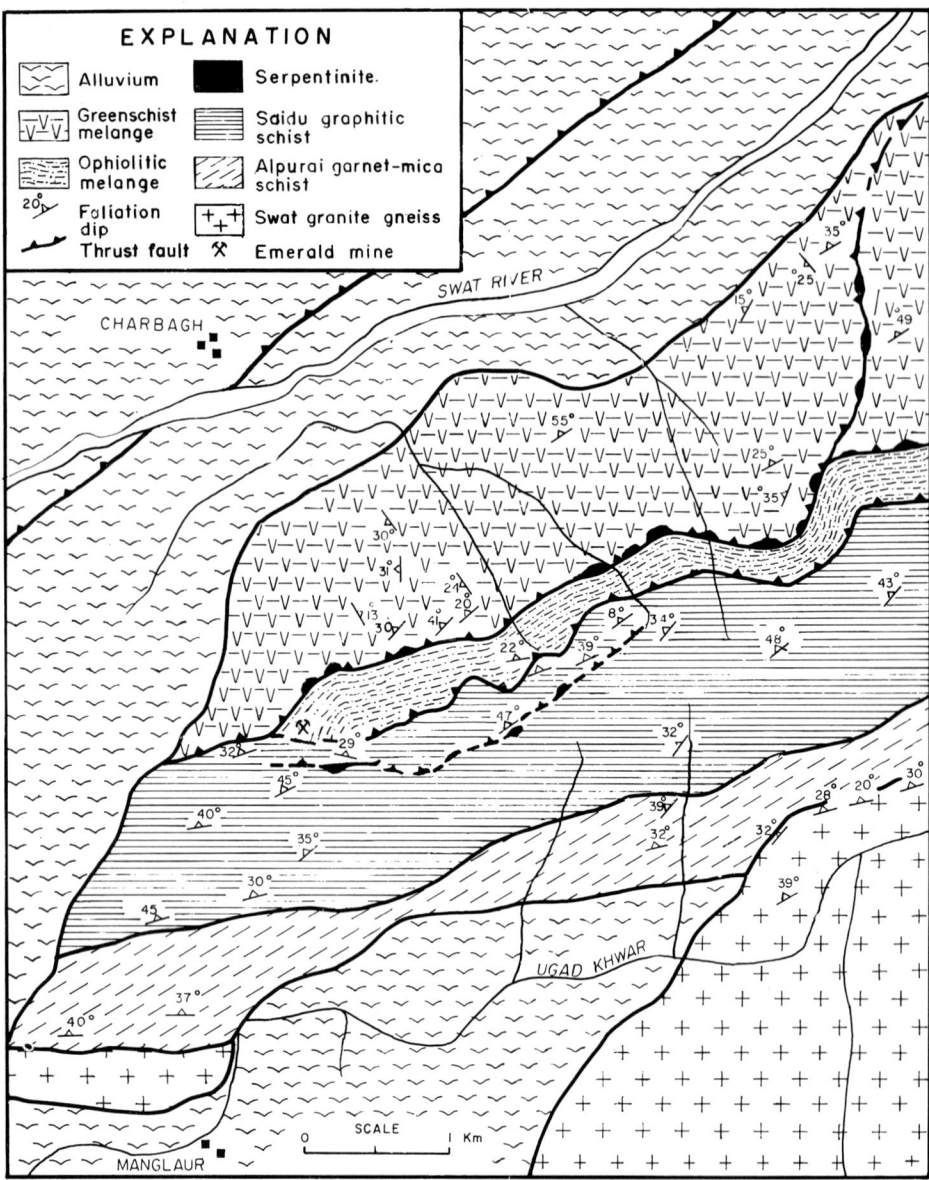

Figure 3.9. Geological map of Charbagh area.

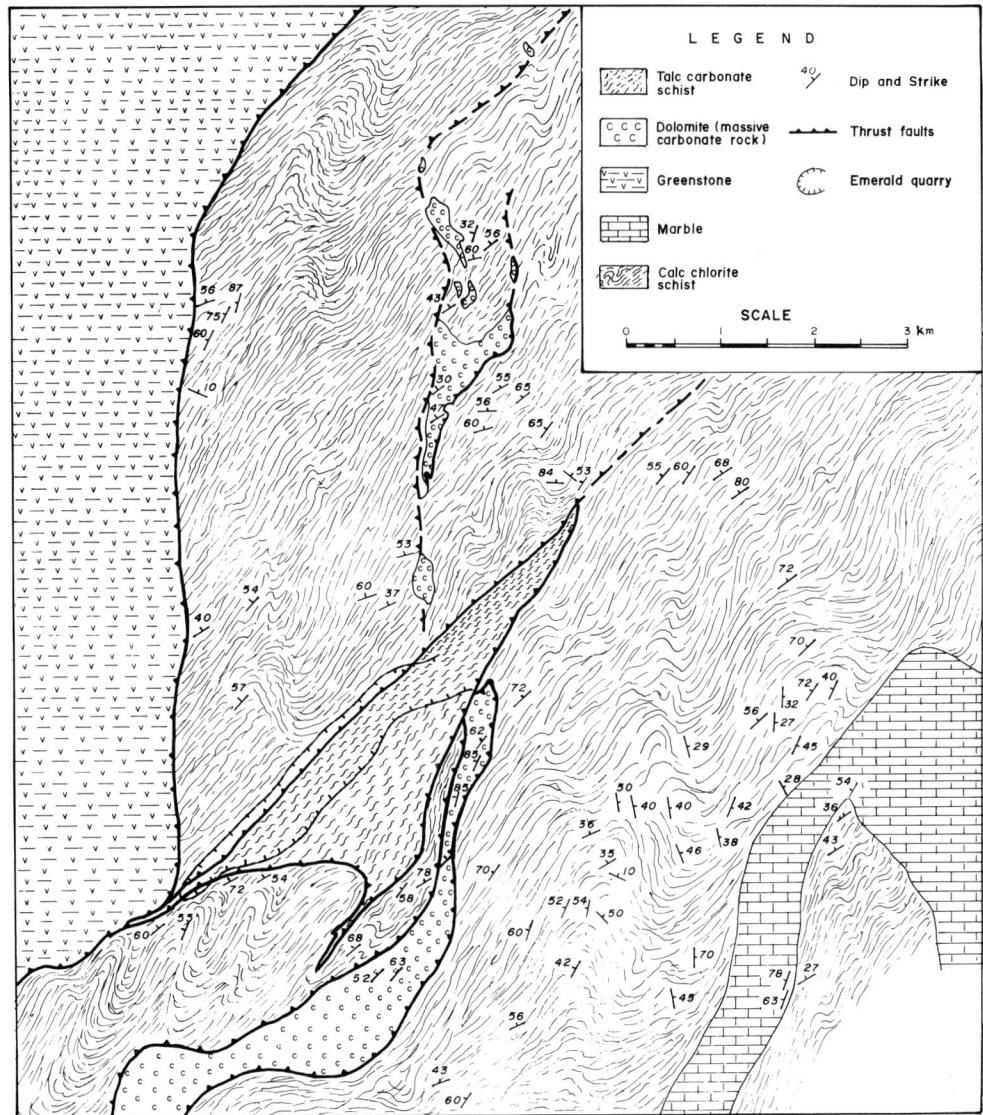

Figure 3.10. Geological map of Charbagh mine.

The Charbagh emeralds occur in the Mingora ophiolitic melange. South of Charbagh this melange is comprised of a 300- to 600-meter-wide east-trending block of chlorite schist and talc-chlorite schist. The dominant rock type however is the rusty brown, greenish gray to whitish green calcareous, fine- to medium-grained, chlorite schist. It is largely comprised of chlorite, mica, calcite, siderite, and quartz, with subordinate amounts of talc and graphite. At places it has a phyllitic appearance.

The chlorite schist contains lenticular clasts of talc-chlorite-dolomite schist (talc carbonates), dolomitic carbonate rock, serpentinites and marbles. The talc-chlorite-dolomite schist and serpentinite lenses have formed along the major thrust zones (fig. 3.10). These bodies are up to 30 meters in length and 0.5 to 8 meters in width and are the alteration products of ultramafic rocks.

Emerald Occurrence

The Charbagh deposits are composed of two lenticular bodies of talc-carbonate schists in which emeralds occur along a northeast-trending zone of thrust faulting within the chlorite schist (fig. 3.10).

The northwestern mineralized zone is comprised of reddish brown to dark-brown talc-chlorite-siderite schist, in which the main constituents are talc, chlorite, siderite, dolomite, and limonite with small amounts of quartz. Quartz lenses and nodules, 2.5 to 10 cm in size are commonly found along the rock foliation. This mineralized zone contains very thin and lean dissemination of emeralds of very poor quality and has not been worked.

The other mineralized zone consists of a lenticular body of light greenish white to brownish white talc-chlorite-dolomite schist, about 2 meters thick and 7 meters long. It occurs along a thrust fault. Light yellowish brown fault gouge has formed along the fault zone between the chlorite schist and the talc-chlorite-dolomite schist. Quartz veins are common in the mineralized zone. Emerald crystals are found disseminated in the talc-chlorite-dolomite schist as well as in the fault gouge.

Quality of Charbagh Emeralds

The Charbagh emeralds are largely found as relatively small crystals commonly in the 0.5 to 2 carat size-range. They are largely transparent and of light green color but most contain dark colored inclusions. Most of the stones are therefore not of a high quality. However, a small percentage of the production is cuttable and the Gemstone Corporation of Pakistan has produced some good facetted gems from the mine (photo. 3.21).

Mining and Production

Despite the extremely small nature of the deposit, open cast mining has been carried out intermittently by the local people and more recently during the past four years by the Gemstone Corporation of Pakistan. The GEMCP has produced about 12,500 carats of gem-grade stones from this mine.

MAKHAD EMERALD DEPOSIT

The Makhad emerald deposit occurs near the Makhad village at an altitude of about 1230 meters above sea level. Makhad is situated about 3.2 km north of Telegram village, which is on the metalled Mingora-Malam Jabba road, about 19 km northeast of Mingora (fig. 3.11). Makhad emeralds have been intermittently mined by the local tribesmen as well as by the Gemstone Corporation of Pakistan (photo. 3.22).

Geology

The geological setting of the Makhad emerald deposit is somewhat similar to the Charbagh deposit inasmuch as this deposit also occurs within the Mingora ophiolitic melange and it is also located in the vicinity of the Makhad thrust fault, east of Charbagh (fig. 2.4). In this area, the rocks have a general east-west strike and dip northward. The Charbagh greenschist has been thrust over the Mingora ophiolitic melange, which in turn has been thrust southward on the Saidu graphitic schist.

Photo.3.22. A view of Makhad emerald mine, Swat, 1981 (photo *A. H. Kazmi*).

Photo.3.23. A view of Khaltaro emerald pegmatite (photo *Khalid Aziz*).

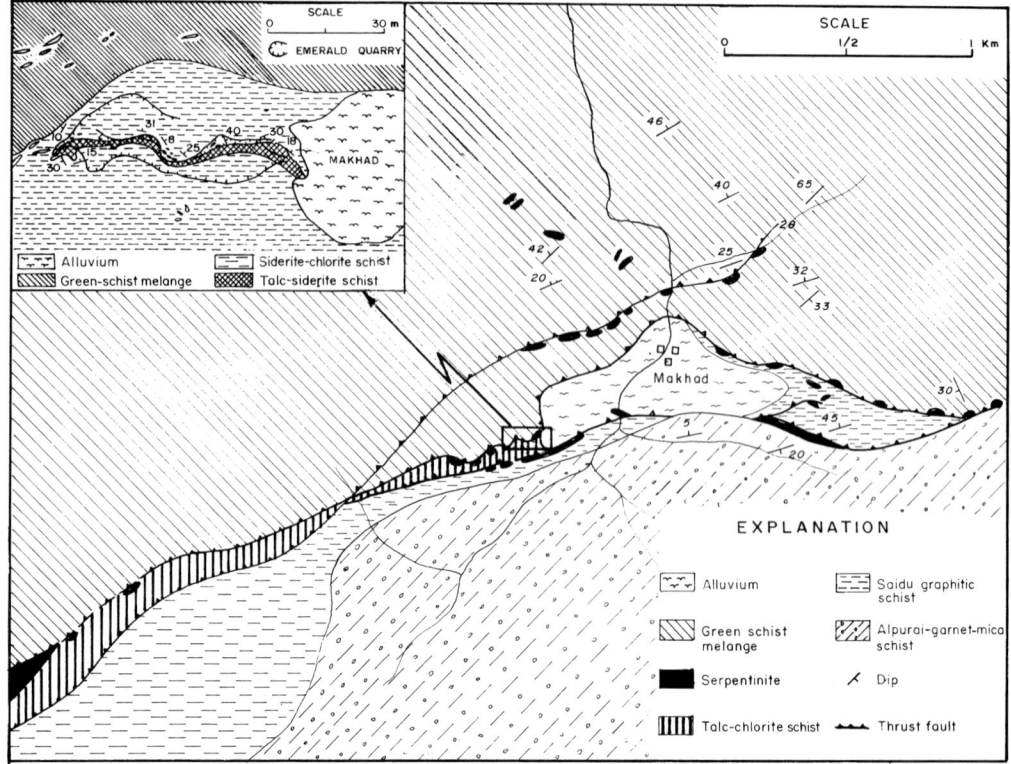

Figure 3.11. Geological map of Makhad area.

In the vicinity of the Makhad emerald mines, the Mingora ophiolitic melange is largely comprised of siderite-talc-quartz-mica schist and contains clasts of talc-carbonate schist, graphitic schist, marble, and quartzite. Close to the Makhad village a series of small en echelon tectonized lenticular bodies of talc-carbonate schist occur (fig. 3.11). The emeralds are found in a small lenticular outcrop of talc-carbonate schist, about 0.5 km west of the Makhad village. This outcrop is about 2.5 meter wide and 70 meter long. It has an eastwest strike and dips at 15° to 35° to the northwest. However, it contains numerous crenulate folds.

The talc-carbonate schist is largely comprised of talc, siderite, magnesite, and quartz. Talc occurs as flakes or as minute lenses and veinlets. Siderite, calcite, and quartz also occur in the form of nodules and veins.

Emerald Occurrence

The entire talc-carbonate body west of Makhad is mineralized and emeralds occur largely as disseminated euhedral crystals. The host rock has been highly weathered into a soft talcose limonitic powdery or clayey mass in which the emeralds are embedded.

Quality of Makhad Emeralds

Makhad emeralds commonly occur as fair size euhedral crystals mostly in the size range of 10 carats or above. However they are of a very dark green color. The crystals are largely opaque to translucent and are full of inclusions. Talc and siderite commonly occur as inclusions and give them a pitted appearance. These stones are not of much value as gems though some of the stones when cut and polished into thin slices provide exquisite, clear deep green hexagonal platy tabloids which can be easily used in jewellery rather in an unconventional fashion.

Mining and Production

Local tribesmen and the Gemstone Corporation of Pakistan have sporadically mined this deposit. Initially the deposit was opened with the help of an open trench. Later the GEMCP made a short inclined drift and a few crosscuts to explore the deposit. The GEMCP has mined about 1000 carats of emeralds after which the mines have been closed.

KHALTARO EMERALD DEPOSIT

Khaltar is a small narrow valley in the Haramosh Range, located about 70 km east-southeast of Gilgit. The Khaltar stream follows a southerly course and joins the Indus River near Sassi village. At this point, the Indus makes an inverted U bend, between Bunji and Shengus villages. The discovery of emeralds in the Khaltar valley was made in 1985 by the Gemstone Corporation of Pakistan in the course of geological exploration of gem pegmatites of the Haramosh region.

In the Khaltar valley emeralds occur near the Khaltaro village, at a locality known as Rayjud, about 32 km north of Sassi. Rayjud is about 3500 meters above sea level (about 2000 m above Sassi) and the emerald deposits occur 3 to 5 km farther upstream, at an altitude of about 4,500 meters above sea level (fig. 3.12). The terrain is extremely rugged and each year the deposits are covered with snow from October until June.

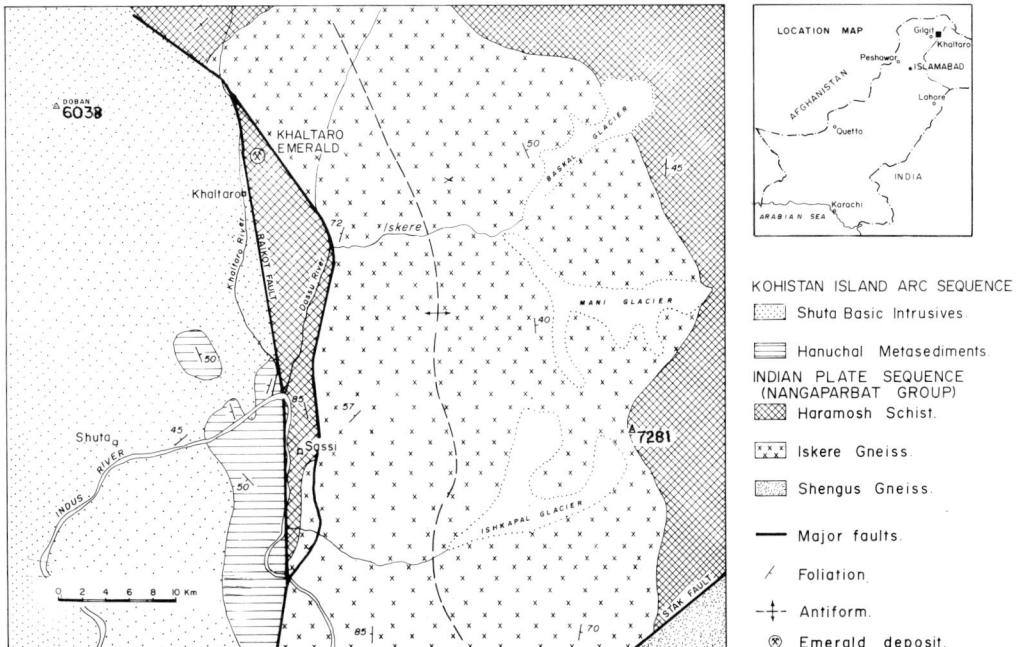

Figure 3.12. Geological sketch map of Khaltaro area (Gilgit District)

Geology

The Khaltaro emerald deposit is located in the Haramosh Range near the snout of the Haramosh glacier at the northern margin of the Nanga Parbat-Haramosh massif (fig. 3.12). This massif is an unique structural and topographic high in the northwestern corner of the Himalayan collision zone (fig. 2.1). It is the northernmost exposure of the Indian continental basement and forms the core of the westernmost syntaxis of the Central Himalayas.

The Nanga Parbat-Haramosh massif is the focal point of one of the most extraordinary and facinating geological settings in the world. Its western margin is bounded by the deep gorge of the Indus River. Between the Nanga Parbat peak (8125 m) and the river bed (1200 m) the topography reveals a rugged and bizzare terrain with the maximum relief (over 7000 m) seen anywhere in the world. This is because the Nanga Parbat-Haramosh massif has been rapidly rising at the rate of 5 to 10 mm/year in Recent and Late Quaternary times and the mountain has undergone an uplift of over 20 km during the past 7 million years (Zeitler, 1985). Even more surprising is the fact that this extreme upsurge has been confined to a 10-20-km-wide zone. This extremely localized uplift of the Nanga Parbat-Haramosh massif occurs at the junction of three suture zones, the Indus suture zone, the Tsangpo suture zone and the Karakoram suture zone. It is essentially a fault block surrounded by major faults on three sides. An active dextral reverse fault, the Raikot fault, forms its northwestern margin (Lawrence and Ghauri, 1983a; Madin, 1986). Southward this fault intersects the MMT near its eastern end and it has been the vehicle for the main upthrust. Previously the Raikot fault was mistaken to be a continuation of the MMT. To the north, the Nanga Parbat-Haramosh massif is bounded by the MKT, whereas to the east it is flanked by the Stak fault (Verplanck, 1987), which truncates the Indus suture zone east of the massif (fig. 3.12).

The Nanga Parbat-Haramosh massif is comprised of Precambrian metasediments which have been variously described as Salkhalas (Wadia, 1933), Nangaparbat gneiss (Misch, 1949), Precambrian-Cambrian basement (Coward and others, 1982) and Nanga Parbat gneisses (Tahirkheli, 1983). Madin (1986) has proposed the name Nanga Parbat group for these rocks.

According to Madin (1986) and Verplanck (1987) the Nanga Parbat group comprises the following three lithostratigraphic units (listed from structurally lowest to the highest).

Shengus gneiss: It is fine-grained, finely laminated amphibolite-grade pelitic to psammitic gneiss, with intercalations of amphibolites and calcsilicate gneisses. The Shengus gneiss is over 5 km thick and has formed as a result of metamorphism of a vast pile of shale, marl, arkosic sandstone, and limestone.

Iskere gneiss: It is coarse-grained, banded, amphibolite-grade biotite gneiss, intercalated with biotite schist, amphibolite and calcsilicate gneiss. The Iskere gneiss has formed as a result of the metamorphism of plutonic rocks of intermediate composition which had intruded a thick sequence of sandstone with minor amounts of pelitic and calcareous rocks. The Iskere gneiss is over 8 km thick. The base of the gneiss is not exposed and its actual thickness is thus likely to be considerably more. Metaplutons in the Iskere gneiss are about 1.8 b.y. old (U/Pb date by Zartmann, quoted in Madin, 1986).

Haramosh schist: It is medium- to coarse-grained amphibolite-grade biotite schist and gneiss with marble, calcsilicate gneiss and subordinate amphibolite. The Haramosh schist may be differentiated from Iskere gneiss by its lack of coarse biotite orthogneiss. The protolith of the Haramosh schist has been interpreted as a sequence of arkosic and greywacke sandstones with minor marl and limestone (Madin, 1986).

The rocks of the Nanga Parbat group have been extensively intruded by aplite and pegmatite dikes. In the proximity of larger faults such as the Stak fault, the Baroluma fault, and the Raikot fault, these pegmatites are gem-bearing and are locally mined for bi- and tri-color tourmaline, topaz, aquamarine, garnet, and other minerals (Kazmi and others, 1985). In the Khaltar valley in the vicinity of the Raikot fault some of the pegmatites contain emeralds.

Emerald Occurrence

In the Khaltar valley, emeralds are confined to pegmatite dikes which have intruded the Haramosh schist in the vicinity of the Raikot Fault (fig. 3.12).

This is a highly crushed and sheared zone between the Haramosh schist (Indian plate sequence) and the Shuta basic intrusives of Madin (Kohistan island arc sequence). The Shuta basic intrusives are largely comprised of gabbro, norite and diorite with commonly occurring inclusions of amphibolites (Hanuchal amphibolite of Madin, 1986; Kamila amphibolite and pyroxene granulite of Tahirkheli, 1983). Blocks of these mafic and ultramafic rocks along with those of the Haramosh schist are found in the shear zone of the Raikot fault which is at places several hundred meters wide. The fault zone has been hydrothermally altered and it contains a wide but discontinuous zone of mylonite.

At Khaltaro, the country rock in which the emerald pegmatites occur are banded biotite gneisses which strike northward and dip eastward. The rocks have been tightly folded. The gneisses are mainly comprised of mica, feldspar, quartz, garnet, and hornblende.

The emerald-bearing pegmatites cut through slivers of dark olive green mafic to ultramafic rocks which occur randomly in the Raikot fault zone.

At Khaltaro two pegmatite veins have been worked for emeralds by the Gemstone Corporation of Pakistan. These are briefly described below.

Khaltaro I

The outcrop of this pegmatite vein is about 5 meters long and 3 meters wide. It cuts across the biotite gneiss. Three distinct zones are discernible in the pegmatite. Its outer margin is fine grained and it is largely comprised of feldspar, quartz, tourmaline, hornblende, and mica. It is followed by an intermediate zone which is coarse and contains the same mineral assemblage. The third zone comprising the inner part of the pegmatite is blocky and coarse grained. Tourmaline and hornblende crystals are concentrated in the core surrounded by massive feldspar and quartz (fig. 3.13) with lesser amounts of calcite, apatite, fluorite, beryl, and mica. Small crystals of muscovite, biotite, and hornblende, however, are spread throughout the groundmass of the pegmatite (Khan and Aziz, 1985). The beryl is commonly emerald, though rarer crystals of goshenite also occur. Emerald is found embedded in feldspar. The emerald crystals are euhedral, well formed and of light to medium light green color and commonly range to 3 cm in diameter (photo. 3.23).

Khaltaro 2

This pegmatite vein is shaped like an octopus (fig. 3.14) and cuts across coarse- grained green- colored mafic (gabbroic?) rock (Khan and Aziz, 1985). It is about 3 to 4 meters thick and is mainly comprised of feldspar, quartz, mica, tourmaline, calcite, fluorite, and apatite, with lesser amounts of transparent to semitransparent euhedral green crystals of emerald and goshenite.

In Pakistan, Khaltaro emeralds are the only ones which occur in pegmatites. Nevertheless,

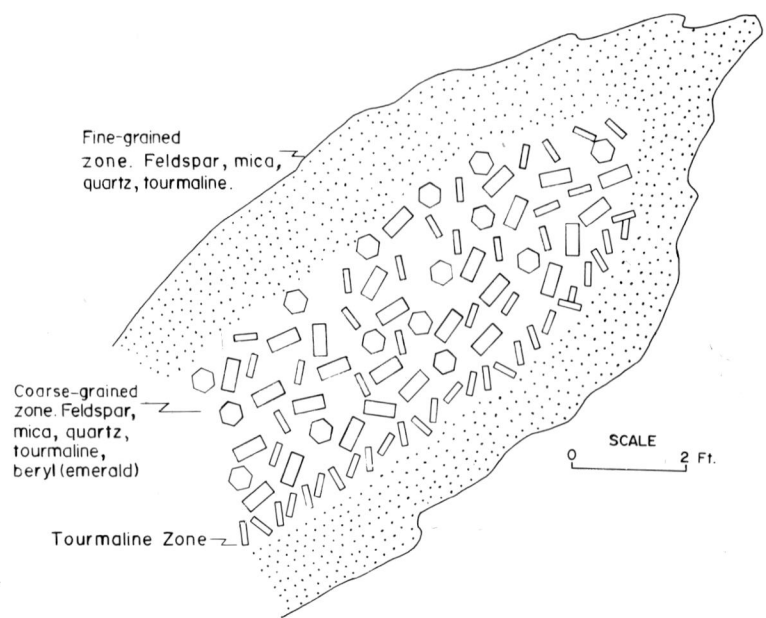

Fine-grained
zone. Feldspar, mica,
quartz, tourmaline.

Coarse-grained
zone. Feldspar,
mica, quartz,
tourmaline,
beryl (emerald)

Tourmaline Zone

SCALE
0   2 Ft.

Figure 3.13. Sketch showing geological section of the emerald-bearing Khaltaro I pegmatite vein (after Khan and Aziz, 1985).

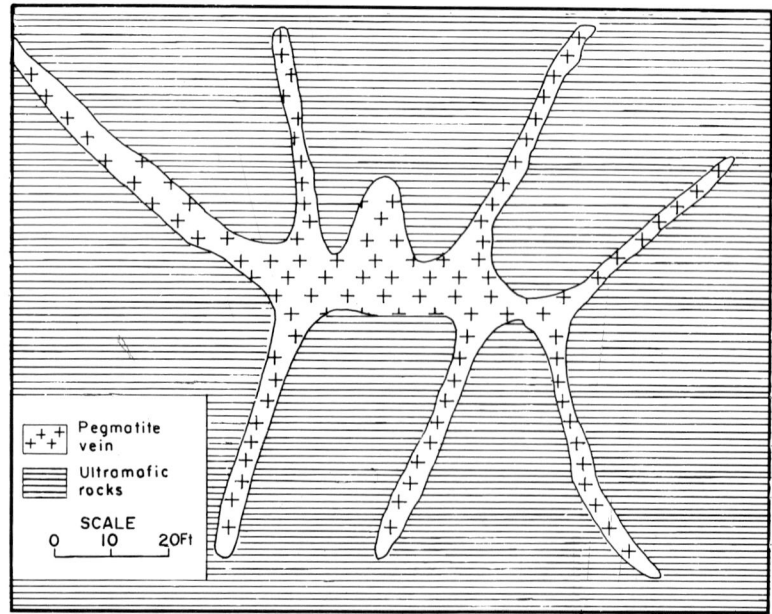

Pegmatite
vein

Ultramofic
rocks

SCALE
0   10   20Ft

Figure 3.14. Sketch showing geological section of the emerald-bearing Khaltaro 2 pegmatite vein (after Khan and Aziz, 1985).

like the other emerald deposits of Pakistan, the Khaltaro emeralds are also located along a major suture zone, although at this point the suture has been displaced by an active transform fault, the Raikot fault. The sources of Be and Cr in the Khaltaro emeralds are likely the result of the fortuitous juxtaposition of two distinctly different rock types. The Cenozoic granitic intrusion in this region and in close proximity to the Nanga Parbat-Haramosh massif undoubtedly provided the Be whereas their late magmatic hydrothermal solutions, while traversing the mafic and ultramafic rocks in the faulted suture zone, picked up Cr.

Quality of Khaltaro Emeralds

The Khaltaro emeralds are commonly of light to medium green color. They occur as well-formed euhedral prismatic crystals. The crystals are largely 1 to 3 cm in diameter, though larger crystals are not uncommon. These stones generally contain inclusions and fine cracks, though some very fine clear stones also have been formed (photo. 3.24).

Mining and Production

Due to extremely rugged terrain and high altitude where mining is possible only for about three months in a year the Khaltaro deposits have not yet been either fully explored or developed. Exploratory mining was attempted by Gemstone Corporation of Pakistan at two points and a small production of about 600 carats of gem-grade stones was obtained.

OTHER PAKISTANI EMERALD OCCURRENCES

Emerald showings have been noted at Gandao, Nawe Dand (Ahmed, 1966) and Bucha (Rafiq and Jan, 1985; Hussain and others, 1984) in Mohmand Agency, at Pranghar and Khanori (Kot) in Malakand Agency (Hussain and others, 1984) and at Aman Kot and Maimola in Bajaur Agency (fig. 2.1). There are unconfirmed reports that emerald has been found and mined in Alai Kohistan, east of the Indus River. However, all these localities are inaccessible and have not been geologically investigated though some of the deposits have been visited by geologists who have made brief comments about them. The available information is summarized below.

Gandao Green Beryl

This deposit is located about 43 km northwest of Peshawar. Gandao green beryl has been mined and sold as emerald, although recent electron microprobe analyses by Hammerstrom (chapter 6, this volume) and geochemistry by Snee and others (chapter 5, this volume) on some samples obtained from local gem merchants indicate that the green color is due to vanadium substitution for aluminium in the beryl structure. Thus the Gandao beryl is not emerald but is green vanadian beryl. The Gandao green beryl commonly occurs in dolomite (photo. 3.25).

The Gandao site was visited by Mohammad Aslam, geologist, Geological Survey of Pakistan, several years ago and according to him (personal commun., 1985) mineralization has taken place in quartz veins in dolomite. The dolomite forms tight folds in the hillocks southeast of Tora Tigga village. The dolomite is overlaid by greenschist and underlaid by undifferentiated schist and limestone. According to Aslam at the time of his visit, emerald (green beryl) was being mined in a haphazard manner but the linear arrangement of some of the workings showed that these were located on a particular quartz vein. Most of the workings were on the

Photo.3.24. Emerald crystals in Khaltaro pegmatite. Length of the pegmatite specimen approximately 25 cm (photo *A. H. Kazmi*).

Photo.3.25. Gandao green beryl crystals in dolomite matrix (photo *A. H. Kazmi*).

southwestern slope of the Tora Tigga hill. There were 75 workings in a distance of 1000 meters and 100 people were engaged in mining.

Barang (Aman Kot and Maimola) Emerald

These deposits are located 80 kilometers north of Peshawar in Bajaur Agency and were visited by Shahid Hussain, Geologist, Gemstone Corporation of Pakistan in 1983. According to him (personal commun., 1983), the emeralds occur in talc-chlorite schist in a melange zone.

Tribal people have sporadically mined these deposits and the emeralds have been sold in Swat and Peshawar under the name of "Barang" emeralds. The authors have seen some of the rough emeralds from this area. They are commonly in the 0.5—5.0 ct or larger size range and are well formed, euhedral, clear, transparent and of attractive medium green color. It appears that after the Mingora and Gujarkili emeralds the Barang emeralds might well be the next best Pakistani emeralds.

Nawe Dand Emerald

This deposit is located on a small hillock south of Nawe Dand village, 40 kilometers north of Peshawar. The emeralds are found in quartz-feldspar veins in talc schist at its contact with hornblende-chlorite schist. The emerald-bearing veins are mainly composed of quartz, calcite and feldspar and they generally follow the foliation planes. These veins are spread over an area of over 5 square kilometers (Ahmed, 1966). This deposit has been sporadically mined in the past.

REFERENCES

Ahmed, W., 1966, A short note on emerald deposits of Nawe Dand, Mohmand Agency, Peshawar Division: Geological Survey of Pakistan, pre-publication Issue, No. 29, 4p.

Bowersox, G.W., and Anwar, J., 1989, The Gujar Killi emerald deposit, Northwest Frontier Province, Pakistan: Gems and Gemology, v. 25, p. 16-24.

Coward, M.P., Jan, M.Q., Rex, D., Tarney, J., Thirlwall, M., and Windley, B.F., 1982, Geotectonic framework of the Himalaya of N. Pakistan: Geological Society of London, Journal v. 139, p. 299-308.

Davies, R.G., 1962, A green beryl (emerald) near Mingora, Swat State: Panjab University, Geological Bulletin, v. 2, p. 51-52.

Gübelin, E.J., 1968, Gemmologische Beobachtungen am neun Smaragd aus Pakistan: Der Aufschluss, Sonderheft 18, p. 110-116.

Gübelin, E.J., 1982, Gemstones of Pakistan: emerald, ruby and spinel: Gems and Gemology, v. 18, p. 123-139.

Hammarstrom, J.M., 1989, Mineral chemistry of emeralds and some associated minerals from Pakistan and Afghanistan: an electron microprobe study: *in* A.H. Kazmi and L.W. Snee, editors, *Emeralds of Pakistan: Geology, Gemology* and *Genesis;* Van Nostrand Reinhold, New York, p. 125-150.

Herzberg, C., 1981, Pakistan debuts first official gemstone collection: National Jeweller (New York), April 16, 1981.

Hussain, S.S., Khan, T., Dawood, H., and Khan, I., 1984, A note on Kot-Prang Ghar melange and associated mineral occurrences: University of Peshawar, Geological Bulletin, v. 17, p. 61-68.

Jan, M.Q., 1968, Petrography of the emerald-bearing rocks of Mingora (Swat State) and Prang Ghar (Mohmand Agency), West Pakistan: University of Peshawar, Geological Bulletin, v. 3, p. 10-11.

Jan, M.Q., Kamal, M., and Khan, M.I., 1981, Tectonic control over emerald mineralization in Swat: University of Peshawar, Geological Bulletin, v. 14, p. 101-109.

Kazmi, A.H., Peters, J.J., and Obodda, H., 1985, Gem pegmatites of Shingus-Dusso area, Gilgit, Pakistan: Mineralogical Record, v. 16, p. 393-411.

Kazmi, A.H., Lawrence, R.D., Anwar, J., Snee, L.W., and Hussain, S., 1986, Mingora emerald deposits (Pakistan): suture-associated gem mineralization: Economic Geology, v. 81, p. 2022-2028.

Khan, I.A., and Aziz, K., 1985, A brief note on the beryl and emerald-bearing pegmatites of Khaltaro area, Gilgit: Gemstone Corporation of Pakistan, unpublished report, 4 p.

Lawrence, R.D., and Ghauri, A.A.K., 1983, Evidence of active faulting in Chilas district, northern Pakistan: University of Peshawar, Geological Bulletin, v. 16, p. 185-186.

Lawrence, R.D., Kazmi, A.H., and Snee, L.W., 1989, Geological setting of the emerald deposits: *in* A.H. Kazmi and L.W. Snee, editors, *Emeralds of Pakistan: Geology, Gemology and Genesis,* Van Nostrand Reinhold, New York, p. 13-38.

Madin, I.P., 1986, Geology and neotectonics of northwestern Nanga Parbat-Haramosh massif: M.S. thesis, Oregon State University, Corvallis, Oregon, 99 p.

Misch, P., 1949, Metasomatic granitization of batholithic dimensions: American Journal Science, v. 247, p. 209-245.

Rafiq, M., and Jan, M.Q., 1985, Emerald and green beryl from Bucha, Mohmand Agency, NW Pakistan: Journal of Gemmology, v. 19, no. 5, p. 404-411.

Snee, L.W., Foord, E.E., Hill, B., and Carter, S.J., 1989, Regional chemical differences among emeralds and host rocks of Pakistan and Afghanistan: implications for the origin of emerald: *in* A.H. Kazmi and L.W. Snee, editors, *Emeralds of Pakistan: Geology, Gemology and Genesis,* Von Nostrand Reinhold, New York, p. 93-124.

Tahirkheli, R.A.K., 1983, Geological evolution of Kohistan Island Arc: The southern flank of the Karakoram-Hindu Kush in Pakistan: Bolletino Li Geofisica Teorica Applicata, v. 25, No. 99-100.

Verplanck, P.L., 1987, A field and geochemical study of the boundary between the Nanga Parbat-Haramosh massif and the Ladakh arc terrane, Northern Pakistan: M.S. thesis, Oregon State University, Corvallis, Oregon, 136 p.

Wadia, D.N., 1933, Note on the geology of Nanga Parbat (Mt. Diamir) and adjoining portions of Chilas, Gilgit District, Kashmir: Geological Survey of India, Records, v. 66, p. 212-234.

Zeitler, P.K., 1985, Cooling history of the northwestern Himalaya, Pakistan: Tectonics, v. 4, p. 127-151.

Gemological Characteristics of Pakistani Emeralds

Edward J. Gübelin

INTRODUCTION

Emeralds and their deposits have been described time and again ever since antiquity. One piece of very early information has been handed down to us by a bas-relief on the wall of a temple east of Edfu (Egypt), which depicts the geographic situation of the so-called "Cleopatra's Emerald Mines" in the barren desert Wadi of Sikeit, near Ras Banas on the coast of the Red Sea. The literary references are of a quite different kind—either simple travel reports, descriptions by prospectors, appraisals by jewellers or investigations by gemologists. Through the ages, the accuracy of such reports kept pace with the increase of knowledge and the progress of science. The most welcome of these have always been those which describe a newly discovered deposit or introduced a new gemstone, but also those which shed further light upon a known source or a known gem by means of a more reliable method of research. This chapter is devoted to a precise account of the gemologically ascertainable diagnostic properties of the emeralds from the various occurrences in Pakistan.

For this examination the author received a parcel of five to eight specimens from each of the following deposits, ranging from east to west: Khaltaro, Gujarkili, Makhad, Charbagh, and Swat Mines 1 and 2; as well as 70 specimens from the Mingora area. The green beryls from Gandao cannot be regarded as emeralds because their green color is not imparted by Cr_2O_3 but exclusively by V_2O_3 (see chapters 5 and 6).

COLOR AND APPEARANCE

The specimens ranged between a low quality of pale green to grayish green hue marred by numerous inclusions rendering them translucent rather than transparent, and a very fine quality of an exquisite bluish or yellowish green shade, highly transparent with only very few inclusions. The majority were in a rough state, with one or two polished prism faces so that the refractive indices could be measured and their interior studied under the microscope. The finest emeralds were faceted in baguette or emerald cuts. Of the cut gems, the best were of good to excellent quality; outstanding in terms of transparency with a vivid, saturated hue. The best Swat emeralds were comparable to fine Sandawana emeralds, while the poorer quality were rather reminiscent of the lifeless, cloudy emeralds from the Transvaal (RSA). Cut gems were favored for a more exact determination of the coloration range. Referring to the DIN color chart 6164, the best colors of the lighter and darker hues were found to be:
$X_c15.8$; $Y_c25.7$; $Z_c14.0$, to $X_c16.2$; $Y_c25.1$; $Z_c21.2$ while the less desirable colors concurred with the following marks:
from $X_c36.8$; $Y_c43.0$; $Z_c40.3$; to $X_c32.4$; $Y_c41.3$; $Z_c37.4$, and $X_c16.2$; $Y_c25.1$; $Z_c21.2$; to $X_c10.8$; $Y_c16.6$; $Z_c14.1$. The color is in some cases marred by the multitude of inclusions in general, and of *included chromite* in particular (photo. 4.1).

PHYSICAL PROPERTIES

Examinations to establish the physical properties were carried out on all the samples, of which 70 originated from the Mingora mines, while five to eight polished crystals originated from each of the other smaller mines. The results tallied well with observations made on the cut gems which were preselected for detailed examinations. The latter ranged in weight from 0.51 to 2.34 ct. and were consistently outstanding in clarity and color.

Refractive indices were determined with sodium light (589.3 nm) on two different gemological refractometers (Eickhorst LED and LED Dialdex) and a Rayner with spinel prism and extended scale allowing an exact reading to the third decimal place with a deviation of ± 0.0005. All the emeralds thus measured supplied mean constants of n_e= 1.583 and n_o=1.590. The birefringence of $\triangle n$=0.007 proved to be remarkably consistent in value and rather high for emeralds of gem quality, but in agreement with emeralds with a high percentage of Fe.

The density was determined in distilled water at room temperature, using a hydrostatic balance. Values between 2.66 and 2.78 were obtained with an arithmetic median of 2.71 g/cm³.

A comparison of the data in table 1, which presents the values for the emeralds from the various mines in Pakistan, confirms that the refractive indices, birefringence, and density vary somewhat, yet not sufficiently to serve as a reliable means of discrimination between the emeralds from the different deposits (with the exception perhaps of those from the Swat mine 2, which possess the highest constants). Checking the data of table 2 and collating them with each other, it becomes apparent that the emeralds from Pakistan figure among those with the highest constants—especially the emeralds from the Swat mine 2. Yet even so, the frequent overlapping of values does not encourage a reliable dependence upon the accidental differences in the measurable properties of emeralds from the listed sources.

The other optical properties not provided in tables 1 and 2 are discussed below. These do not differ appreciably, but are characteristic of emeralds from any deposit (with the exception of the absorption band at 420-430 nm). The *dichroism* is distinct yet not intensive, and alternates between bluish green along the c-axis and yellowish green at right angles to the c-axis. In general the *absorption spectrum* (fig. 4.1), which may be scrutinized easily with the optical spectroscope, manifests the normal chromium (Cr^{+3}) absorption lines with various intensities (the weak ones sometimes missing) at the red end,

for o: 685, 680, 637 and 604,

for e: 683, 680, 662, 646 and 630

as well as iron (Fe^{+3}) absorption lines in the blue section, for o: 477.4, 445, and 372.5—also with changing intensities. However, the unusually high iron content of most of these emeralds from Pakistan results in a broad absorption band in the blue from 420 nm for e to 430 nm for o, with an absorption maximum at 427 nm. Since this absorption band was consistently present in the Pakistani emeralds tested, it is a welcome additional means of identification. The emeralds from Pakistan appear light red under the *color filter* and glow red to orange in the *Stokes fluoroscope* (double filter method). They fail to respond to either short or long wave *ultraviolet rays,* and the shortwave ultraviolet radiation (253 nm) is completely absorbed from 330 nm downwards (fig. 4.1). The lack of luminescence and the shortwave absorption is due to the high iron content.

INCLUSIONS

The inclusion scenes in the emeralds from Pakistan present features which are primarily of diagnostic value in enabling a distinction of these gems from synthetic counterparts, but in

Table 4.1. Varying physical properties of emeralds from Pakistan

Deposits	Refractive indices for		Birefringence $\triangle n$	Frequency median for		Density g/cm³	Density median
	n_e	n_o		n_e	n_o		
Swat Mine-1 (Mingora)	1.578—1.586	1.584—1.594	-0.006—0.008	1.582 (14 measurements)	1.589	2.72—2.77	2.76
Swat Mine-2 (Mingora)	1.583—1.591	1.591—1.600	-0.008—0.009	1.586 (9 measurements)	1.594	2.70—2.78	2.74
Charbagh	1.577—1.585	1.584—1.592	-0.007—0.009	1.585 (21 measurements)	1.592	2.68—2.70	2.69
Gujarkili	1.582—1.593	1.589—1.600	-0.007—0.009	1.583 (14 measurements)	1.591	2.69—2.74	2.73
Makhad	1.579—1.587	1.586—1.595	-0.007—0.008	1.583 (15 measurements)	1.590	2.74—2.76	2.75
Khaltaro	1.581—1.582	1.589—1.590	-0.008	1.581 (5 measurements)	1.589	2.66—2.72	2.69
Sarkan Mine	1.575—1.590	1.582—1.599	-0.007—0.009	1.584 (5 measurements)	1.591	2.68—2.73	2.71
Arithmetic medians of 7 different occurrences	1.583	1.590	—0.007				2.71

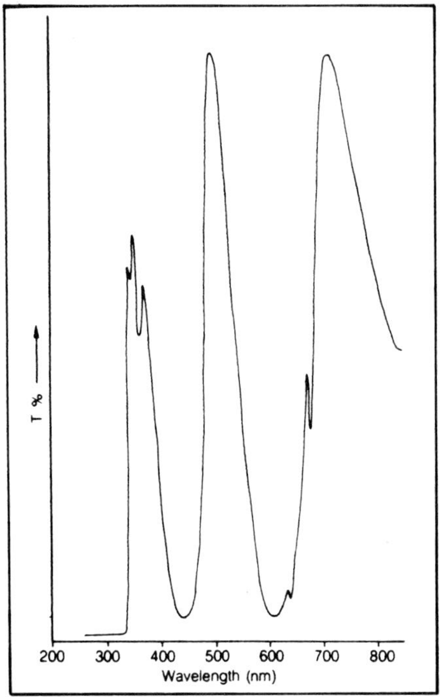

Figure 4.1. Absorption curve of Pakistan emerald plotted by a Beckman Spectral Photometer at room temperature. Note the conspicuous amplitude between 425 and 430 nm.

Table 4.2. Optical constants and densities of emeralds from various world deposits

Country and deposit	Refractive indices		Birefringence $\triangle n$	Density g/cm3
	n_e	n_O		
AUSTRALIA				
Poona	1.572	1.578	-0.005—0.007	2.693
BRAZIL				
Bahia				
Brumado	1.573	1.579	-0.005—0.006	2.682
Carnaiba	1.583	1.588	-0.006—0.007	2.72
Salininha	1.583	1.589	—0.006	2.70
Socotó	1.577	1.583	—0.006	2.71
Goiaś				
Santa Terezinha	1.585	1.592	—0.006	2.752
	1.587	1.595	—0.008	2.764
Minas Gerais				
Belmont (Itabira)	1.576	1.581	—0.005	2.728
	1.582	1.590	—0.008	2.742
INDIA				
Ajmer	1.585	1.595	—0.007	2.735
COLOMBIA				
Borbur	1.569	1.576	—0.007	2.704
Chivor	1.570	1.579	-0.005—0.006	2.688
Muzo	1.570	1.580	-0.005—0.006	2.698
MADAGASCAR				
Ankadilalana	1.581	1.589	—0.008	2.727
	1.585	1.591	—0.006	
MOZAMBIQUE				
Maria III	1.585	1.591	—0.006	2.73
Melela (Morrua)	1.585	1.593	—0.008	2.73
NORWAY				
Eidsvol (Minnesund)	1.583	1.590	—0.007	2.759
AUSTRIA				
Habachtal	1.584	1.591	—0.007	2.734
PAKISTAN				
(medians of 7 deposits)	1.583	1.590	—0.007	2.71
ZAMBIA				
Kafubu	1.592	1.602	—0.010	2.77
Kitwe	1.580	1.586	—0.006	2.794
Miku	1.582	1.589	-0.007—0.008	2.738
ZIMBABWE				
Mayfield	1.584	1.590	—0.006	2.72
Sandawana	1.584	1.590	—0.006	2.75
TANZANIA				
Lake Manyara	1.578	1.585	—0.006	2.72
REPUBLIC OF SOUTH AFRICA				
Cobra Mine (Transvaal)	1.583	1.594	-0.006—0.007	2.75
USSR				
Tokowaya (Sverdlovsk)	1.580	1.588	-0.006—0.007	2.74
USA				
Hiddenite (N.C.)	1.581	1.588	—0.007	2.73

* These data represent arithmetic medians of the examined specimens.

some cases they also grant an indication as to the place of origin.

To the unaided eye, the general appearance of the emeralds from Pakistan agrees with the so-called "jardin" in natural emeralds, yet the wispy "feathers" and "veils" somewhat resemble the lace-like wavy "banners" in most synthetic emeralds. These distinguish themselves from the "feathers" by a granular appearance both in synthetic flux emeralds from Russia and in the hydrothermal synthetic emeralds by Biron.

Under the microscope however, a host of manifold inclusions becomes discernible which immediately challenges the inquisitive mind of the keen observer. Partially healed fractures distinctly marked by fluid inclusions of either secondary or pseudosecondary formation encompass the majority of the internal features. They are accompanied by primary fluid inclusions as well as by solid inclusions of protogenetic and syngenetic formation (photo. 4.2).

The *primary fluid inclusions* are of two types: Type one occurs as bevies of narrow rectilinear filaments in parallel alignment and elongated along the c-axis. They commonly arise from tiny crystalline growth obstacles (mineral inclusions), and in some cases they form such dense masses that a cat's-eye effect would result if the host emerald were cut as a cabochon (photos. 4.3 and 4.4). These are aqueous two-phase inclusions composed of a watery solution with its vapor bubble. Type two is most remarkable, yet somewhat bewildering, as it encompasses pronged three-phase inclusions oriented parallel to the c-axis, such as gemologists are very familiar with in Colombian emeralds (photo. 4.5). These—just as the Colombian ones—are hydrous and saline, containing cubic crystals of either halite or sylvite and occasionally some other dark daughter crystals. These jagged inclusions, which so far have only been observed in the emeralds from the Mingora mines, represent natural primary syngenetic growth defects which trapped part of the nutrient solution during crystal formation. If such fluid is chemically pure, upon condensation resulting from lower temperature and pressure these inclusions become two-phase: liquid and gas. However, if the solution is saline, solid phases in the form of minute daughter crystals may precipitate (photo. 4.6).

The secondary fluid inclusions are also of two kinds: Either they line irregular fractures criss-crossing the emerald, or they mark cleavage planes parallel to the basal pinacoid (0001). The partly healed fractures hold fluid droplets which determine the pattern by their arrangement and their rounded, hose-like, elongated, drop-like or free-form shapes (photo. 4.7). They form webs over flat planes or follow irregularly curved internal surfaces which conform with the course of former cracks (photo. 4.8). Directional forces during the healing process of such fractures must have had an influence, because, interestingly enough, the droplets not only display irregular but also elongated tube-like forms closely paralleling crystallographic directions. Despite this orientation, they neither lie along the prism faces nor do they follow the c-axis (photo. 4.9). In this respect they decisively differ from the jagged and filamental growth defects manifested by the primary inclusions. As far as their origin and composition are concerned, they are two- and three-phase inclusions consisting either of one solution with its vapor bubble or two immiscible liquids with one gas bubble representing the lixiviated remnants of the therapeutic fluid that entered the fractures and partly repaired them (photo. 4.10).

Completely different in appearance from these "fingerprint" inclusions which settled in former fractures, are the flat film-like, two-phase inclusions the contours of which are normally irregularly rounded-off, but occasionally also subhexagonal (photo. 4.11). Under oblique illumination they glow with brilliant interference colors which vary depending upon the thickness of the droplets. The gas libellae usually show a complementary color to the one of the embracing liquid (photo. 4.11). This second category of secondary fluid inclusions is absolutely

Photo. 4.1. Photomicrograph depicting a *general view* of a dense inclusion scene with numerous fluid and chromite inclusions in an emerald from Pakistan. Darkfield. 20x (photo *E. Gübelin*).

Photo. 4.2. *General view* of various inclusions (primary and secondary fluid inclusions accompanied by dolomite rhombohedra, growth tubes, and flat parallel fissures). Emerald from the Mingora area. Darkfield. 20x (photo *E. Gübelin*).

Photo. 4.3. Parallel *growth tubes* with two-phase (l-g) fillings. Near the bottom, secondary fluid (l-g and l-1g) two-phase and multi-phase inclusions marking partly healed fractures. Emerald from the Mingora area. Darkfield. 25x (photo *E. Gübelin*).

Photo. 4.4. A very dense concentration of parallel filamental *growth tubes* interspersed by larger two-phase (l-g and three-phase (l-g-s) inclusions cause chatoyancy on this cabochon-cut emerald from the Mingora area. Darkfield. 32x (photo *E. Gübelin*).

Photo. 4.5. Sharply jagged, primary two-phase (l-g) and *three-phase (l-g-s) inclusions* following the direction of the c-axis in an emerald from the Mingora area. Darkfield. 32x (photo *E. Gübelin*).

Photo. 4.6. An individual primary *three-phase (l-g-s-) inclusion* orientated parallel to the c-axis of an emerald from the Mingora area. Darkfield. 100x (photo *E. Gübelin*).

Photo. 4.7. Oddly shaped secondary monophase (l) and two-phase (l-g) *fluid inclusions* designating a partly healed fracture in an emerald from the Mingora area. Darkfield. 40x (photo *E. Gübelin*).

Photo. 4.8. Primary fluid *three-phase (l-g-s) inclusions and secondary fluid two-phase (l-g) inclusions* (some of the latter slightly curved) prevail in the typical endogenesis of this emerald from the Mingora area. Darkfield. 20x (photo *E. Gübelin*).

4.1

4.2

4.3

4.4

4.5

4.6

4.7

4.8

Photo. 4.9. Primary and secondary *fluid inclusions* in an emerald from the Mingora area. Note the conspicuously curved arrangement of the secondary inclusions. Such a curved configuration is highly characteristic of the emeralds from Pakistan. Darkfield. 35x (photo *E. Gübelin*).

Photo. 4.10. Very typically curved distribution of *secondary fluid two-phase* (l-g) inclusions which populate the basal plane and reflect its hexagonal contour in an emerald from the Mingora area. Darkfield. 32x (photo *E. Gübelin*).

Photo. 4.11. Numerous tiny *fluid two-phase (l-g) films* populating a cleavage plane parallel to the basal face in an emerald from the Mingora area. Under inclined illumination these fluid films reflect with interference colors. Note dolomite crystal above fluid films. 40x (photo *E. Gübelin*).

Photo. 4.12. Ultra-thin *fluid films* expanding on a basal plane display a vivid array of interference colors in an emerald from the Mingora area when illuminated vertically. 50x (photo *E. Gübelin*).

Photo. 4.13. One section featured by elongated and triangular *two-phase (l-g) inclusions* is separated from another section devoid of inclusions by a tooth-like *growth edge* in an emerald from the Mingora area. Darkfield. 80x (photo *E. Gübelin).*

Photo. 4.14. *Zig-zag growth seams* emphasized by parallel swirl marks suggest irregular growth successions which may often be observed in emeralds from Pakistan. Darkfield. 25x (photo *E. Gübelin*).

Photo. 4.15. The central core of this emerald from Swat Mine 2 in the Mingora area consists of an almost colorless, *hexagonal column,* and thus betrays that this crystal first grew without incorporating the coloring pigment chromium. Darkfield. 20x (photo *E. Gübelin*).

Photo. 4.16. Cross section of an emerald from Khaltaro, subtly marked by countless *albite inclusions* in a pale green body which is wrapped in a darker green coat. Darkfield. 10x (photo *E. Gübelin*).

4.9

4.10

4.11

4.12

4.13

4.14

4.15

4.16

Photo. 4.17. Side view of an emerald crystal from Khaltaro. It is accompanied by *satellite emeralds* which grew along its prisms. It is also capped by a cluster of *albite feldspar,* and in some of the fractures, dendritic *limonite* has formed, rendering this emerald crystal a particularly interesting object for investigation. Darkfield. 15x (photo *E. Gübelin*).

Photo. 4.18. Shattered black grain of pyrrhotite amongst *secondary fluid two-phase (1-g) inclusions* and subdued accumulation actinolite fibres in an emerald from Charbagh. Darkfield. 45x (photo *E. Gübelin*).

Photo. 4.19. Group of brownish green euhedral prismatic crystals and greenish black fragments of *actinolite* above a strongly corroded crystal of *albite* in an emerald from Charbagh. Darkfield. 25x (photo *E. Gübelin*).

Photo. 4.20. Individual brownish green slightly corroded *actinolite* rod in contact with greyish black cubes of gersdorffite in an emerald from Khaltaro. Darkfield. 40x (photo *E. Gübelin*).

Photo. 4.21. SEM portrait of a prismatic crystal of actinolite, distinctly discloses the morphology of this crystal inclusion. 15000x (photo *E. Gübelin*).

Photo. 4.22. A well preserved black crystal of *chromite* accomodated by a relatively clear emerald from the Mingora area. Darkfield. 40x (photo *E. Gübelin).*

Photo. 4.23. *Scattered black grains of chrome chlorite* surrounded by indistinct strongly corroded magnesite crystals mar the interior of an emerald from Charbagh. Darkfield. 20x (photo *E. Gübelin*).

Photo. 4.24. A small group of well shaped rhombohedral *dolomite* crystals hovers amongst primary two-phase (1-g) inclusions and rectilinear growth tubes. Darkfield. 50x (photo *E. Gübelin*).

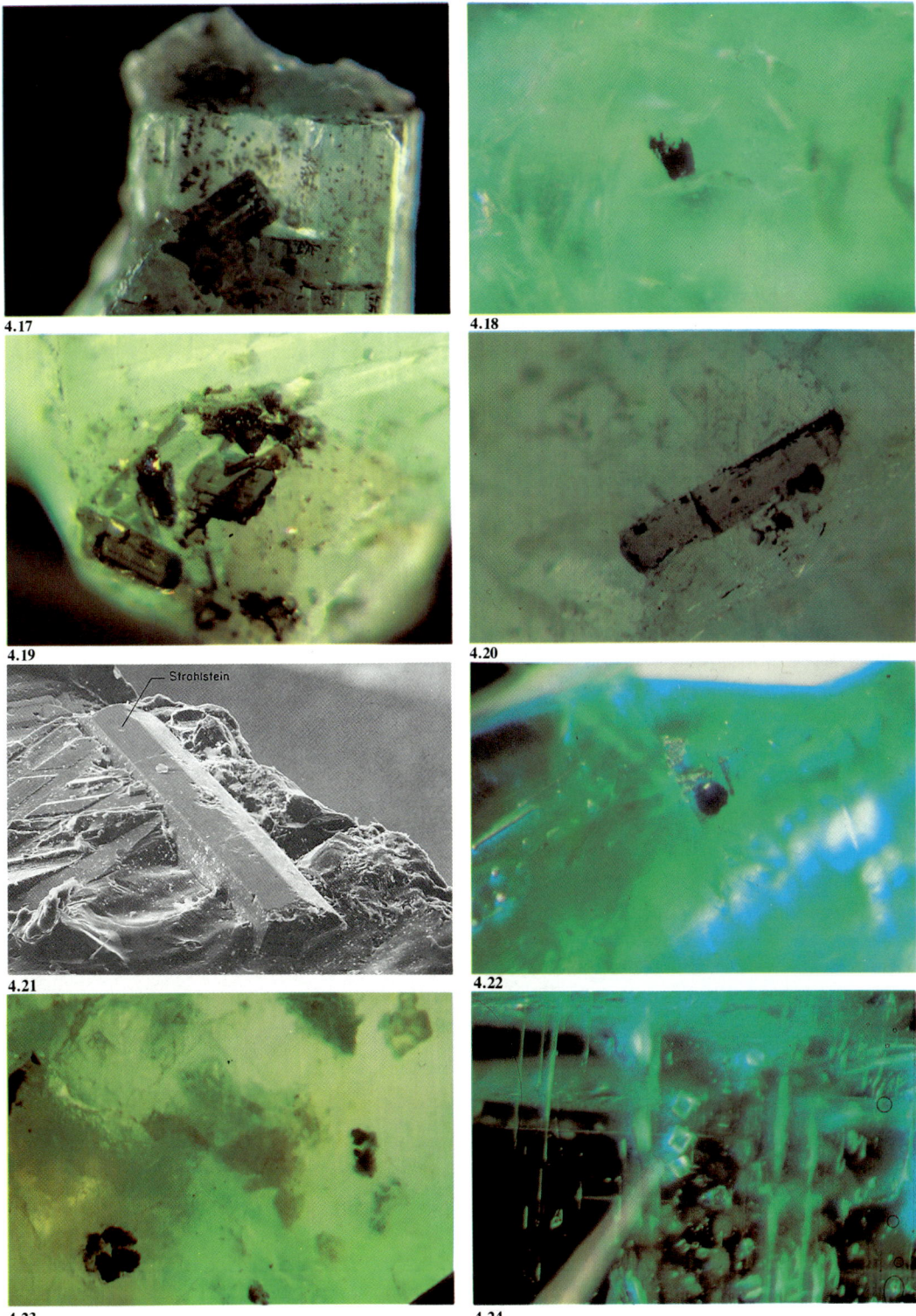

4.17

4.18

4.19

4.20

Strohlstein

4.21

4.22

4.23

4.24

Photo. 4.25. Group of distinctly shaped brownish, iron-stained *dolomite* rhombohedra represent an essential element of the interior of an emerald from the Mingora area. Darkfield. 50x (photo *E. Gübelin*).

Photo. 4.26. A clearly rhombohedral *dolomite* crystal suspended above a mass of *fluid films* in an emerald from the Mingora area. Darkfield. 40x (photo *E. Gübelin*).

Photo. 4.27. Large black patches of chromite inclusions escort strongly corroded yellowish *enstatite* crystals in an emerald from the Swat 1 Mine in the Mingora area. Darkfield. 25x (photo *E. Gübelin*).

Photo. 4.28. A tabular *albite feldspar* set apart from larger, scattered *chromite* grains as well as from tiny dots of chromite lined up in straight rows marking growth periods, in an emerald from Charbagh. Darkfield. 50x (photo *E. Gübelin*).

Photo. 4.29. Numerous *albite feldspar* and rhombohedral *calcite* crystals govern the inclusion scenery inside this emerald from Khaltaro. Darkfield. 50x (photo *E. Gübelin*).

Photo. 4.30. Two intergrown, well shaped cubic crystals of *gersdorffite* are identified for the first time as mineral inclusions in emerald from Charbagh. Note the growth marks on the greyish reflecting cube face. Darkfield. 50x. (photo *E. Gübelin*).

Photo. 4.31. *Gersdorffite* crystals with sharply edged cubic forms are surrounded by numerous rounded grains of chromite and shredded fragments of pyrrhotite in an emerald from Charbagh. Darkfield. 25x (photo *E. Gübelin*).

Photo. 4.32. *Gersdorffite* crystals were also identified amongst black masses of minute dots and larger patches as well as distinctly outlined crystals of chromite, in an emerald from Makhad. Darkfield. 15x (photo *E. Gübelin*).

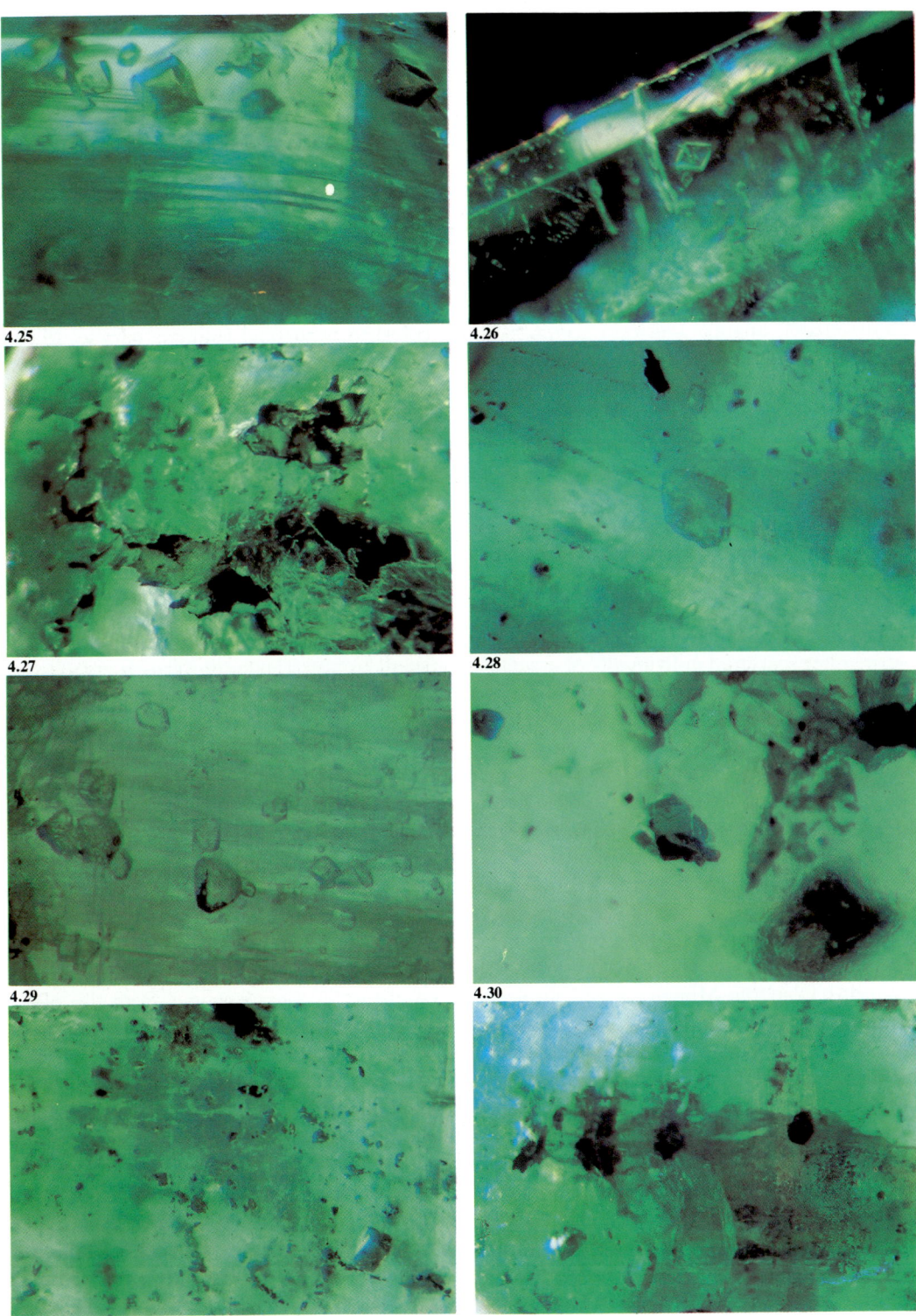

4.25

4.26

4.27

4.28

4.29

4.30

4.31

4.32

diagnostic for beryl from several localities, and when present, distinguishes natural emeralds (photo. 4.12) easily from their synthetic fakes.

A certain resemblance with regard to appearance and composition of the fluid inclusions in Pakistan emeralds to those in emeralds from Colombia (Chivor as well as Muzo) is quite evident and suggests hydrothermal influence during the growth process of the emeralds from Pakistan (at least those from the Mingora area).

More detailed information on fluid inclusions is presented in chapter 7 by R.R. Seal, II, who carried out a meticulous investigation of the fluid inclusions of the emeralds from the Swat district.

In addition to the fluid inclusions described above (which are indeed inclusions in the true sense of the word), zoned color banding (similar to that observed in many emeralds from other occurrences), angular (zig-zag) accretion steps, colorless hexagonal columns parallel to the c-axis and crystals with a pronounced green coat over a paler core may be perceived (photos. 4.13 to 4.17). These indicate spasmodic growth conditions and irregular coloration on account of a varying concentration of the coloring pigment chromium.

Some of the *mineral inclusions* emphatically reflect the mineral suite of the mother rocks, while others have been trapped by chance as accessory minerals of the external paragenesis. Thus some guest minerals (photo. 4.18) are protogenetic while others are syngenetic and might be of original formation or of secondary origin (reformed after antecedent dissolution). For the sake of a more distinct survey, the mineral inclusions will be arranged hereinafter in alphabetical order:

Actinolite [Ca_2 (Mg, Fe)$_5$ (OH$_2$) Si$_8O_{22}$] was observed only in emeralds from Charbagh and Khaltaro, in which it exhibited well-shaped, rod-like crystals of brownish-green color. They were either scattered individually or assembled as small groups. The latter were observed to be in close contact with an irregular whitish mass of albite. Although they were analyzed by micro X-ray spectroscopy, they may easily be recognized through the microscope (photos. 4.19 to 4.21), thanks to their monoclinic morphology, the striated prism faces, the presssure cracks parallel to the basal plane and the oblique extinction (approx. 10-20°).

Chromite [Fe Cr$_2O_4$] is the most ubiquitous guest mineral in Pakistan emeralds from all sources. It occurs as discrete, fairly well-shaped crystals, or as irregular grains of black color either in metallic or dull lustre (photo. 4.22). It may also occur as the filling of fractures and holes (photo. 4.1). Originating from ultramafic rocks, it is a protogenetic inclusion and indicates the participation of these rocks in the formation of the emeralds. In one emerald from Charbagh, blackish green aggregates of *chrome chlorite* were identified (photo. 4.23).

Dolomite [Ca Mg (Co$_3$)$_2$] crystals dominate the internal paragenesis of most of the emeralds from the Mingora area, but seem to be absent in the specimens from the other localities. They develop more or less euhedrally-shaped rhombohedra with relatively few faces (photo. 4.24). Occasionally they are accompanied by calcite rhombohedra (CaCO$_3$), which may be distinguished from dolomite because of the twin lamellae parallel to the long diagonal cleavage rhombohedron, as these occur more frequently in calcite. Dolomite as well as calcite usually occurs as colorless crystals, but in places dolomite is stained brown by traces of iron (photo. 4.25). At the time of this book's publication, only two emerald occurrences are known in the world whose emeralds contain dolomite inclusions: those from the Mingora area and those from Santa Terezinha de Goiás (Hanni and Kerez, 1983). As opposed to those from Santa Terezinha de Goiás, the emeralds from Mingora contain neither aggregated clusters of dolomite crystals nor pyrite inclusions. Furthermore, the chromite inclusions dotting the interior of the emeralds from Santa Terezinha de Goiás are minute grains densely disseminated throughout the whole gemstone. The presence of dolomite inclusions in emeralds from Mingora is a logical consequence for which the host rocks are responsible (photo. 4.26).

Enstatite [$Mg_2Si_2O_6$] was identified in an emerald from Swat Mine 1 and in one from Charbagh (photo. 4.27). Since no pyroxene was found during the chemical analyses of the rock, it may be an accidental secondary mineral which could form readily under the prevailing conditions and in face of an abundance of the necessary constituents Mg and Si.

Feldspar has been definitely determined as *albite* [$NaSi_3AlO_8$] in emeralds from Khaltaro, where it shares hospitality with calcite. Both these guest minerals display more or less well-shaped crystals and, as may be expected, they are colorless. Their presence concurs excellently with the late pegmatitic formation of the emeralds from Khaltaro (photos. 4.28 and 4.29).

Gersdorffite [$NiAsS$] was found to be present in emeralds from three mines at Charbagh, Makhad and Gujarkili and this came as a great surprise (photo. 4.30). Belonging to the same geological zone as Gujarkili, whose rocks proved to be rich with Ni, this trace element could be expected. Gersdorffite crystallizes in the isometric system as cubes, and with its dark gray color and metallic luster it might be mistaken for galena. This is the first time ever that gersdorffite has been discovered as an inclusion in any gemstone (photos. 4.31 and 4.32).

Magnesite [$MgCO_3$] may, in face of the abundance of Mg in these host rocks as well as in the emeralds from these deposits, be expected to be accomodated as a guest mineral in the emeralds from the Swat region (photo. 4.33). It is a component of regional metamorphic rocks, a replacement product of calcite and dolomite, or a decomposition mineral of silicates and rocks rich with Mg. As a guest mineral in the emeralds from the Swat mines, it forms strongly corroded grains and aggregates of whitish to yellowish color (photo. 4.34). In an emerald from Charbagh, magnesite was accompanied also by strongly corroded, rounded grains of *pentlandite* [$(Fe, Ni)_9S_8$], which also concurs with the chemistry of the mother rocks. In small quantities it is encountered with pyrrhotite in basic plutonites (photo. 4.35).

Mica was observed as the chrome-bearing, green variety *fuchsite* [$Na (Al, Cr)_2 (OH)_2Si_3AlO_{10}$] in emeralds from the Mingora district, but it could not be chemically analyzed. Belonging to the mineral assembly of the emerald-bearing rocks of the Mingora district and crystallizing simultaneously with the emeralds there, fuchsite may well be expected to round off the internal paragenesis of their host gems.

Pyrrhotite [$Fe_{1-x}S$] seems to be a rare inclusion in the emeralds from Pakistan, because it was only noticed in three specimens, originating from Charbagh. In these specimens it was scattered randomly throughout the gems, either in the form of discrete grains, or concentrated in small groups. Just as in emeralds from other sources, some of the grains displayed orbicular shapes—some closed, some half open—creating a leopard skin pattern. This is the pattern by which pyrrhotite may readily be recognized (photos. 4.34 and 4.36).

The three sulphide mineral inclusions gersdorffite, pentlandite and pyrrhotite unerringly indicate the presence of sulphur in those rocks at Charbagh which took a share in the formation of emeralds.

CONCLUSION

It may be stated that in general the emeralds from Pakistan may be identified and separated from those of other sources by the sum of their properties and especially by their inclusions—on condition that more than one kind be present. With regard to this statement it can be added that they may be distinguished with absolute certainty from any synthetic emerald!

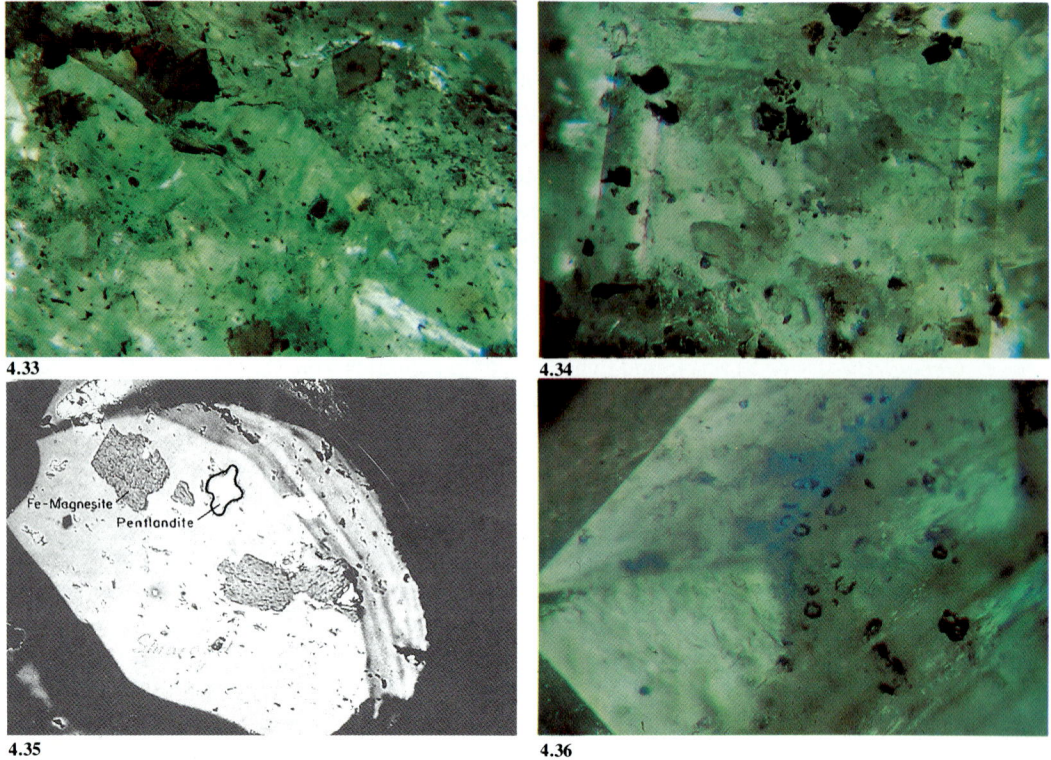

Photo 4.33. Brownish yellow, well shaped rhombohedra of *magnesite* and one distorted cube of *gersdorffite* have associated with dense accumulations of *chromite* inclusions to create a rich internal paragenesis in this emerald from Gujarkili. Darkfield. 15x. (photo *E. Gübelin*).

Photo 4.34. Translucent, whitish, and strongly corroded *magnesite* crystals have combined with rounded grains of chromite and disrupted fragments of *pyrrhotite* to build an interesting mineral association inside an emerald from Charbagh. Darkfield. 20x. (photo *E. Gübelin*).

Photo 4.35. *Pentlandite* (Fe, Ni)$_9$ S$_8$ was encountered once only in an emerald from Charbagh by means of the S.E.M. It exhibited corroded orbital shapes, and was accompanied by irregular crystals of *magnesite*. S.E.M. 15000x. (photo *E. Gübelin*).

Photo 4.36. Highly characteristic leopard-skin-like pattern is often designed by rugged crystals of *pyrrhotite* in emeralds, as is well manifested here within an emerald from Charbagh. Darkfield. 25x (photo *E. Gübelin*).

The combination of curved secondary fluid (1-g or 1-1-g) inclusions, jagged three-phase (1-g-s) primary inclusions and either well-shaped dolomite crystals plus chromite and/or fuchsite, would, for instance, suggest an emerald from the Mingora area. Curved fluid (1-g or 1-1-g) inclusions accompanied by gersdorffite and/or magnesite and/or pentlandite, might hint at an emerald from Charbagh. Primary and secondary fluid inclusions forming an endogenesis with albite, calcite and actinolite could indicate an origin at Khaltaro. Yet in face of the few really significant inclusions and the relatively few samples which qualified for inclusion research, *no* absolutely certain and reliable conclusion may be drawn concerning the exact origin of these emeralds from Pakistan. Nevertheless, they can readily be distinguished from synthetic emeralds of any brand.

ACKNOWLEDGMENTS

The author wishes to extend his gratitude to Brig. Kaleem ur Rahman Mirza for inviting him to visit the various emerald mines of the Swat region (Swat 1,2 and 3; Charbagh; Gujarkili and Makhad) and for entrusting him with a great number of emeralds from the various deposits for carrying out the above study. He is equally grateful to his friend A.H. Kazmi for accompanying him to the deposits and explaining them to him as well as inspiring him to compose this chapter on the gemological characteristics of the emeralds from Pakistan. Last but not least, the author wishes to thank Prof. Dr. M. Weibel of the Institute for Crystallography and Petrology at the Federal Institute of Technology, Zürich, Switzerland, for executing the non-destructive analyses of the mineral inclusions.

REFERENCES

Hammarstrom, J.M., 1989, Mineral chemistry of emeralds and some associated minerals from Pakistan and Afghanistan: an electron microprobe study: *in* A.H. Kazmi and L.W. Snee, editors, *Emeralds of Pakistan: Geology, Gemology and Genesis,* Van Nostrand Reinhold, New York, p. 125-150.

Hanni, H.A., and Kerez, C.J., 1983, Neues vom Smaragdvorkommon von Santa Terezinha de Goias, Goias Brasilien: Deutsche Gemologischen Gesellschaft, Zeitschrift, v. 32, p. 50-58.

Seal, R.R., II, 1989, A reconnaissance study of the fluid inclusion geochemistry of the emerald deposits of Pakistan and Afghanistan: *in* A.H. Kazmi and L.W. Snee, editors, *Emeralds of Pakistan: Geology, Gemology and Genesis,* Van Nostrand Reinhold, New York, p. 151-164.

Snee, L.W., Foord, E.E., Hill, B., and Carter, S.J., 1989, Regional chemical differences among emeralds and host rocks of Pakistan and Afghanistan: implications for the origin of emeralds: *in* A.H. Kazmi and L.W. Snee, editors, *Emeralds of Pakistan: Geology, Gemology and Genesis,* Van Nostrand Reinhold, New York, p. 93-124.

Regional Chemical Differences Among Emeralds and Host Rocks of Pakistan and Afghanistan: Implications for the Origin of Emerald

Lawrence W. Snee, Eugene E. Foord, Brittain Hill, and Stephen J. Carter

INTRODUCTION

Emerald is the exquisite result of the chemical substitution of a small amount of chromium (or vanadium) for aluminum in $Be_3Al_2Si_6O_{18}$, the mineral beryl (Deer and others, 1986). On the world gem market, if the green coloration of emerald is caused by any element other than chromium, it is not called emerald according to standards adopted by the London Chamber of Commerce and Industry's Precious Stone Trade Section (Farm, 1975). The coexistence in nature of chromium and beryllium is unusual and thus emerald is one of the rarest and most precious gemstones. The general incompatibility of beryllium and chromium is the result of their different geochemical behavior as clearly demonstrated in the compilation of chemical data for igneous rocks by Wedepohl (1978). Beryl, the most common beryllium mineral, forms during late-stage (pneumatolytic and hydrothermal) igneous activity. Because Be^{2+} has a small ionic radius (approximately 0.3 Å; Shannon and Prewitt, 1969), it is excluded from the crystal structure of most minerals. In addition, because of the small average continental crustal abundance of beryllium (less than 5 ppm; Wedepohl, 1978), beryllium is not concentrated enough to form beryl until late stages of igneous activity, i.e., during the formation of pegmatites or hydrothermal deposits. In contrast, chromium, also a trace constituent in the Earth's crust (averaging approximately 50 ppm; Wedepohl, 1978), precipitates early during igneous fractionation in the mineral chromite, $FeCr_2O_4$, or it substitutes for aluminum, iron, magnesium, or titanium in early igneous phases, such as pyroxene, amphibole, spinel, or garnet. Thus, the greatest abundance of chromium is found in ultramafic and mafic rocks. So even though chromium, which has an ionic radius of 0.62 A (Cr^{3+}; Shannon and Prewitt, 1969), will readily substitute for aluminum, which has an ionic radius of 0.54 Å (Al^{3+}; Shannon and Prewitt, 1969), in the beryl crystal structure, chromium is not expected in the late-stage liquid or vapor from which beryl may be formed. Special circumstances are necessary to bring chromium and beryllium together; as a result, emerald is extremely rare and is found only where geologic processes have been favorable for its formation.

One of the places on the Earth where the geologic environment is appropriate for emerald occurrence is the area of northern Pakistan and Afghanistan. In this region, oceanic and island arc rocks that are rich in chromium are juxtaposed against felsic, beryllium-bearing continental rocks of the Indian plate. Within this geologic setting, emeralds are found in two very different geologic environments. In the economically most important deposits, emeralds formed from hydrothermal fluids that penetrated along shear zones into talc-carbonate schists within melange zones. In the other setting, the emeralds are in pegmatites or quartz-ankerite veins which intruded schist, gneiss, and amphibolite. In both cases, the beryl apparently precipitated from a late-stage beryllium-carrying hydrothermal or pneumatolytic fluid which had scavenged chromium from the host rock.

Even though the causal relationship among the occurrence of emerald, type of host rock, and presence of hydrothermal veins or pegmatites has been recognized in Pakistan for some

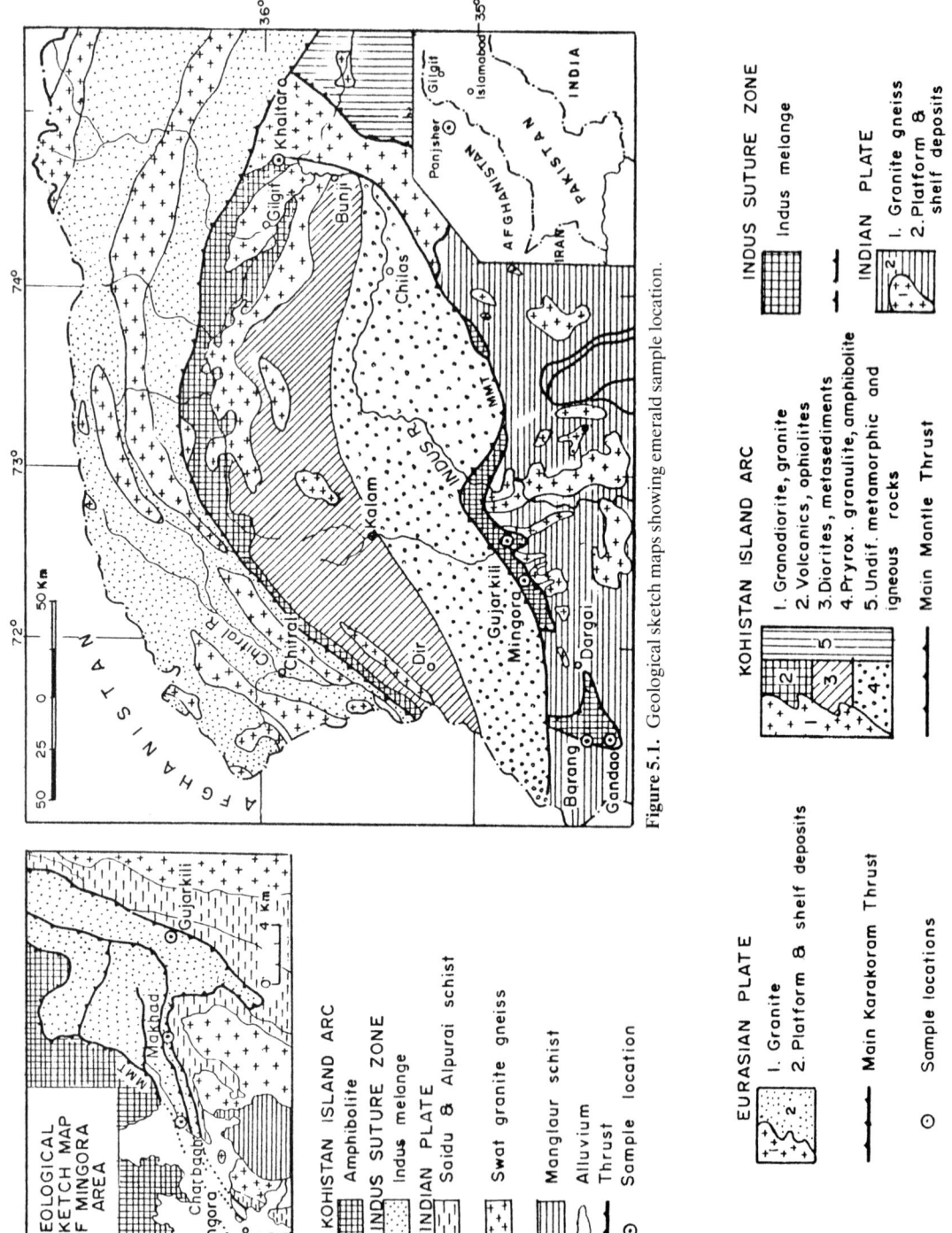

Figure 5.1. Geological sketch maps showing emerald sample location.

time (Kazmi and others, 1984, 1986), the details of the chemistry of the emeralds, mineralized zones, and host rocks remain unknown. As in other emeralds of the world, chromium is the suspected coloring agent of the Pakistan emeralds, but without chemical data this has not yet been proven. Similarly, without data on the beryllium and chromium contents in rocks near the emerald deposits, the mechanism for emerald formation is not certain. Some areas that are apparently geologically favorable for emerald mineralization are devoid of emeralds. Therefore, a chemical study of samples of emeralds, mineralized zones, unmineralized host rocks, and major lithologies of the region near the important emerald occurrences was conducted to help provide constraints on the formation of emeralds in Pakistan and Afghanistan. These data are combined in a later chapter with other chemical, microprobe, field, and fluid inclusion results to produce a genetic model for the origin of the emerald deposits of Pakistan and Afghanistan.

SAMPLES AND SAMPLE LOCATIONS

Nine emerald samples from seven different localities representing four geographic and two geologic settings were irradiated. Emerald mine locations are shown on figure 5.1. Aspects of regional geology that are important for understanding the location of emerald mineralization are discussed by Lawrence and others in chapter 2 of this volume; geologic details of each mine are outlined by Kazmi and others in chapter 3. Of the nine emerald samples, four are from Swat, Pakistan—one from Gujarkili, one from Makhad, and two from Mine 1 and Mine 2 at Mingora. All four Swat samples come from shear zones cutting talc-chlorite-dolomite schists in the Mingora ophiolite melange. Besides the Swat emeralds, two samples are from Barang and Gandao. The Barang and Gandao localities are in the Mohmand area about 50 km southwest of Mingora. In Mohmand, emeralds were formed in shear zones in talc-carbonate-quartz rock. This setting is similar to that at Mingora and the host rock may be correlative to the Mingora ophiolite melange. The emeralds from Panjsher, Afghanistan and Khaltaro, Pakistan represent samples from geologic environments different from the Swat and Mohmand emeralds. The Panjsher emerald comes from quartz-ankerite and dolomite veins along shear zones in altered dioritic plutons. The Khaltaro sample is from a relatively new emerald find in pegmatite that cuts Nanga Parbat gneiss east of Gilgit.

Forty-nine whole rock samples were also irradiated. Samples included mineralized host rock from the Mingora and Gujarkili mines and representative rocks from the region (fig. 5.1). The locations within the Mingora emerald mine of sample suites from Mines 1, 2, and 3 are shown on figure 5.2 and the distribution of these samples at each site is shown on figures 5.3, 5.4, 5.5., and 5.6. Locations of Gujarkili samples are shown on figure 5.7. In addition to the host rock samples, six samples of major lithologies, including Swat granite gneiss, Manglaur amphibolite, tourmaline granite, post-tectonic tonalite, Manglaur schist, and Manglaur graphitic quartzite, exposed in the Indian plate in Swat were analyzed; locations of these samples are shown on figure 5.8.

ANALYTICAL METHODS

The chemical data presented herein were obtained through the application of three analytical techniques—instrumental neutron activation (INAA), induction-coupled argon plasma-atomic emission spectrometry (ICAP-AES), and X-ray fluorescence (XRF). All three methods provide precise geochemical data from relatively small samples. INAA and XRF are

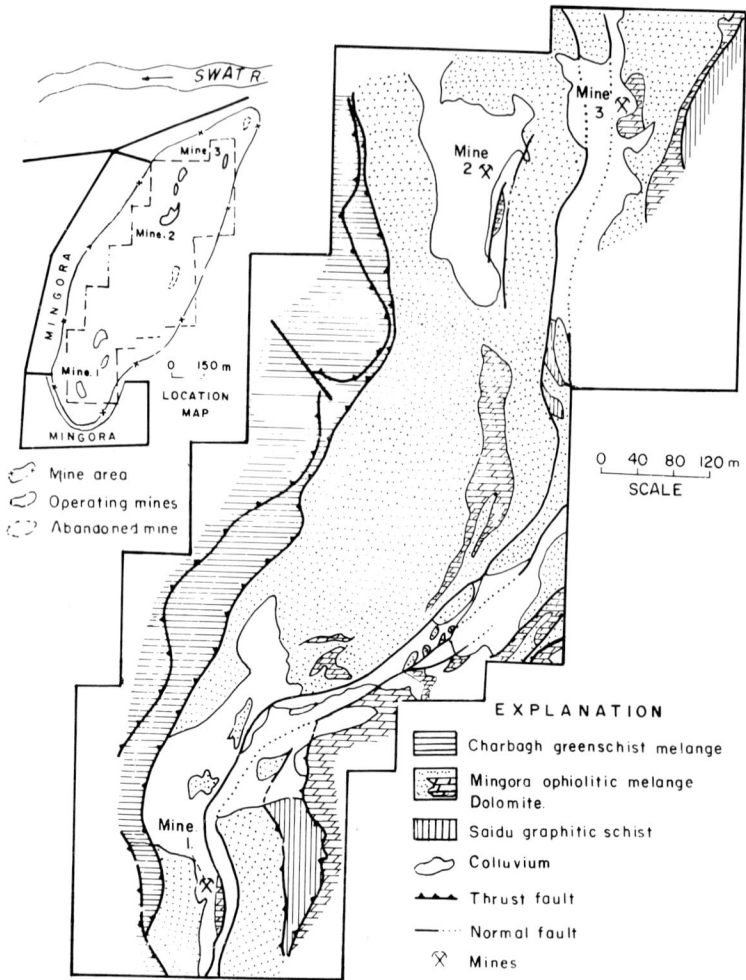

Figure 5.2. Geological sketch map of Mingora emerald mines and locations of Mines 1, 2 and 3.

non-destructive techniques whereas ICAP-AES is destructive. Rocks and emeralds were analyzed for 46 major, minor, and trace elements at the Oregon State University Radiation Center and the Analytical Laboratories of the U. S. Geological Survey, Denver, Colorado.

In the discussion that follows, the use of trade names is for descriptive purposes only and does not constitute endorsement by the U.S. Geological Survey.

Sample Preparation

Due to the small sample volume and the high precision of the analytical techniques, care must be taken to ensure sample purity. To avoid contamination, samples were collected from the freshest possible exposures and were not trimmed on a rock-saw. During sample preparation, all equipment was cleaned after each sample was processed to avoid contamination.

Samples of mineralized and non-mineralized rocks were placed on a clean steel plate and carefully broken into centimeter-sized chips with a clean rock hammer. The chips were then reduced to millimeter-sized particles in an alumina-plate pulverizer, and powdered to <200 mesh (<0.13 mm) in an alumina-plate mill. Samples were homogenized and aliquants were separated and weighed for either INAA or XRF.

Emerald samples were cleaned with purified acetone to remove adhering rock particles and hand-picked to ensure purity. Samples for INAA were weighed, then sealed in irradiation vials. ICAP-AES samples were pulverized using an agate mortar and pestle.

Standards

Elemental abundances for all three methods are determined by comparing the activity produced in a standard of known composition with the activity of the sample. The standards that were used in this study include U.S. Geological Survey rock standards BCR-1 (basalt), PCC-1 (peridotite), BHVO (basalt), and G-2 (granite), National Bureau of Standards sample SRM-1633a (coal fly ash), and Oregon State University sample CRB-1 (basalt). Appropriate standards were chosen to optimize conditions for determination of the abundance of each element. For example, PCC-1 was used for determining the abundances of MgO, Cr, Co, and Ni because of the high concentrations of these elements and lack of interferences in this standard.

Experimental Procedure

Instrumental Neutron Activation Analysis

Seven emeralds and 49 whole rock samples were analyzed by this technique, which involves bombarding samples and standards with thermal neutrons in a research nuclear reactor, measuring the activity produced by gamma decay, and comparing the activity between samples and standards to obtain elemental abundances. In this study, samples were irradiated under two different conditions—a short irradiation of 3 minutes under a relatively low neutron flux of 3×10^{10} neutrons per square centimeter per second and a longer irradiation of four hours under a neutron flux of 3×10^{12} neutrons per square centimeter per second. The short, low intensity irradiation provides optimum conditions for analysis of Ti, Al, Mg, Ca, Na, V, and Mn; the longer, higher intensity irradiation is best for Fe and the majority of the trace elements listed in the tables. For the first irradiation, samples were sealed in polyethylene vials and pneumatically transferred to the reactor. Samples were allowed to decay for 10 to 15 minutes after irradiation before the gamma-ray spectra were collected for 400 seconds each on a Ge(Li) detector coupled to a 2048-channel pulse-height analyzer. For the second, longer irradiation, samples were placed in a rotating rack to ensure equal neutron flux. Samples were repackaged after irradiation and allowed to decay for seven to ten days before each gamma-ray spectrum was collected on the detector for 8000 seconds; after an additional 30 days, gamma-ray spectra of the samples were analyzed for 16000 to 20000 seconds. Additional details of the irradiation procedures and data reduction are discussed in the appendix of this chapter.

Induction-coupled Argon Plasma—Atomic Emission Spectrometry

This technique was employed only for selected emerald samples and is described in detail by Crock and others (1983). All ICAP-AES measurements were performed on a 63-channel

Jarrell-Ash model 1160 Atom Comp ICAP emission spectrometer with some in-house modifications. These modifications include a Matheson mass-flow controller, model 8249, for controlling the sample and plasma-gas flow-rates, a Gilson Minipulse II peristaltic pump to deliver the sample to a modified Babington nebulizer, an autoprofiler, a water saturation system for the nebulizer argon-gas flow, and a Perkin-Elmer model 100 autosampler. A Digital Equipment Corporation PDP 11-34 minicomputer controls the ICAP emission spectrometer and is interfaced to a Hewlett-Packard 1000 computer for data reporting. Operating conditions, analytical wavelengths, and background corrections have been described by Crock and others (1983).

Samples for ICAP-AES analysis are normally prepared by a multi-acid digestion; details of this sample-preparation procedure have been described by Crock and others (1983). Because some common rock-forming minerals, such as beryl, chromite, ilmenite, monazite, sphene, topaz, tourmaline, and zircon, are not readily dissolved using the standard multi-acid decomposition procedure, a sodium peroxide sinter method is used in their digestion. In the sinter digestion, up to 100 mg of sample and 500 mg of sodium peroxide are fluxed at 420°C in a zirconium or vitreous carbon crucible and then transferred into a teflon crucible and dissolved in 6N hydrochloric acid. Besides the ability of the sinter method to digest beryl, it also allows the use of smaller sample sizes and lower temperatures during dissolution.

X-ray Fluorescence Spectroscopy

Six rock samples were analyzed for major elements by X-ray fluorescence spectroscopy using a Phillips PW1600 simultaneous wavelength-dispersive spectrometer according to the method of Taggart and others (1981). Samples were incorporated in lithium tetraborate fusion discs prepared by the method of Taggart and Wahlberg (1980a, 1980b). In summary, for each sample, approximately 0.8 gram of rock powder (<100 mesh) was weighed into an ignited, tared, platinum-gold (95:5) crucible. The samples were then ignited for 20 minutes in a muffle furnace at 925°C, cooled in a desiccator, and reweighed. To each of the samples was added 8 grams of lithium tetraborate and the mixture was reheated to 1120°C for 17 minutes. During the reheating the mixture was in constant motion to ensure homogeneity of the mixture. The resultant glass discs are analyzed on the X-ray spectrometer, which is on-line with a computer.

GEOCHEMICAL DATA

Geochemical data for this study are given in tables 5.1—5.6. Major element concentrations are given in weight percent (wt. %), trace element concentrations in parts per million (ppm). In the context of this paper, the term major element is used for chemical constituents whose abundance in common rocks is normally greater than 0.1 wt. % (except BeO in table 5.1); trace element is used for an element whose abundance in common rocks is less than 0.1 wt. %.

DISCUSSION

Emerald Chemistry

Partial chemical data are presented in table 5.1 for nine emerald samples. Both INAA and ICAP-AES techniques were employed to provide maximum elemental coverage. Many elements are detectable by both methods and in general the overall agreement is very good. The abundances of some elements are not the same by the two techniques. This may be due to real chemical differences and variability between sample aliquants or it may be due to inaccuracies in either or both techniques. Because ICAP-AES analyses were conducted nearly two years after INAA on different mineral grains from each location's sample set, we suspect that real sample chemical variability is displayed by the data. However, no attempt is made in this paper to evaluate the differences between methods. An exception is the careful evaluation of the ICAP-AES data for sample K-1, which yielded an anomalously high CaO content (6.4 wt. %). Based on X-ray diffraction analysis, optical microscopy, and energy dispersive data collected during scanning electron microscopic study, it was discovered that sample K-1 included 10% fluorite (CaF_2) as inclusions within the emerald. This included fluorite would cause a 10% dilution effect on the total chemical composition. Therefore, the data for this sample were corrected by subtracting 10 wt. % CaF_2 from the overall data and proportionally correcting the abundances of other elements found in the sample. Both corrected and uncorrected values are reported in table 5.1.

Of the analyzed major elements, the concentration of TiO_2 is generally below detection limit, although the Panjsher and Mohmand areas contain a detectable amount. Trace amounts of CaO are present in most emeralds. (As noted above, the anomalously high CaO content of sample K-1 is due to the admixture of 10% fluorite). SiO_2 ranges from 63.1 to 67.0 wt. %. Total FeO ranges from 0.4 to 1.6 wt. % and MgO ranges from 0.6 to 3.4 wt. %. For both total FeO and MgO, the lowest values are found in Panjsher and Khaltaro samples and there appears to be a trend in MgO from lowest concentrations in Panjsher and Khaltaro emeralds, to intermediate values in Mohmand samples, to highest values in Swat emeralds. Total Al_2O_3 ranges from 12.0 to 16.4 wt. % and again there appears to be differences among areas; highest values are from Panjsher and Khaltaro, intermediate abundances from Mohmand, and lowest concentrations from Swat. Na_2O ranges from 0.99 to 2.49 wt. % with lowest values from Panjsher, intermediate values from Swat, and highest values from Mohmand. In addition, BeO content seems to show a geographical variation with highest values from Panjsher and Khaltaro emeralds; overall, BeO ranges from 9.5 to 12.0 wt. %.

The emeralds contain significant concentrations of the trace elements Sc, V, and Cr and moderate amounts of Mn, Ni, Sn, Cs. The anomalously high Cr concentration in sample M-2 reflects the presence of small chromite inclusions. Co, Ba, and Sr are present in low abundance; Zr and Rb may be present but the data are inconclusive because of the large associated analytical errors. Zr, Hf, Ta, Th, U, and rare earth elements are virtually absent. In contrast to the major element data, no clear geographical relationships are exhibited by any of the trace element data.

Significance of Chemical Variations and Trends

Because the amount of chemical data of this study is limited, caution should be used when evaluating chemical differences and apparent geographical trends among emeralds. However, microprobe data summarized by Hammarstrom (chapter 6, this volume) on additional emerald samples from these and other areas display similar variations and differences. Hammarstrom also noted that Swat and Mohmand emeralds have lower Al_2O_3 but higher MgO and Na_2O contents than Panjsher and Khaltaro emeralds. Hammarstrom's observation that Swat emeralds have more Cr and less V than Mohmand emeralds is supported by the INAA data. Hammarstrom also found that Sc is higher in Gandao emeralds than Mingora; our data support this. Thus, the combined data, although somewhat limited, reinforce these observations and point toward the need for a comprehensive study of not only Pakistan and Afghanistan emeralds

Table 5.1. Partial chemical analyses of beryls from Pakistan and Afghanistan

	Afghanistn SEM-1 Panjsher		Pakistan. Pegmatite K-1 Khaltaro[5]	Pakistan, Mohmand SEM-2 Barang	SEM-4 Barang		SEM-3 Gandao	
	INAA	ICAP-AES	ICAP-AES	INAA	INAA	ICAP-AES	INAA	ICAP-AES
TiO_2 (wt. %)	0.21±0.10	<0.04	<0.05	0.12±0.03	0.13±0.08	<0.05	<0.02	<0.07
SiO_2	—	65.5	58.6 (64.5)	—	—	63.1	—	63.8
Al_2O_3	16.4±0.3	16.2	16.1 (17.7)	14.5±0.2	14.7±0.2	12.7	14.8±0.3	14.5
FeO^*	0.61±0.02	0.62	0.44 (0.5)	—	—	1.2	1.6±0.2	1.4
MgO	0.7±0.1	1.03	0.73 (0.8)	2.0±0.3	2.5±0.7	4.8	2.4±0.7	2.2
CaO	0.07±0.07	0.52	6.4 (0)	0.3±0.1	<0.3	0.13	<0.04	0.53
Na_2O	0.99±0.02	—	—	2.30±0.02	2.22±0.06	—	2.49±0.03	—
K_2O	—	<0.5	<0.4	—	—	<0.4	—	<0.5
BeO	—	12.04	10.80 (11.9)	—	—	10.96	—	11.41
P_2O_3	—	<0.07	0.11	—	—	—	—	<0.09
Li (ppm)	—	80	700 (770)	—	—	80	—	120
B	—	<30	30	—	—	<30	—	<30
Sc	109.1±0.1	320	70	—	—	2100	1180±10	1400
V	690±20	1000	60	69±4	169±14	390	13300±300	5300
Cr	19180±60	3700	1900 (2090)	—	—	4500	340±80	280
Mn	8.5±0.8	<30	60	260±10	380±34	40	37±2	40
Co	0.3±0.1	<7	6	—	—	12	1.7±0.9	<7
Ni	<120	70	40	—	—	230	<70	50
Cu	—	<7	8	—	—	<7	—	8
Zn	—	<7	20	—	—	<10	—	60
Rb	24±7	—	—	—	—	—	40±40	—
Sr	60±40	<10	10	—	—	<10	3000±2000	20
Cs	22±2	—	—	—	—	—	42±7	—
Ba	50±40	21	7	—	—	<7	1300±1000	300
La	0.04±0.03	<10	<10	—	—	<10	<0.3	<10
Ce	1.2±0.9	<30	<20	—	—	<30	35±13	<30
Nd	2±2	<30	<20	—	—	<30	<30	<30
Sm	0.02±0.01	—	—	—	—	—	0.14±0.09	—
Eu	<0.05	<10	<10	—	—	<10	<0.1	<10
Tb	<0.2	—	—	—	—	—	<1	—
Yb	<0.1	<7	<5	—	—	<7	0.9±0.6	<7
Lu	<0.2	—	—	—	—	—	<10	—
Sn	—	—	—	—	—	—	—	—
Zr	140±90	—	—	—	—	—	<300	—
Hf	0.1±0.1	—	—	—	—	—	0.4±0.4	—
Ta	<0.1	—	—	—	—	—	<0.1	—
Th	<1	<30	<20	—	—	<30	<1	<30
U	<1	<700	<500	—	—	<700	<1	<700

Table 5.1. (Continued)

			Pakistan, Swat			
SEM-5 Makhad	GK-21 Gujarkili		M-1 Mine 1, Mingora		M-2 Mine 2, Mingora	
ICAP-AES	INAA	ICAP-AES	INAA	ICAP-AES	INAA	ICAP-AES
<0.05	<0.02	<0.07	<0.2	<0.07	<0.02	<0.05
64.8	—	65.0	—	67.0	—	64.6
14.4	12.0±0.2	13.4	13.7±0.5	13.4	13.0±0.2	12.5
0.49	—	1.5	0.8±0.1	1.3	1.6±0.1	1.7
2.8	3.1±0.8	3.0	3.4±0.4	3.6	3.0±0.4	4.1
0.27	<0.3	0.36	<0.04	0.29	<0.04	0.20
—	1.67±0.05	—	2.02±0.03	—	2.07±0.03	—
<0.4	—	<0.5	—	<0.5	—	<0.4
11.54	—	11.63	—	10.49	—	9.49
<0.07	—	<0.09	—	<0.09	—	<0.07
200	—	170	—	200	—	140
<30	—	100	—	<30	—	<30
950	—	990	359±1	1100	404±1	1600
590	359±15	170	510±20	220	100±30	240
3900	—	11000	9750±70	5100	55300±200	8000
<30	60±20	<30	11±1	<30	3±1	30
9	—	10	4±1	8	4±1	9
130	—	190	<300	110	<200	210
10	—	20	—	<7	—	13
<7	—	9	—	<10	—	<10
—	—	—	<20	—	20±10	—
10	—	30	100±100	20	<100	<10
—	—	—	72±8	—	240±30	—
9	—	14	180±120	9	50±50	10
<10	—	<20	<0.3	<10	<0.3	<10
<30	—	<30	<30	<30	<3	<30
<30	—	<30	<10	<30	<10	<30
—	—	—	<0.06	—	<0.06	—
<10	—	<20	0.2<0.1	<10	<0.1	<10
—	—	—	<0.2	—	<0.2	—
<7	—	<8	<0.1	<7	<0.1	<7
—	—	—	<0.2	—	<0.2	—
—	—	—	—	—	—	—
—	—	—	200±200	—	200±200	—
—	—	—	0.6±0.2	—	<0.7	—
—	—	—	0.1±0.1	—	<0.1	—
<30	—	<30	<1	<30	<0.1	<30
<700	—	<800	<1	<700	<1	<700

Notes: 1. Other elements analyzed by ICAP-AES but not detected at respective limits of determination are As (<50 ppm), Ga (<30 ppm), Ho (<30 ppm), Mo (<10 ppm), Nb (<30 ppm), Pb (<30 ppm), and Y (<10 ppm).
2. By ICAP-AES, Sn content of all emeralds is between approximately 70 to 90 ppm.
3. Analytical precision on ICAP-AES analyses is variable depending on element but ranges from 1 to 5%, relative.
4. No attempt is made in this paper to evaluate differences between INAA and ICAP-AES methods.
5. Based on semiquantitative energy dispersive analysis on a scanning electron microscope, optical microscopy, and x-ray diffraction analysis, sample K-1 contains 10% fluorite contamination. Values in parentheses are calculated by removing 10% CaF_2.

but emeralds from elsewhere to determine if similar chemical differences are common and whether these differences can be correlated to geologic environment. The control of geologic environment on the formation of emeralds is discussed in the final chapter of this book.

An additional interesting result of the chemical data is the discovery of the very high vanadium concentration in Gandao beryls (up to 13,300 ppm). In fact, the very high vanadium and low chromium contents (300 ppm) of the Gandao beryl appear to indicate that the green coloration of Gandao beryls is due to vanadium instead of chromium. Synthetic and natural beryls with less than about 0.1 wt. % Cr_2O_3 and containing comparable amounts of other chromophores are colorless or nearly so (unpublished data, E.E. Foord). The fact that vanadium provides the color in the Gandao beryl is important because, as pointed out by Anderson (1966), vanadium has been the "mystery" element in beryl coloration. Because the coloration is due to vanadium, the Gandao beryls are not emeralds according to the standards of the London Chamber of Commerce and Industry's Precious Stone Trade Section (Farm, 1975) but should be referred to as green vanadian beryl.

An added important aspect of the beryl data is the unexpected discovery of fluorite in the Khaltaro emerald sample. As noted above, about 10% fluorite was found within the emerald crystals. The presence of fluorite is important because it provides direct evidence that fluorine was present in the fluids from which the emeralds crystallized. An important requirement for the formation of emeralds is the existence of a transporting and (or) concentrating mechanism for beryllium. Although the Khaltaro emerald occurs within pegmatite, which may be enriched in beryllium by fractionation, fluoride-ion (as well as chloride- and hydroxyl- ion; Renders and Anderson, 1987) is a known complexing agent for beryllium. Thus, beryllium either could have remained in solution or it was transported to the crystallization site by complexes formed with fluorine.

Comparison to other Emeralds

Complete chemical data for other emeralds are also limited but some data on natural emeralds are compiled in Sinkankas (1981), Graziani and others (1983), Gübelin (1958 and 1982), Rafiq and Jan (1985), Bank (1974), Metson and Taylor (1977), Hanni (1982), Schrader (1983), and Stockton (1984); the last three articles discuss chemical differences between natural and synthetic emeralds. Based on these studies, all samples of our study clearly fall within the compositional range of natural emeralds; the presence of numerous trace components makes them distinct from synthetic stones. Compared to other natural emeralds of the world, the Swat and Mohmand emeralds are chemically distinct, whereas the compositions of Panjsher and Khaltaro emeralds are indistinguishable. Although the chemical ranges overlap, Swat and Mohmand emeralds are generally higher in iron, magnesium, and sodium and lower in aluminum than other natural emeralds. (Compare table 6.7 of chapter 6 by Hammarstrom with table 5.1 of this chapter.) In addition, chromium, vanadium, and scandium are relatively high. Natural emeralds have been reported to contain between 0.004 and 0.6 wt. % Cr_2O_3 In comparison, synthetic flux-grown emeralds contain between 0.1 and 2.7 wt. % Cr_2O_3 (e.g., Rogers and Sperisen, 1942; Foord and Mills, 1978). Our data show that Cr_2O_3 for all analyzed Pakistan and Afghanistan emeralds ranges from 0.28 to 2.81 wt. % (excluding the Gandao beryl and Mingora Mine 2 sample).

Host Rock Chemistry

Host rocks from three mines at Mingora and from Gujarkili were analyzed to determine whether the chemical constituents of the emeralds were present in the host prior to mineralization or whether some or all of the constituents were derived from an outside source during mineralization. In addition, in some locations, samples were collected from unaltered host rock, altered host rock, and the zone of mineralization to evaluate any chemical gradients that may have formed during mineralization. These data are discussed below according to location and/or individual mine.

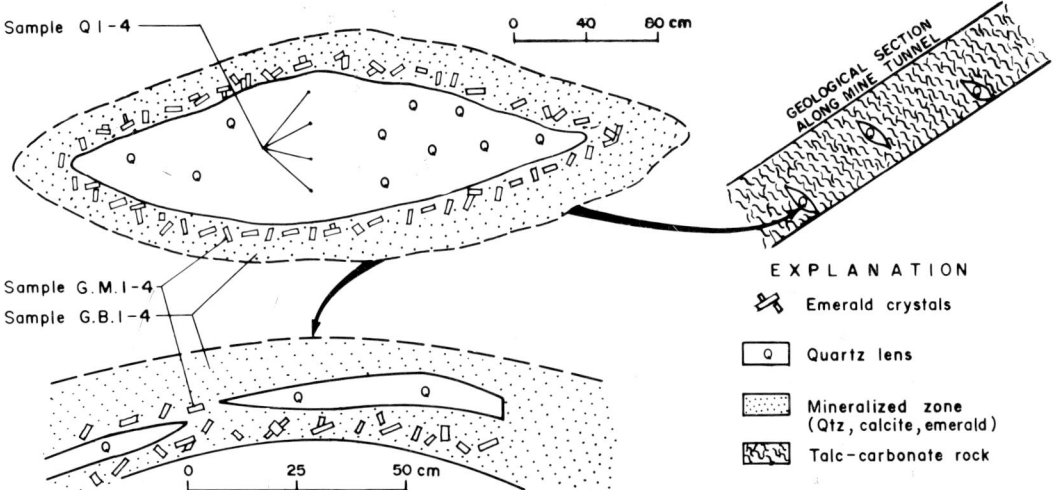

Figure 5.3 Sketch showing mode of emerald mineralization at Mine 1, Mingora, with general locations of samples.

Mingora, Mine 1

Emerald mineralization at Mine 1, one of several mines at Mingora (Fig. 5.2), occurs in talc-carbonate schists adjacent to quartz lenses (fig. 5.3). The emeralds are found in a 15-cm-wide band around each lens; altered host rock that is devoid of emeralds extends an additional 15 cm away from the lenses. Twelve samples were collected; four each from a representative quartz lens (Q1-Q4), the emerald-bearing mineralized zone around that lens (GM1-GM4), and the nearby emerald-barren unaltered host rock (GB1-GB4). Partial chemical analyses of the samples are presented in table 5.2.

Detectable amounts of analyzed elements except Al_2O_3, MgO, Cr, and Co are virtually non-existent in representative samples of the quartz lens. Traces of Al_2O_3 (0.64—0.68 wt. %) were found in all four samples. Sample Q1 contains 0.66 wt. % MgO. Notably, all four samples contain some Cr with amounts ranging from 1.2 to 6.6 ppm; the presence of these significant amounts of chromium in nearly pure quartz is unusual. Detectable Co was found in all samples; Q2 contains the highest abundance—0.33 ppm. (It is unlikely that any of the detected MgO, Cr, or Co was contributed to the samples during sample preparation but some Al_2O_3 may have been added from the alumina plates. However, we feel most, if not all, of the Al_2O_3 is from the

Table 5.2. INAA data for host rocks from Mine 1, Mingora, Pakistan.

	GM-1	GM-2	GM-3	GM-4	GB-1	GB-2	GB-3	GB-4	Q-1	Q-2	Q-3	Q-4
TiO_2 (wt. %)	<0.02	<0.02	<0.02	<0.02	0.04±0.02	0.04±0.02	0.05±0.02	<0.04	<0.04	0.06±0.03	0.05±0.04	<0.05
Al_2O_3	2.02±0.04	1.79±0.04	3.00±0.05	5.23±0.06	1.83±0.04	1.45±0.04	1.49±0.03	3.91±0.06	0.64±0.01	0.68±0.01	0.68±0.01	0.64±0.01
FeO^*	4.81±0.02	4.73±0.02	5.48±0.02	9.54±0.03	7.21±0.02	6.31±0.02	4.67±0.02	5.82±0.02	0.002±0.001	0.01±0.01	0.01±0.01	0.01±0.01
MgO	23.6±0.4	26.2±0.05	23.4±0.5	21.4±0.5	26.7±0.5	26.5±0.5	26.9±0.5	26.9±0.5	0.66±0.01	0.07±0.02	0.05±0.02	0.03±0.02
CaO	17.1±0.8	9.1±0.5	15.4±0.8	17.1±0.8	5.5±0.4	8.4±0.5	6.3±0.4	8.0±0.4	0.01±0.01	0.13±0.02	0.01±0.01	0.04±0.01
Na_2O	0.023±0.001	0.023±0.001	0.020±0.001	0.019±0.001	0.034±0.001	0.035±0.001	0.022±0.001	0.020±0.001	0.002±0.001	0.003±0.001	0.001±0.001	0.002±0.001
Sc (ppm)	13.2±0.1	11.0±0.01	13.8±0.1	13.3±0.1	10.5±0.1	10.7±0.1	6.4±0.1	12.2±0.1	0.01±0.01	0.02±0.01	0.01±0.01	0.01±0.01
V	32±2	31±3	47±3	66±3	34±2	33±2	27±2	56±3	<0.3	0.2±0.1	<0.3	<0.3
Cr	1600±3	1615±3	2090±10	1580±3	1100±10	1951±3	2306±4	930±3	1.8±0.1	6.6±0.2	1.2±0.1	1.3±0.1
Mn	1590±10	1075±3	2400±10	1890±10	1140±5	965±5	1163±5	1616±6	0.9±0.1	9±11	0.5±0.1	2.0±0.1
Co	76.8±0.2	54.8±0.2	32.9±0.1	92.4±0.2	75.8±0.2	80.8±0.2	60.5±0.2	75.0±0.1	0.04±0.01	0.33±0.01	0.07±0.01	0.05±0.01
Ni	1310±20	1330±10	980±10	1980±20	1690±20	1640±20	1370±10	1680±20	<1	1.5±0.3	<1	<1
Rb	<1	<1	<1	2±1	<1	<1	2±1	<1	<0.1	<0.1	<0.1	<0.1
Sr	650±60	200±20	580±60	100±10	250±30	200±30	100±20	280±30	0.7±0.3	7±1	0.8±.3	2.0±0.5
Cs	0.61±0.06	0.06±0.04	0.09±0.04	0.12±0.04	0.04±0.02	0.12±0.03	0.03±0.02	0.05±0.02	0.007±0.002	0.009±0.002	0.006±0.002	0.007±0.002
Ba	30±10	<10	30±10	40±20	10±10	10±10	<10	<10	2±1	1±1	<1	<1
La	1.82±0.03	0.42±0.02	0.83±0.02	0.22±0.02	0.09±0.01	0.09±0.01	0.35±0.02	0.36±0.02	0.001±0.001	0.003±0.001	0.003±0.001	<0.001
Ce	4.3±0.2	1.7±0.3	2.9±0.3	1.8±0.3	1.4±0.3	1.1±0.3	1.5±0.3	1.4±0.3	<0.01	<0.01	<0.01	<0.01
Nd	2±1	<1	2±1	2±1	1±1	<2	<2	<2	0.1±0.1	0.1±0.1	0.1±0.1	<0.1
Sm	1.16±0.02	0.48±0.01	1.30±0.01	0.24±0.01	0.16±0.01	0.13±0.01	0.29±0.01	0.65±0.01	0.001±0.001	0.003±0.001	<0.001	0.001±0.000
Eu	0.52±0.01	0.18±0.01	0.47±0.01	0.08±0.01	0.07±0.01	0.07±0.01	0.11±0.01	0.19±0.01	<0.01	<0.01	<0.01	<0.01
Tb	0.25±0.03	0.10±0.02	0.34±0.05	0.07±0.03	0.06±0.03	0.14±0.03	0.09±0.02	<0.2	<0.01	<0.01	<0.01	<0.01
Yb	0.55±0.06	0.31±0.04	0.72±0.06	<0.09	0.20±0.05	0.10±0.04	0.31±0.03	0.44±0.04	<0.01	<0.01	<0.01	<0.01
Lu	0.14±0.02	0.05±0.01	0.12±0.01	0.07±0.01	0.02±0.01	<0.02	<0.02	<0.03	<0.1	<0.01	<0.01	<0.01
Zr	10±10	<10	20±10	<10	10±10	<10	<10	<10	<1	<1	1±1	1±1
Hf	<0.03	<0.03	<0.03	0.04±0.02	<0.03	<0.03	<0.03	<0.03	0.005±0.002	0.003±0.001	0.004±0.001	0.003±0.001
Ta	0.01±0.01	<0.01	<0.01	<0.01	<0.01	0.01±0.01	<0.01	<0.01	<0.01	<0.01	<0.001	<0.01
Th	<0.1	<0.1	<0.1	<0.1	<0.1	<0.1	<0.1	<0.1	<0.001	<0.001	<0.002	<0.001
U	<0.2	<0.2	<0.1	<0.2	<0.2	<0.1	<0.1	<0.1	<0.01	<0.01	<0.01	<0.01

samples). Because there is no field evidence indicating that the quartz lens was leached after its formation, the present composition of this lens represents the original fluid that formed it. Even though an anomalous amount of Cr was carried in this SiO_2-rich solution, the absence of other elements present in the Mingora emeralds is strong evidence that the quartz lens-forming solution was not the source for the formation of emeralds in the host rock. The unusually large amount of chromium in the quartz lens is probably a result of interaction between a normally chromium-poor siliceous fluid and the chromium-rich wall rock.

Samples from the mineralized zone and from the barren zone of Mine 1 are geochemically complex. The host rock is talc-carbonate and within the mineralized zone the talc-carbonate host contains quartz, calcite, and emerald. Except for a few differences, the eight samples are chemically indistinguishable. All eight samples contain appreciable amounts of Al_2O_3, FeO, MgO, and CaO. TiO_2 and Na_2O are virtually absent. High amounts of Sc, V, Cr, Mn, Co, Ni, and Sr are present. Rb, Cs, Ba, Zr, Hf, Ta, Th, and U are absent. Rare earth elements are present in amounts ranging from one to 12 times normal carbonaceous chondritic abundance. The major chemical differences between mineralized and non-mineralized samples are in relative abundances of Al_2O_3, CaO, Sc, Cs, and REEs. Samples from the mineralized zone are distinctly enriched in CaO and, on the average, are enriched in the other four, although the abundances of Al_2O_3, Sc, Cs, and REEs for single samples from the barren zone may overlap those of samples from the mineralized zone.

The chemical difference between mineralized and barren host rock shows that the mineralized rock was enriched in Al_2O_3, CaO, Sc, Cs, REEs, and probably BeO and SiO_2 by some process. This enrichment probably occurred during fluid movement through the volumes now occupied by quartz lenses because chemical enrichment is symmetrically disposed around the lenses. The absence of significant amounts of any element except Cr in the nearly pure SiO_2 lens is strong evidence that the fluid that formed the quartz lens did not form the emeralds. Thus, the enrichment of the mineralized zone probably occurred before quartz precipitation. If this is correct, at least two fluids of distinctly different compositions affected the host rock. The first was a chemically rich hydrothermal fluid and it was followed by a solution of nearly pure SiO_2. This conclusion is supported by fluid inclusion results of Seal (chapter 7). In that study, Seal shows that at least two different fluids were associated with emerald and quartz deposition and the earlier fluid, which formed the emeralds, had a higher temperature and salinity than the later, quartz-forming fluid. In addition, the chemical enrichment of the mineralized zone in REEs, Al_2O_3, BeO, and SiO_2 is strong evidence that the fluids had a continental chemical affinity. Thus, it is likely that mineralizing fluids were derived during some process that affected nearby continental crustal rocks and then migrated through shear zones in the chromium-rich melange rocks. The high salinity of the fluid inclusions as determined by Seal in chapter 7 is consistent with a hydrothermal fluid-transport of beryllium, which chemically complexes with chloride-, fluoride-, or hydroxyl-ion, in slightly acid waters (Renders and Anderson, 1987). As pointed out by Renders and Anderson (1987), an increase in pH of the hydrothermal fluids as a result of wall rock alteration, or CO_2 loss by boiling, could cause deposition of emerald. However, Seal (chapter 7, this volume) found no evidence for boiling in the fluid inclusions of Pakistani emeralds.

Mingora, Mine 2

Emerald mineralization at Mine 2 occurs in talc-chlorite-dolomite schist (figs. 5.4 and 5.5.). In the central quarry, from which our samples were taken, emeralds are found along a

Table 5.3. INAA data for host rocks from Mine 2, Mingora, Pakistan.

	M-1	M-3	M-5	M-7	M-9	H-1	H-3	H-5
TiO_2 (wt. %)	0.05±0.03	...	0.02±0.01	...	<0.02	0.02±0.01	...	0.02±0.01
Al_2O_3	0.24±0.01	...	0.36±0.01	...	0.57±0.01	0.04±0.01	...	0.38±0.01
FeO*	7.08±0.03	6.33±0.02	6.98±0.03	6.74±0.03	5.63±0.03	6.30±0.03	4.18±0.03	6.57±0.03
MgO	33.5±0.4	...	34.4±0.4	...	31.6±0.4	35.0±0.4	...	32.2±0.4
CaO	0.37±0.05	...	<0.1	...	4.4±0.4	0.19±0.04	...	<0.2
Na_2O	0.018±0.002	0.018±0.002	0.018±0.002	0.018±0.002	0.016±0.02	0.018±0.002	0.018±0.002	0.018±0.002
Sc (ppm)	3.96±0.01	4.77±0.02	4.60±0.02	2.92±0.01	7.82±0.01	4.20±0.01	5.11±0.02	4.00±0.01
V	17±	...	27±1	...	29±1	27±1	...	12±1
Cr	2104±3	1660±2	2522±3	2158±3	1703±3	2218±2	2197±3	1432±2
Mn	1050±3	...	980±10	...	840±10	910±10	...	890±10
Co	180.2±0.4	94.4±0.2	106.9±0.3	2553±5	2029±4	2233±4	2255±5	2177±5
Ni	1881±16	2060±20	2040±20	1962±16	1619±16	1890±15	1769±16	1770±15
Rb	2±1	4±2	<2	<2	2±1	<2	1±1	1±1
Sr	<10	<50	<50	<10	<10	<10	<10	<10
Cs	0.05±0.02	<0.04	<0.04	<0.03	0.04±0.02	<0.03	0.03±0.01	0.03±0.01
Ba	<20	<30	<30	<20	<10	9±9	13±10	15±10
La	<0.01	<0.01	<0.01	0.009±0.003	<0.01	<0.01	<0.01	<0.01
Ce	<0.2	0.5±0.3	1.7±0.5	1.0±0.3	<0.2	0.5±0.2	<0.3	0.7±0.2
Nd	<0.1	<0.1	<0.1	<1	<1	<1	<1	<1
Sm	0.01±0.01	±0.002	<0.01	<.01	0.016±0.002	<.003	<0.004	0.023±0.006
Eu	<0.01	<0.01	<0.01	<0.01	0.01±0.01	0.01±0.01	0.01±0.01	0.01±0.01
Tb	<0.02	0.04±0.01	0.03±0.01	0.01±0.01	0.04±0.02	<0.02	0.02±0.02	0.02±0.02
Yb	<0.01	<0.01	<0.01	0.03±0.01	0.08±0.02	0.02±0.01	0.02±0.01	0.07±0.02
Lu	<0.07	<0.03	<0.03	<0.07	0.04±0.02	<0.05	<0.05	<0.05
Zr	<10	20±10	20±10	20±10	<10	<20	<20	10<10
Hf	0.01±0.02	<0.03	<0.03	<0.01	<0.01	0.03±0.01	<0.02	0.03±0.02
Ta	0.02±0.01	<0.02	<0.02	<0.02	<0.02	0.01±0.01	<0.1	<0.1
Th	<0.2	<0.1	<0.1	<0.1	<0.1	<0.2	<0.1	<0.1
U	<0.2	<0.1	<0.2	<0.1	<0.1	<0.1	<0.1	<0.2

Table 5.3. (continued).

	H-7	H-9	F-1	F-3	F-5	F-7	F-9
TiO_2 (wt. %)	...	<0.02	<0.02	...	<.02	...	<0.02
Al_2O_3	...	0.47±0.01	0.18±0.01	...	0.42±0.01	...	0.32±0.01
FeO*	6.26±0.03	4.69±0.03	7.25±0.03	7.40±0.03	6.83±0.03	6.18±0.03	6.06±0.03
MgO	...	30.3±0.4	34.8±0.4	...	33.9±0.4	...	32.9±0.4
CaO	...	4.0±0.4	0.18±0.03	...	0.14±0.03	...	1.4±0.2
Na_2O	0.018±0.002	0.016±0.002	0.016±0.002	0.018±0.002	0.018±0.002	0.018±0.002	0.016±0.002
Sc (ppm)	4.43±0.01	4.64±0.01	3.65±0.01	4.05±0.01	4.46±0.01	4.84±0.01	3.62±0.01
V	...	14±1	13±1	...	27±2	...	22±1
Cr	2447±3	2216±3	1579±2	1989±3	1686±2	2145±3	2468±3
Mn	...	680±10	1000±10	...	775±10	...	740±10
Co	2397±5	2015±4	2478±5	2590±5	2485±5	2254±5	2348±5
Ni	1914±16	1280±12	2070±17	2025±17	1953±16	1856±15	2036±16
Rb	<1	1±1	<2	2±1	2±1	<2	1±1
Sr	<10	<10	<10	<10	<10	<10	<10
Cs	<0.03	0.05±0.02	<0.03	<0.03	0.03±0.01	<0.03	<0.03
Ba	10±10	10±9	<10	<10	<10	10±10	<10
La	<0.01	<0.01	<0.02	<0.01	<0.01	<0.02	<0.01
Ce	<0.3	0.5±0.2	<0.3	0.5±0.2	<0.4	0.3±0.2	<0.04
Nd	<1	<1	<1	<2	<1	<2	<1
Sm	<0.003	0.033±0.004	0.008±0.005	0.005±0.003	<0.004	<0.002	0.005±0.002
Eu	0.01±0.01	0.01±0.01	<0.01	0.01±0.01	<0.01	<0.01	<0.01
Tb	0.02±.01	<0.02	<0.02	0.03±0.02	0.02±0.02	<0.02	<0.02
Yb	0.01±0.01	0.05±0.02	0.03±0.02	0.02±0.01	0.02±0.01	0.02±0.01	0.03±0.01
Lu	<0.04	<0.04	<0.06	<0.06	0.01±0.01	<0.05	0.02±0.03
Zr	<20	<20	<20	30±10	<20	10±10	<20
Hf	0.03±0.01	0.03±0.02	<0.01	0.02±0.01	0.02±0.02	0.02±0.02	<0.01
Ta	<0.1	0.01±0.01	0.01±0.01	0.02±0.01	<0.02	<0.02	0.02±0.01
Th	<0.1	<0.1	<0.1	<0.2	<0.1	<0.1	<0.1
U	<0.1	<0.1	<0.1	<0.1	<0.2	<0.1	<0.1

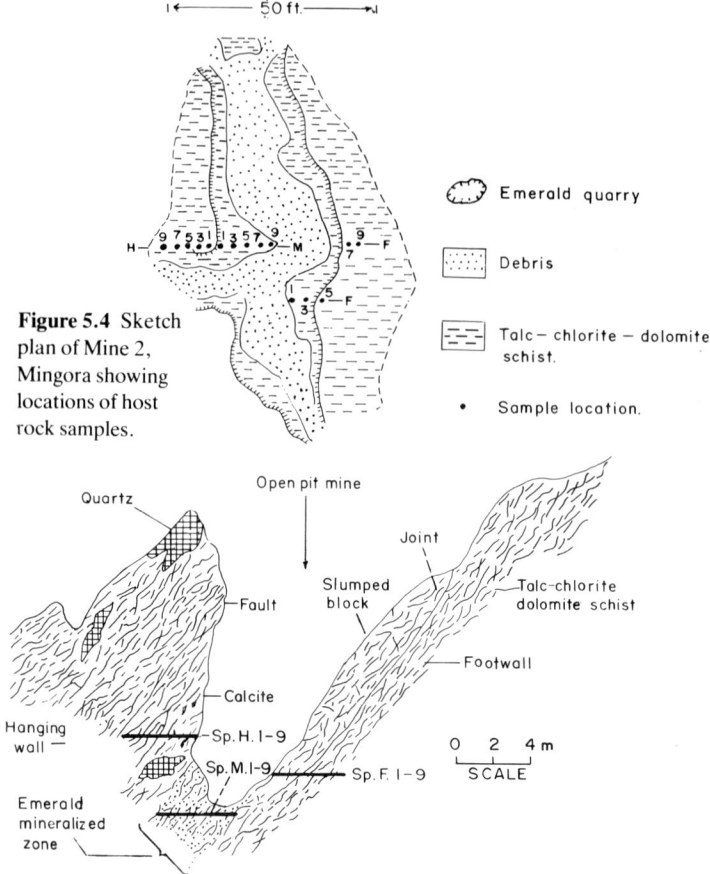

Figure 5.4 Sketch plan of Mine 2, Mingora showing locations of host rock samples.

Figure 5.5. Geological cross-section of Mine 2, Mingora with locations of host rock samples.

well-defined shear zone. Emeralds are disseminated throughout the zone and are commonly associated with fuchsite (a chromium-bearing muscovite) and tourmaline. Barren quartz and calcite veinlets are found in and near the shear zone. In the northern and southern quarries of Mine 2, emeralds are also associated with quartz and calcite. Fifteen analyzed samples were collected from hanging wall (H1, H3, H5, H7, H9), footwall (F1, F3, F5, F7, F9), and mineralized zone (M1, M3, M5, M7, M9) of the central quarry of Mine 2. Partial chemical data are shown in table 5.3.

No detectable chemical differences are exhibited within the INAA data of the 15 samples from Mine 2. Of the analyzed major elements, the samples are high in MgO and FeO and low in TiO_2, Al_2O_3, and Na_2O; CaO varies from 0 to 4.4 wt. %. Of the trace elements, Cr, Co, Mn, and Ni occur in significant amounts and Sc and V are slightly enriched. The other trace elements are undetected.

Significant chemical differences are exhibited between the host rock samples of Mine 1 and Mine 2. Mine 2 has higher MgO and lower CaO and Al_2O_3 than Mine 1. With respect to trace elements relative to Mine 1, Mine 2 is greatly enriched in Co and slightly enriched in Ni but Mine 2 has lower Sc, Mn, Sr, and REEs. It is likely that most of these differences reflect the

Table 5.4. INAA data for host rocks from Mine 3, Mingora, Pakistan.

	M-3A	M-3B	M-3C	M-3D
TiO_2 (wt. %)	0.03±0.03	<0.04	<0.04	0.08±0.05
Al_2O_3	0.98±0.02	1.15±0.02	1.15±0.02	1.23±0.04
FeO*	7.10±0.02	7.48±0.02	7.45±0.02	5.15±0.02
MgO	40.5±0.5	38.5±0.5	38.6±0.05	32.5±0.5
CaO	0.42±0.06	0.48±0.06	0.07±0.04	2.6±0.2
Na_2O	0.059±0.004	0.043±0.003	0.040±0.003	0.077±0.005
Sc (ppm)	6.0±0.1	5.8±0.1	7.6±0.1	4.4±0.1
V	28±1	34±2	35±5	25±4
Cr	2049±4	2134±4	1612±3	1570±3
Mn	818±2	779±3	1186±4	819±3
Co	105.7±0.2	103.2±0.2	87.9±0.2	73.7±0.2
Ni	2010±20	1830±20	1620±20	1370±10
Rb	12±2	15±3	13±3	8±2
Sr	<100	<100	<100	<100
Cs	0.67±0.07	1.00±0.09	0.83±0.08	0.40±0.05
Ba	40±10	60±20	40±10	30±10
La	0.07±0.01	0.06±0.01	0.05±0.01	0.11±0.01
Ce	1.0±0.3	0.8±0.3	0.5±0.2	1.5±0.3
Nd	<1	<1	<1	<1
Sm	0.04±0.01	0.03±0.01	0.03±0.01	0.11±0.01
Eu	0.01±0.01	0.01±0.01	<0.01	0.05±0.01
Tb	<0.05	0.05±0.03	<0.03	<0.05
Yb	0.05±0.03	0.16±0.04	0.12±0.04	0.09±0.03
Lu	<0.02	<0.03	0.03±0.01	<0.02
Zr	<10	<10	<10	10±10
Hf	0.04±0.01	<0.03	0.04±0.01	0.04±0.01
Ta	<0.01	<0.01	<0.01	0.01±0.01
Th	<0.1	0.13±0.03	<0.1	0.10±0.03
U	<0.1	<0.1	<0.1	<0.1

Table 5.5. INAA data for host rocks from the Gujarkili Mine, Pakistan.

	GK-6	GK-7	GK-8	GK-9	GK-10	GK-11	GK-12	GK-13
TiO_2 (wt. %)	<0.02	<0.02	...	<0.02	<0.02	...	...	...
Al_2O_3	0.42±0.01	0.49±0.01	...	0.67±0.01	2.22±0.03	...	...	...
FeO*	4.30±0.02	3.89±0.02	4.90±0.02	5.76±0.02	9.07±0.02	4.87±0.02	2.37±0.01	5.33±0.02
MgO	30.2±0.4	32.5±0.4	...	31.9±0.4	9.4±0.3	...	...	...
CaO	0.9±0.1	0.08±0.02	1.5±0.6	0.60±0.08	<0.08	<0.08	<0.1	<0.1
Na_2O	0.023±0.002	0.02±0.001	0.022±0.001	0.035±0.002	0.028±0.001	7.55±0.02	0.009±0.001	0.038±0.002
Sc (ppm)	8.15±0.03	9.82±0.04	10.00±0.04	7.14±0.02	9.24±0.03	14.4±0.1	1.54±0.1	5.44±0.02
V	16±1	17±1	...	27±1	30±2	108.0±0.8	...	...
Cr	2340±180	2330±170	1840±140	1540±2	1846±3	...	1036±1	2035±2
Mn	604±8	370±8		403±8	1160±10			
Co	59.9	60.0±0.3	90.8±0.04	113.6±0.3	136.9±0.3	14.5±0.1	37.8±0.1	80.5±0.2
Ni	1200±500	900±400	1100±500	2340±20	2255±20	90±10	955±11	1910±20
Rb	<2	<2	2±2	2±2	3±2	3±1	<2	1±1
Sr	<60	<30	<30	<60	<60	50±10	<50	<50
Cs	<0.1	0.38±0.04	0.56±0.05	1.1±0.1	0.2±0.1	0.9±0.1	0.12±0.04	2.0±0.2
Ba	<20	<20	<20	<30	<30	80±30	<30	<30
La	0.07±0.01	0.01±0.01	0.02±0.01	0.12±0.01	1.6±.1	33.7±0.2	0.07±0.01	0.15±0.02
Ce	<0.3	<0.3	<0.3	0.4±0.2	2.6±0.4	71.4±0.6	0.1±0.1	0.5±0.2
Nd	<2	<2	<1	<3	<3	27±1	<2	<2
Sm	0.026±0.001	0.013±0.03	0.006±0.002	0.064±0.005	0.86±0.01	5.37±0.01	0.028±0.003	0.12±0.01
Eu	<0.01	<0.01	0.02±0.01	0.03±0.01	0.52±0.01	0.91±0.01	0.013±0.003	0.05±0.01
Tb	<0.02	<0.02	<0.02	0.02±0.01	0.18±0.04	0.89±0.06	0.02±0.01	0.06±0.03
Yb	0.03±0.01	<0.01	0.03±0.02	0.04±0.02	0.41±0.06	2.6±0.1	0.04±0.02	0.03±0.02
Lu	<0.03	<0.04	<0.03	0.03±0.02	<0.04	0.26±0.04	<0.04	<0.03
Zr	<40	<40	<40	<40	<40	160±40	<40	<40
Hf	<0.03	<0.03	<0.03	<0.03	0.06±0.05	5.0±0.1	<0.03	<0.03
Ta	<0.02	0.02±0.01	<0.02	0.03±0.01	<0.02	1.1±0.1	<0.02	<0.02
Th	<0.1	<0.1	<0.1	0.2±0.1	<0.3	13.7±0.3	<0.1	<0.1
U	<0.1	<0.2	<0.1	<0.1	<0.1	2.4±0.3	<0.1	0.3±0.1

Table 5.5. (Continued)

	GK-17	GK-18	GK-19	GK-20
TiO$_2$ (wt. %)	0.02±0.01	...	<0.02	0.03±01
Al$_2$O$_3$	0.80±0.01	...	0.36±0.01	1.19±0.02
FeO*	5.42±0.02	6.88±0.02	5.47±02	4.95±0.02
MgO	29.5±0.4	...	27.0±0.4	29.5±0.4
CaO	0.56±0.07	...	0.19±0.04	1.6±0.2
Na$_2$O	0.070±0.001	0.034±0.002	0.016±0.001	0.031±0.002
Sc (ppm)	4.51±0.02	5.28±0.02	4.56±0.02	4.52±0.02
V	15±1	...	12±1	18±1
Cr	1960±2	1510±2	1259±2	1160±2
Mn	600±10	...	1460±10	440±10
Co	114.5±0.3	60.4±0.2	71.4±0.2	180.3±0.4
Ni	2130±20	1970±20	1640±20	2540±20
Rb	<2	4±3	5±3	<2
Sr	<50	<50	<50	<50
Cs	0.6±0.1	<0.1	1.0±0.1	1.6±0.2
Ba	<30	<30	<30	<30
La	0.19±0.01	0.12±0.01	0.09±0.01	0.27±0.01
Ce	<0.2	<0.3	0.2±0.1	1.5±0.3
Nd	<1	<1	<1	<1
Sm	0.094±0.003	0.09±0.01	0.034±0.004	0.143±0.002
Eu	0.04±0.01	0.05±0.01	<0.01	0.05±0.01
Tb	0.03±0.02	0.02±0.001	<0.02	0.01±0.01
Yb	0.11±0.04	0.09±0.02	<0.04	0.08±0.01
Lu	<0.03	0.02±0.02	0.02±0.02	0.02±0.01
Zr	<40	20±10	<40	<40
Hf	<0.03	<0.03	<0.03	<0.003
Ta	<0.02	<0.02	0.01±0.01	<0.02
Th	<0.1	<0.1	<0.2	<0.2
U	<0.1	<0.2	<0.1	0.28±0.03

pre-emerald mineralization lithology of the host rock but differences in Al_2O_3 and REEs also in part may be due to alteration.

Mingora, Mine 3

Emerald mineralization at Mine 3 occurs in intensely fractured dolomite adjacent to talc-chlorite-dolomite schist. The emeralds are commonly associated with chlorite and fuchsite in stockwork quartz veins (fig. 5.6). Four samples of dolomitic host rock (M3A-M3D) were analyzed; results are shown in table 5.4.

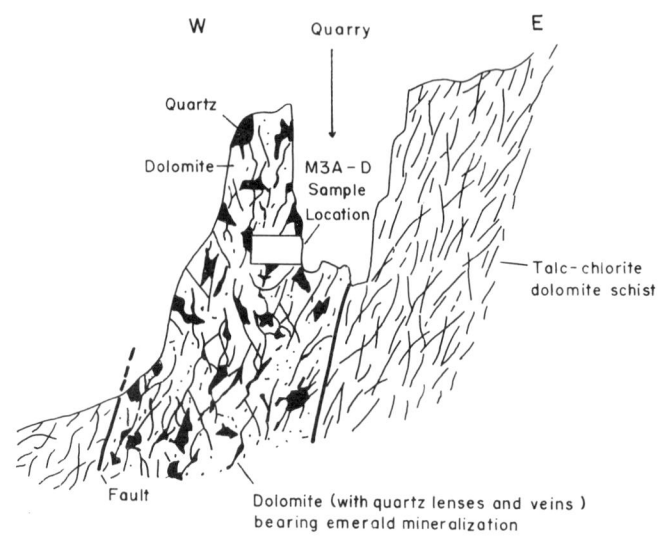

Figure 5.6. Geological cross-section of Mine 3, Mingora showing locations of host rock samles.

Except for some minor variations, the four samples of Mine 3 are chemically very similar. All are high in MgO and FeO (32.5 to 40.5 and 5.2 to 7.5 wt. %, respectively) and have small amounts of Al_2O_3 and CaO (1.0—1.2 and 0.1—2.6 wt. %, respectively); TiO_2 and Na_2O are virtually absent. Trace elements Sc, V, Cr, Mn, Co, and Ni are abundant and Rb, Cs, Ba, and REEs are present in small, but measurable amounts. Sr. Zr. Hf, Ta, Th, and U were not detected. Chemically these samples are similar to samples from the central quarry of Mine 2.

Gujarkili

Emerald mineralization at Gujarkili occurs at intersecting shear zones cutting talc schist, talc carbonate, and carbonate (fig. 5.7). The emeralds are commonly associated with talc, calcite, and quartz. Twelve host rock samples (GK-6 through GK-13 and GK-17 through GK-20) were analyzed; the results are presented in table 5.5. Five samples, GK-8, 11, 12, 13, and 18, were irradiated only once for 4 hours at 1 megawatt and as a result TiO_2, Al_2O_3, MgO, V, and Mn abundances could not be determined.

Except for samples GK-10 and GK-11, the Gujarkili samples are chemically very similar. They are rich in MgO, FeO, Sc, V, Cr, Mn, Co, and Ni. Furthermore, these samples are generally indistinguishable from talc-dolomite-chlorite schists of Mingora, Mine 2. In contrast, the chemistry of sample GK-10 shows the effect of high iron oxide content; the origin of the higher iron content is unclear. Sample GK-11 is a metapelitic rock which has preserved its pre-metamorphic chemical signature; this is well-demonstrated by light rare earth element enrichment, which is very similar to that exhibited by typical marine mud, i.e., North American shale composite (Haskin and others, 1968).

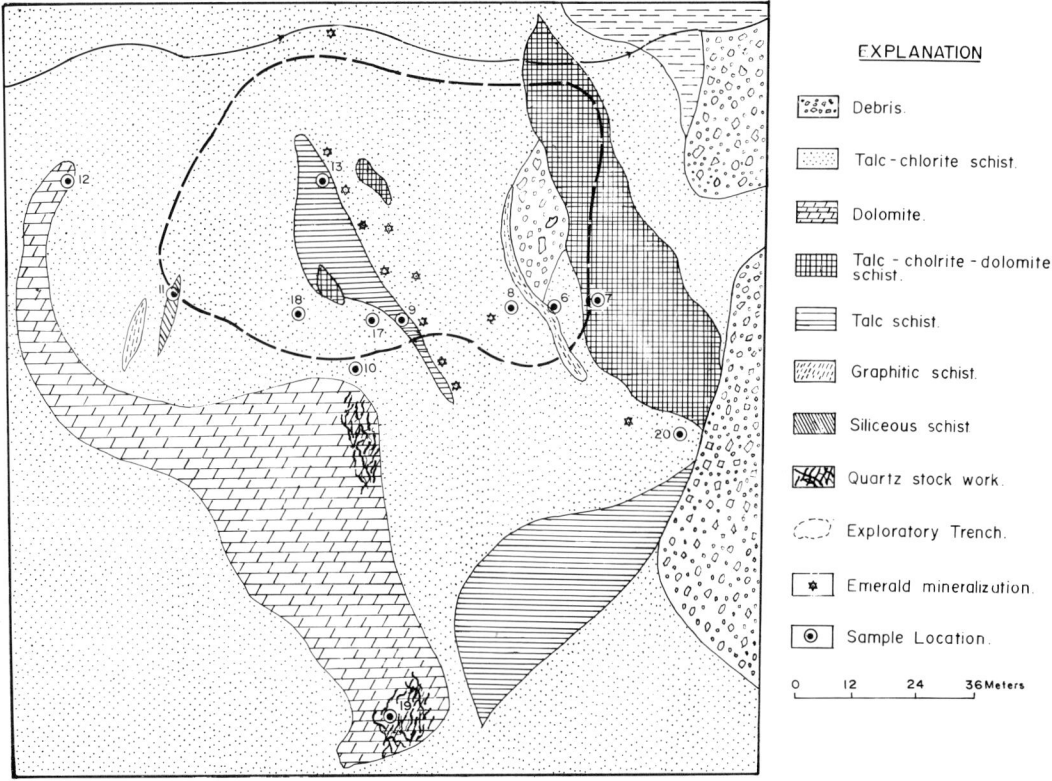

Figure 5.7. Sketch map of Gujarkili mine showing location of host rock samples.

Summary of Host Rock Chemistry

The chemistry displayed by the host rocks is typical of ultramafic, altered ultramafic, and dolomitic rocks of ocean floor and upper oceanic mantle origin. This is consistent with the interpretation presented by Lawrence and others in chapter 2 that the host rocks are part of an altered melange, which includes obducted ocean floor. The high concentrations of MgO, Cr, Mn, Co, and Ni are similar to those found in ultramafic rocks. With the exception of local apparent enrichment of Al_2O_3, CaO, FeO, Sc, Cs, REEs, SiO_2, and BeO in altered rocks, the chemistry of the melange host rocks was relatively unaffected by mineralizing activity associated with emerald formation. However, the localized enrichment of Al_2O_3, REEs, SiO_2, and BeO in the host rock near the emeralds is important because it is an indication that at least part of the emerald-mineralizing fluids had a continental component.

Table 5.6. Partial chemical analyses of Indian Plate rocks near Mingora, Pakistan.

	D-21 Swat granite gneiss	SW-62* Tourmaline granite	SW-68a* Manglaur schist	SW-73* Post-tectonic biotite granite	SW-74* Manglaur amphibolite	SW-202* Manglaur graphitic quartzite
TiO_2 (wt. %)	0.71±0.06	0.03	0.49	0.92	2.51	0.66
Al_2O_3	11.5±0.2	14.3	14.3	14.4	15.3	13.2
FeO*	3.67±0.01	0.89	7.42	4.32	8.58	2.15
MgO	4.2±0.3	0.16	0.97	1.48	3.75	1.74
CaO	2.5±0.3	0.40	0.24	2.72	6.47	4.25
Na_2O	2.61±0.01	3.26	1.71	2.62	3.28	2.54
Sc (ppm)	10.6±0.1	8±0.1	11±0.1	10±0.1	17±0.1	14±0.1
V	66±4	...	...	...	...	...
Cr	24±1	<1	44±1	21±2	13±1	49±2
Mn	565±5	542	387	620	1239	232
Co	5.7±0.1	0.4±0.1	10.3±0.1	6.8±0.1	18.4±0.1	9.9±0.1
Ni	<20	...	...	...	...	...
Rb	190±20	360±30	220±10	230±10	140±10	150±10
Sr	260±30	10±20	400±70	400±60	1480±140	420±60
Cs	6.7±0.6	35.5±1.1	9.7±0.3	3.6±0.2	4.4±0.2	8.7±0.3
Ba	560±40	30±30	460±40	830±30	1080±60	2560±160
La	61.5±0.4	2.2±0.2	41.1±0.6	62.9±0.7	84.2±0.5	43.6±0.4
Ce	113±1	3±2	96±1	123±2	209±2	90±2
Nd	49±2	6±5	42±10	50±8	10±16	40±20
Sm	8.44±0.03	0.85±0.01	7.86±0.04	9.37±0.03	9.70±0.05	7.10±0.06
Eu	1.46±0.01	0.08±0.01	1.30±0.03	1.87±0.03	3.68±0.06	1.1±0.03
Tb	1.35±0.08	0.22±.01	2.23±0.02	1.43±0.10	0.57±0.03	0.35±.02
Yb	3.9±0.1	1.4±0.2	4.5±0.4	4.0±0.2	3.8±0.3	4.1±0.5
Lu	0.62±0.03	0.14±0.02	0.48±0.03	0.48±0.02	0.38±0.05	0.66±0.07
Zr	280±40	10±10	60±10	320±50	120±20	30±10
Hf	8.0±0.1	1.1±0.1	8.2±0.2	6.8±0.3	10.9±0.3	5.2±0.3
Ta	3.2±0.1	5.9±0.2	1.3±0.1	2.1±0.2	4±0.2	1.3±0.1
Th	22.0±0.3	1.8±0.3	13.6±0.2	22.0±0.3	15.3±0.2	18.9±0.1
U	1±1	1±1	3±1	1±1	1±1	1±1

* TiO_2, Al_2O_3, FeO*, MgO, CaO, Na_2O, and Mn, analyzed by XRF (J.E. Taggart, A.J. Bartel, and K. Stewart, analysts); all other elements by INAA.

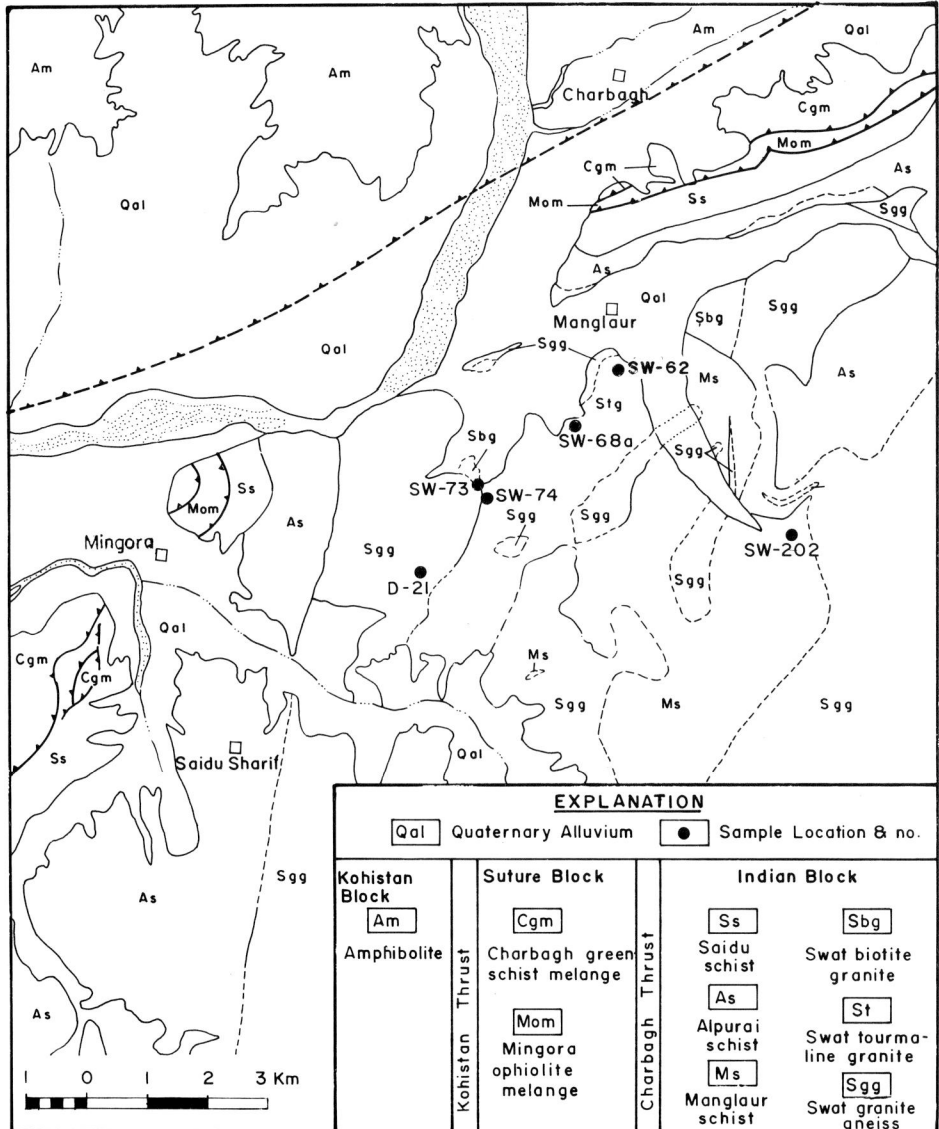

Figure 5.8. Geological map of Manglaur area, Swat District, northern Pakistan, showing locations of representative continental crustal samples analyzed in this study.

Rocks of the Indian Plate

Six representative samples of Indian continental crustal rocks from the Mingora area were selected for analysis. The samples are augen gneiss (D-21), tourmaline granite (SW-62), garnet-mica schist (SW-68a), post-kinematic tonalite (SW-73), amphibolite (SW-74), and graphitic quartzite (SW-202). Sample locations are show on figure 5.8 and the data are presented in table 5.6.

The chemistry of these samples is typical of continental rocks and is strikingly different from that of the melange rocks in which the emeralds are developed. Of particular importance is the relatively low concentrations of Cr, Mn, Co, and Ni and the high relative abundance of Rb, Cs, Ba, Zr, Hf, Ta, Th, U, and REEs compared to melange rocks. Like BeO, the enriched elements are typically incompatible and are normally excluded from "primitive" rocks such as ocean floor ultramafics. The presence of beryl in the melange rocks and the local enrichment of REEs, Al_2O_3, and SiO_2 near emerald occurrences are strong evidence that the solutions from which the beryl precipitated were influenced by, if not derived from, a continental source.

Although not revealed by our chemical analyses, it is important to mention that postorogenic granites, which intruded the Indian plate rocks, contain fluorite. In addition, fluorite deposits are common in shear zones of southern Swat and in Mohmand. These shear zones apparently served as pathways for migrating hydrothermal fluids that carried the fluorite; the presence of fluorite in postorogenic granites suggests that these granites were the source of the fluorine-bearing fluids. Thus, since fluorine is a complexing agent of beryllium, it is possible that the beryllium of Swat and Mohmand emeralds was transported to sites of crystallization within shear zones of the melange terrane by these same fluorine-rich fluids. However, it is also possible that beryllium transport could have occurred as chloride or hydroxy-complexes.

CONCLUSION

Like other emeralds of the world, the emeralds of Pakistan are green because of the substitution of the transition element chromium for aluminum in the beryl crystal structure. (The one exception to this is the Gandao beryl which contains a large amount of vanadium and a small chromium content.) Pakistan emeralds are unusual in that they contain large amounts of iron, magnesium, and sodium. In addition, distinct chemical differences among Pakistan and Afghanistan emeralds act as fingerprints that reflect the chemical differences of their host rocks. The oceanic rocks within the melange that are the host for the emeralds are the likely source for the chromium. Other trace elements in the emeralds were also derived from the host; these include magnesium, iron, manganese, and cobalt. The beryllium, sodium, cesium, rubidium, aluminum, and silicon present in the emeralds were probably derived from mineralizing fluids that had a continental source. Scandium and vanadium in the emeralds may have been derived from either the source of the mineralizing fluid or the host; whatever the source of scandium and vanadium, the emeralds are greatly enriched in both compared to analyzed rocks of this study.

Juxtaposition of chromium-bearing oceanic rocks against beryllium-bearing continental rocks during the suturing of India to Asia provided the geologic setting that was favorable for emerald formation. Hydrothermal fluids gained access to the melange host along shear zones. The mechanism by which continentally derived hydrothermal fluids transported beryllium to chromium-rich melange rocks is unknown but the presence of fluorite in Khaltaro emeralds, postorogenic granites, and shear zones in Swat and Mohmand indicates that fluorine was available to act as a complexing agent. Transport of beryllium by chloride- or hydroxy-complexes is also possible. Emerald formation at Swat and Mohmand probably occurred when the pH of these fluids increased due to wall rock alteration. Thus, even though the entire suture zone is geologically favorable for emerald formation, emeralds are found only where shear zones provided pathways for continentally derived hydrothermal fluids to come in contact with oceanic rocks in the melange along the suture.

Emeralds from different areas in Pakistan and Afghanistan are chemically different. Swat, Mohmand, Khaltaro, and Panjsher emeralds show small but distinct differences in aluminum, magnesium, sodium, chromium, vanadium, and scandium. Emeralds are clearly a reflection of the chemistry of both host rocks and mineralizing fluid. Because chemical differences among emeralds of the world are common, it would be timely to evaluate those differences with respect to geologic setting of each occurrence. Emeralds may hold more magic than just that associated with their beauty and rarity.

ACKNOWLEDGMENTS

The authors would like to thank Scott Hughes and Roman Schmidt, Oregon State University, for their assistance with INAA. James G. Crock, U.S. Geological Survey, performed the ICAP-AES analyses of eight beryls and Joseph E. Taggart of the U.S.G.S. analyzed six whole rock samples by XRF. Ali H. Kazmi supplied the samples used in this study. Robert D. Lawrence provided much appreciated discussion and guidance during the course of this research. This research was supported in part by a Fulbright Research Grant to Ali H. Kazmi during his stay at Oregon State in 1984 and in part by National Science Foundation Grant 81-18403 to Robert D. Lawrence and Robert S. Yeats. Critical reviews by Joan Fitzpatrick and Peter Modreski of the U.S.G.S. greatly improved this manuscript.

REFERENCES

Anderson, B.W., 1966, Chromium as a criterion for emerald: Journal of Gemmology, v. 10, no. 2, p. 41-45.

Bank, H., 1974, The emerald occurrence of Miku, Zambia: Journal of Gemmology, v. 14, no. 1, p. 8-15.

Crock, J.G., Lichte, F.E., and Briggs, P.H., 1983, Determination of elements in National Bureau of Standards' Geological Reference Materials SRM 278 obsidian and SRM 688 basalt by inductively coupled argon plasma—atomic emission spectrometry: Geostandards Newsletter, v. 7, p. 335-340.

Deer, W.A., Howie, R.A., and Zussman, J., 1986, *Rock-forming Minerals*, Volume 1B, Disilicates and ring silicates: Longman Scientific and Technical Limited, London, England, p. 372-409.

Farm, A.E., 1975, Emeralds and beryls: Journal of Gemmology, v. 14, p. 322-323.

Foord, E.E., and Mills, B.A., 1978, Biaxiality in 'isometric' and 'dimetric' crystals: American Mineralogist, 63, p. 315-325.

Graziani, G., Gübelin, E.J., and Lucchesi, S., 1983, The genesis of an emerald from the Kitwe district, Zambia: Neues Jahrbuch für Mineralogie, Monatshefte, v. 4, p. 175-186.

Gübelin, E.J., 1958, Notes on the new emeralds from Sandawana: Gems and Gemology, v. 9, no. 7, p. 195-203.

Gübelin, E.J., 1982, Gemstones of Pakistan: emerald, ruby, and spinel: Gems and Gemology, v. 18, p. 123-139.

Hammarstrom, J.M., 1989, Mineral chemistry of emeralds and some associated minerals from Pakistan and Afghanistan: an electron microprobe study: *in* A.H. Kazmi and L.W. Snee, editors, *Emeralds of Pakistan: Geology, Gemology and Genesis,* Van Nostrand Reinhold, New York, p. 125-150.

Hanni, H.A., 1982, A contribution to the separability of natural and synthetic emeralds: Journal of Gemmology, v. 18, no. 2, p. 138-144.

Haskin, L.A., Frey, F.A., and Wildeman, T.R., 1968, Relative and absolute terrestrial abundances of the rare earths: *in* Ahrens, L.H., editor, *Origin and Distribution of the Elements.* International Series of Monographs on the Earth Sciences 30, p. 889-912.

Kazmi, A.H., Anwar, J., Hussain, S., Khan, T., and Dawood, H., 1989, Emerald deposits of Pakistan: *in* A.H., Kazmi, and L.W., Snee, editors, *Emeralds of Pakistan: Geology, Gemology and Genesis,* Van Nostrand Reinhold, New York, p. 39-74.

Kazmi, A.H., Lawrence, R.D., Anwar, J., Snee, L.W., and Hussain, S., 1986, Mingora emerald deposits (Pakistan): suture-associated gem mineralization: Economic Geology, v. 81, p. 2022-2028.

Kazmi, A.H., Lawrence, R.D., Dawood, H., Snee, L.W., Hussain, S., 1984, Geology of the Indus suture zone in the Mingora-Shangla area of Swat, northern Pakistan: University of Peshawar, Geological Bulletin, v. 17, p. 127-144.

Laul, J.C., 1979, Neutron activation analysis of geological materials: Atomic Energy Reviews, v. 17, p. 603-695.

Lawrence, R.D., Kazmi, A.H., and Snee, L.W., 1989, Geologic setting of the emerald deposits: *in* A.H. Kazmi, and L.W. Snee, editors, *Emeralds of Pakistan: Geology, Gemology and Genesis,* Van Nostrand Reinhold, New York, p. 13-38.

Metson, N.A., and Taylor, A.M., 1977, Observations on some Rhodesian emerald occurrences; Journal of Gemmology, v. 15, no. 8, p. 422-434.

Rafiq, M., and Jan, M.Q., 1985, Emerald and green beryl from Bucha, Mohmand Agency, NW Pakistan: Journal of Gemmology, v. 19, no. 5, p. 404-411.

Renders, P.J., and Anderson, G.M., 1987, Solubility of kaolinite and beryl to 573 K: Applied Geochemistry, v. 2, p. 193-203.

Rogers, A.F., and Sperisen, F.J., 1942, American synthetic emeralds: American Minerologist, v. 27, p. 762-768.

Schrader, H.W., 1983, Contributions to the study of the distinction of natural and synthetic emeralds: Journal of Gemmology, v. 18, no. 6, p. 530-543.

Seal, R.R., II, 1989, A reconnaissance study of the fluid inclusion geochemistry of the emerald deposits of Pakistan and Afghanistan: *in,* A.H. Kazmi and L.W. Snee, editors, *Emeralds of pakistan: Geology, Gemology, and Genesis,* Van Nostrand Reinhold, New York, p. 151-164.

Shannon, R., and Prewitt, C., 1969, Effective ionic radii in oxides and fluorides: Acta Crystallographica, B25, p. 925-946.

Sinkankas, J., 1981, *Emeralds and other beryls:* Chilton Book Co., Radnor, Pennsylvania, USA, 665p.

Stockton, C.M., 1984, The chemical distinction of natural from synthetic emeralds: Gems and Gemology, v. 20, no. 3, p. 141-145.

Taggart, J.E., and Wahlberg, J.S., 1980a, New mold design for casting fused samples: Advances in X-ray Analysis, v. 23, p. 257-261.

Taggart, J.E., and Wahlberg, J.S., 1980b, A new in-muffle automatic fluxer design for casting glass discs for X-ray fluorescence analysis: Federation of Analytical Chemists and Spectroscopy Societies meeting, Philadelphia, Pennsylvania, September, 1980.

Taggart, J.E., Lichte, F.E., and Wahlberg, J.S., 1981, Methods of analysis of samples using X-ray fluorescence and induction-coupled plasma spectroscopy: U.S. Geological Survey, Professional Paper 1250, p. 683-687.

Walker, F.W., Miller, D.G., and Feiner, F., 1983, Chart of the Nuclides, 13th edition: General Electric Company, San Jose, California.

Wedepohl, K.H., 1978, *Handbook of Geochemistry:* Springer-Verlag, Berlin, Germany.

APPENDIX

Instrumental Neutron Activation Analysis

The instrumental neutron activation analytical techniques employed in this study were chosen to provide optimum conditions for analysis of emeralds and whole rock samples. While experimental design is controlled by the types of rocks analyzed, the following INAA procedures provide accurate and precise results for a heterogeneous sample suite. Irradiation and decay times for each sample run are dependent on the half-lives and assumed concentrations of the elements in the sample suite. Table 5.7 lists the elements of interest in this study, along with the appropriate gamma energies and half-lives. It is possible to determine the chemical concentration of additional elements (see Laul, 1979) if those elements occur in abundances greater than the detection limit of the elements in question or if chemical separations are made. The samples in this study were irradiated under two different conditions, in order to effectively determine the abundance of both long-lived (days) and short-lived (minutes) isotopes.

The first irradiation utilized a pneumatic sample-transport system, which forms a two-way path between a clean lab and the reactor. Samples were placed in the system and transported into the reactor in under five seconds, then exposed to a low neutron flux of 3×10^{10} neutrons per square centimeter per second for three minutes. The samples were then pneumatically transferred to the clean lab, measured for safe levels (<1 millicurie) of radioactivity and placed into clean irradiation vials to avoid contamination of the researcher and equipment. Samples were allowed to decay for 10-15 minutes until activity levels were low enough to allow counting with less than 10% dead time on the detector system. The gamma-ray detection system used in this study consisted of a Ge(Li) detector coupled to a 2048-channel pulse-height analyzer. The system was calibrated to 2 keV per channel for the first series of counts, and samples were counted for 400 seconds each.

Figure 5.9 illustrates a gamma-ray spectrum for standard CRB-1 after 15 minutes of decay since end-of-bombardment (E.O.B.). Note the high background levels in the low gamma-energy range, which obscures most peaks. High background levels are primarily the result of ^{28}Al and ^{49}Ca decay. These radioactive isotopes will almost completely decay to stable isotopes within two hours after E.O.B. The samples were thus allowed to decay for an additional 3-6 hours, then counted for 1000 seconds with the detector system calibrated at 0.75 keV per channel. After counting, all samples were resealed in irradiation vials and placed into the rotating rack of the reactor.

For the second activation, samples were irradiated for four hours at 1 megawatt (3×12^{12} neutrons per square centimeter per second) while the rack rotated at approximately 1 revolution per minute; rotation ensures that all samples are exposed to the same neutron flux. The samples were allowed to decay for 7-10 days to reduce background levels, then placed into new irradiation vials. Samples were then counted for 8000 seconds with the detector calibrated at 1 keV per channel. Figure 5.10 illustrates the gamma-ray spectrum of CRB-1 after a seven day decay interval. Note that while other elements are present, only elements with half-lives from 40-100 hours (table 5.7) are analyzed in this count; continued decay will result in a lower background and hence greater precision in counting longer-lived (>100 hours) radioisotopes. After counting for 8000 seconds, the samples were allowed to decay for an additional 30 days. The samples were then counted for 16000 to 20000 seconds with the detector calibrated at 0.75 keV per channel. Figure 5.11 illustrates the gamma-ray spectrum for CRB-1 after 40 days decay. Note the larger and more clearly defined peak areas for the longer-lived elements than observed in figure 5.10.

Table 5.7. Radionuclides used in this study. Half-lives and gamma energies from the Chart of the Nuclides, 13th edition, General Electric Corporation (Walker and others, 1983).

Element	E (KeV) gamma	Half-life	Count time	Decay time
51 Ti	320.1	5.76m	———	———
27 Mg	1014.4	9.45m	800	15
52 V	1434.6	3.76m	seconds	minutes
28 Al	1778.7	2.25m		
49 Ca	3084.4	8.72m	———	———
			———	———
56 Mn	846.8	2.58h	1000	3-6
24 Na	1368.6	14.97h	seconds	hours
			———	———

SHORT IRRADIATIONS: 3 minutes © 20 Kilowatts

LONG IRRADIATIONS: 4 hours © 1 Megawatt

Element	E (KeV)	Half-life	Count time	Decay time
153 Sm	103.2	46.7h	———	———
239 Np(U)	228.2	56.4h	8000	7-10
175 Yb	396.3	100.6h	seconds	days
140 La	1596.2	40.3h	———	———
147 Nd	91.1	11.0d	———	———
141 Ce	145.4	32.5d		
177 Lu	208.4	6.7d		
233 Pa(Th)	311.9	27.0d		
51 Cr	320.1	27.7d		
181 Hf	482.0	42.4d		
131 Ba	496.3	11.8d		
85 Sr	514.0	64.8d	16,000	30
95 Zr	756.7	64.0d	to	to
134 Cs	795.8	754.2d	20,000	50
58 Co(Ni)	810.8	70.9d	seconds	days
160 Tb	879.4	72.4d		
86 Rb	1076.6	18.6d		
46 Sc	1120.5	83.8d		
182 Ta	1221.4	114.5d		
59 Fe	1291.6	44.5d		
60 Co	1332.5	1925.6d		
152 Eu	1408.0	4894.4d	———	———

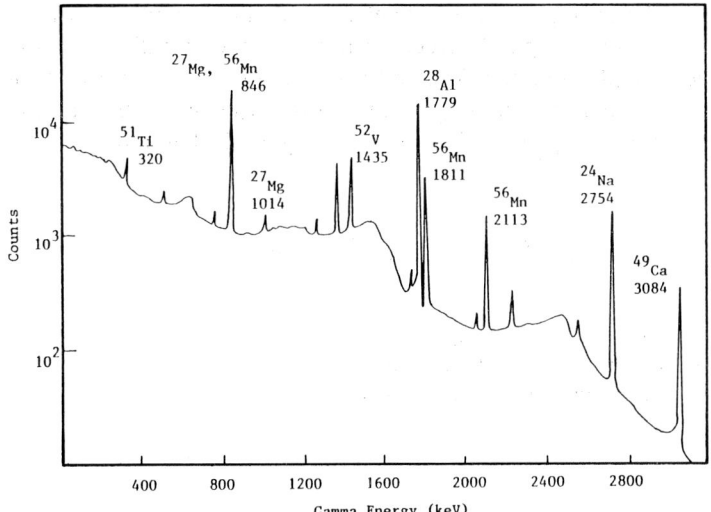

Figure 5.9 Gamma-ray spectrum for standard CRB-1 after 15 minutes of decay since end of bombardment.

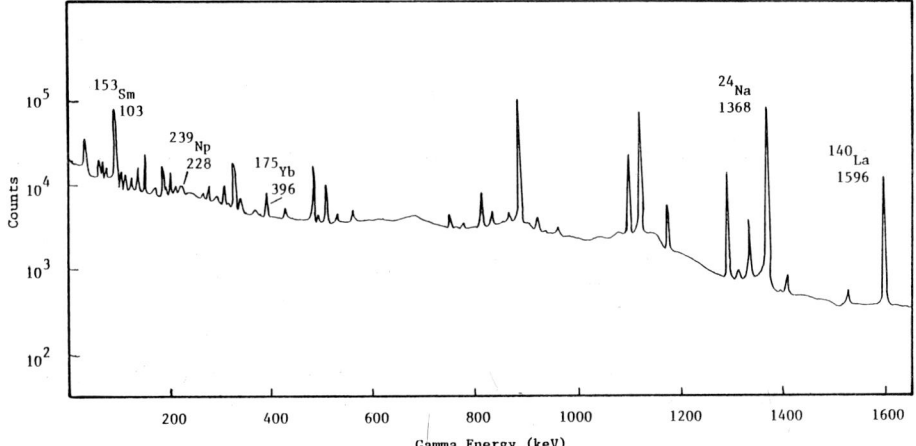

Figure 5.10. Gamma-ray spectrum for CRB-1 after a 7 day decay interval.

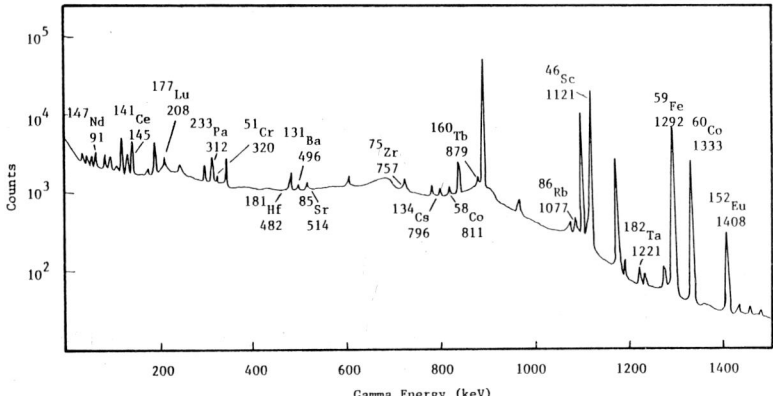

Figure 5.11. Gamma-ray spectrum for CBR-1 after 40 days decay.

Data reduction for INAA involves comparing the net peak area of an element with known concentration to the net peak area of the sample. Corrections must be made for differences in decay and counting times between samples and standards and an estimation of the uncertainty of the analysis must also be calculated. When the counting time is less than one half of the half-life of an element, the net counts (N) under the peak of that element can be calculated by:

$$N = T \ \frac{n}{2} \left(\frac{t_1}{n_1} + \frac{t_2}{n_2} \right)$$

where
$$\begin{aligned}
T &= \text{Total counts in n channels of the peak.} \\
t_1 &= \text{Total counts in } n_1 \text{ channels of left background} \\
t_2 &= \text{Total counts in } n_2 \text{ channels of right background}
\end{aligned}$$

e.g.

```
                            *
                    *            *
                *                     *
            *                              *
    *   *   *                          *   *   *
   |← n₁ →|←            n            →|← n₂ →|
```

The uncertainty associated with the peak area at the 68.3% confidence level (1 sigma) is given by:

$$E = \left\{ \frac{n^2}{4} \left(\frac{t_1}{n_1^2} + \frac{t_2}{n_2^2} \right) + T \right\}^{1/2}$$

In order to arrive at the elemental concentration (ppm) of the sample, differences in decay and count times between samples and standards must be accounted for:

$$\left[\text{ppm} \right] = \frac{N_x e^{L d_x} * \mu gs * C_s}{N_s e^{L d_s} * gx * C_x} \quad \text{and}$$

$$\text{ERROR} = \left[\text{ppm} \right] * \left(\frac{E_x^2}{N_x^2} + \frac{E_x^2}{N_s^2} \right)^{1/2}$$

where
$$\begin{aligned}
N &= \text{Net peak area,} \\
E &= \text{Uncertainty associated with N,} \\
x &= \text{Sample,} \\
s &= \text{Standard,} \\
L &= \text{Decay constant for analyzed element} = \text{ln 2/half-life,} \\
d &= \text{Decay time from some common time to midpoint of counting interval.} \\
ugs &= \text{Micrograms of element in standard,}
\end{aligned}$$

g   = Grams of sample,
and   C   = Counting interval.

These equations can easily be adapted into an interactive computer program, which greatly facilitates data reduction for elements with clearly defined peak areas. When peak areas are not clearly defined, as for low abundance elements, background and peak areas must be calculated manually with the above equations.

It can be shown that for any channel count, the statistical uncertainty (at a 68.3% confidence level) associated with that count is equal to the square root of the count. For elements that do not have a identifiable peak, the best estimate of the elemental abundance can be obtained by taking three times the square root of the average background level, adding that value to the average background level, and using that sum as the net peak area T (n = 1). The calculated abundance will thus represent a 99.5% certainty that the elemental abundance is less than that value, otherwise a peak would be observed.

Mineral Chemistry of Emeralds and Some Associated Minerals from Pakistan and Afghanistan: an Electron Microprobe Study

Jane M. Hammarstrom

INTRODUCTION

An electron microprobe was used to study a suite of nine emerald samples representing four separate occurrences of gem-quality emeralds from northwestern Pakistan and neighboring Afghanistan. The study was undertaken to characterize the chemistry of the emeralds of the region, to compare the emeralds with emeralds from other locations, and to investigate local variations in composition. The results of the study could help unravel the complex and unusual parageneses of these deposits, and ultimately could aid in refinement of exploration techniques. The study also includes data for (1) green vanadian beryls from the Mohmand emerald district, (2) green chromium-rich tourmalines from the Swat emerald district, and (3) minerals in the carbonate host rock at Mingora. The electron microprobe technique allows nondestructive chemical analysis of discrete spots across a single crystal and is useful in the study of minerals. This technique can determine transition metals responsible for color and can identify compositional zoning within single grains. The principal disadvantages of the technique are that elements lighter than fluorine (atomic number 9) cannot be determined and that different oxidation states of elements such as iron cannot be distinguished. Thus, we do not report data for beryllium, a major constituent of emerald; lithium and water, which can occur in emeralds in amounts up to several weight percent; and boron, a major constituent of tourmaline. Snee and others report on the beryllium content of these emeralds in chapter 5 of this volume.

DESCRIPTION OF SAMPLES

Emeralds occur in several geologic settings in northwestern Pakistan and eastern Afghanistan. Samples from four separate gemstone occurrences were examined in this study. Locations are shown on figure 6.1. Samples from Pakistan represent (1) the Swat area near Mingora, where emeralds occur in shear zones in the Mingora ophiolite melange of the Indus suture zone (Kazmi and others, 1984), between the Kohistan andesite arc and the Indian crustal plate, (2) the Mohmand area, located about 128 km southwest of Mingora in a similar setting of ophiolite and volcanics, separated from the Mingora area by platform and shelf deposits, and (3) the Khaltaro area to the north, near Gilgit, where emeralds occur in pegmatites. The Panjsher samples come from the Darkhenj area of Afghanistan, where emeralds occur in quartz-calcite-ankerite veins. Kazmi and others discuss the detailed geology of all these deposits in chapter 3 of this volume.

The Swat samples, from west to east across the area, represent two mines at Mingora (sample SEM51-53/P3 from Mine 1 and sample SDA-25 from Mine 2), the Charbagh Mine (sample C/5C), the Makhad Mine (samples SEM-5 and Makhad No. 6), and the Gujarkili Mine (sample GK21/P3). All the emerald mines occur in shear zones in talc-chlorite-dolomite schists except for Mine 1 at Mingora, where emeralds occur in mineralized zones around quartz lenses

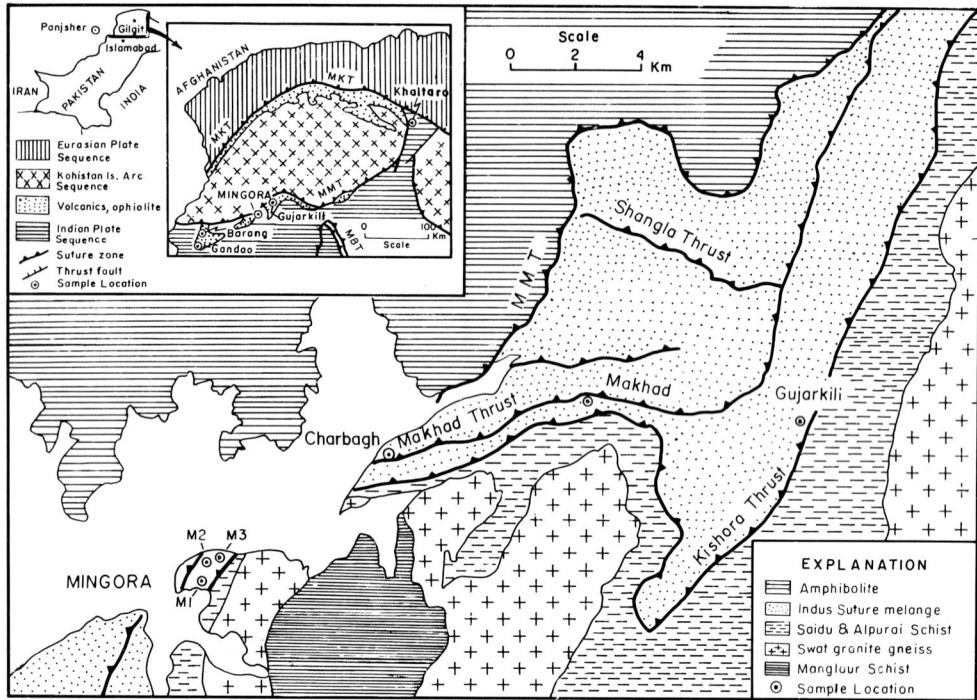

Fig. 6.1. Geological sketch maps showing sample locations.

in talc-carbonate schists. The emeralds vary in size from 1-mm-long lath-shaped grains to 1-cm-long euhedral crystals that have hexagonal outlines. Although all the samples appear green on casual examination, they all exhibited heterogeneous color distribution when cut and polished for microprobe analysis. The color zoning and patchiness of these emeralds allowed us to examine the relationship between color and composition on a small scale and comprise one of the most interesting aspects of this study. Carbonate inclusions are common and most inclusions encountered in the microprobe study proved to be dolomite or quartz. One dark mineral inclusion is the locus of a zone of intense green color in one emerald (see figure 6.2A). This inclusion lies beneath the surface of the emerald and could not be positively identified—but may be chromite, which has been reported as an inclusion in other emeralds (e.g., Gübelin, 1969, 1974; Gübelin and Koivula, 1986; Cassedanne and Sauer, 1984; and see Gübelin, this volume).

Two localities in the Mohmand area are represented: emerald from Barang (sample SEM-4) and green beryl from Gandao (sample SEM-3). Samples from both areas are subhedral grains several mm long and have colors similar to those of the Swat samples. Rafiq and Jan (1985) described emeralds from shear zones in talc-carbonate-quartz rocks that represent hydrothermally altered ultramafics near Bucha, Mohmand. They noted a considerable range in color, transparency, and size for the Bucha emeralds.

The Khaltaro sample is a recent find from a pegmatite that intrudes amphibolites and hornblende-mica schists. The geologic setting at Khaltaro is probably similar to that of the nearby gem pegmatites at Gilgit. Several varieties of beryl are reported from Gilgit (Kazmi and others, 1985), although emeralds have not previously been reported from there. We studied a pale green to almost colorless anhedral grain that is 5 mm in diameter.

The Panjsher emeralds occur in quartz-ankerite and dolomite veins along shear zones in altered diorites and gabbros, dolomitic marble, quartz-biotite schist, and quartz porphyry. We analyzed four emeralds and performed detailed work on a 5-mm-long subhedral crystal that had a dark green 1-mm-wide rim surrounding a pale green to almost colorless core. A 0.1-mm-wide colorless rim is present at the edge of the crystal on one side, and a large 1-mm-long dark brown, triangular-shaped inclusion occurs within the grain. A partial microprobe analysis of this inclusion gave 48.9% CaO, 2.4% MgO, and 0.10% FeO, which suggests that it is a magnesian calcite.

ANALYTICAL METHODS

Principles of electron microprobe analysis

The electron microprobe has become an invaluable tool in studying the chemistry of minerals. Although the equipment is costly and complex, the microprobe is being used more frequently in the study of gemstones and is widely used in petrologic studies. Gübelin (1969) discussed the technique and demonstrated its utility for identifying mineral inclusions in gemstones and more recently, Dunn (1977) reviewed the electron microprobe method of analysis and its applications in gemology. Hanni (1982) and Stockton (1984) analyzed natural and synthetic emeralds from a variety of sources by electron microprobe and established chemical criteria for distinguishing among them. We refer the reader to Smith (1976) for a thorough review of the instrumentation, theory, and applications of microbeam techniques. In brief, a polished thin section or grain mount is mounted on a movable stage and is viewed through a microscope. The section is exposed to an electron beam, which excites x-ray emission from a small volume (to a depth of a few microns) in the sample. Each element present in the sample emits a discontinuous spectrum of x-rays having a very limited number of wavelengths, so that examination of the x-ray spectrum produced by a sample provides compositional information. The relationship between x-ray intensity and concentration of the element giving rise to x-rays of a particular wavelength is roughly linear. By comparing intensities of x-rays generated by samples of unknown composition with intensities of x-rays from samples of known composition (standards) and by applying appropriate corrections to account for a variety of effects that can alter the linearity of this relationship, one can obtain a quantitative analysis of a discrete small volume within a mineral grain. Limiting factors on the range of elements that can be analyzed include the quality of the vacuum inside the chamber, the operating voltage, the range of wavelength-dispersive spectrometers for measuring emitted x-rays, and the availability of suitable standards. Few electron microprobes can analyse elements lighter than fluorine (atomic number 9). Total iron is measured and usually reported as FeO, because the distinction between the ferrous and ferric state cannot be made. The detection limits for elements analyzed by microprobe are generally in the range of hundreds to thousands of parts per million. This range of detection limits is orders of magnitude greater than detection limits for some other analytical techniques, such as instrumental neutron activation (INAA). The detection limit for microprobe analysis of a particular element can be lowered by increasing the counting time or beam current; however, long counting times or high beam currents can result in volatilization of alkali elements. Thus, to determine a variety of elements, the operating procedures for microprobe analysis of minerals must represent a compromise on optimum conditions. The principal advantages of the technique over other methods of analysis are that the sample is not consumed in the process, in situ analyses of small volumes within individual grains are possible, and inclusions of other mineral grains (which are so common in emeralds) can be avoided.

X-ray intensity at a given wavelength is measured as a number of counts. The standard deviation associated with a single measurement based on counting statistics alone is defined as the square root of the count rate. The percentage of this standard deviation relative to the total count rate defines the precision of the measurement. For heterogeneous materials, the cumulative standard deviation associated with an average of several analyses from a grain or from a group of grains will be large relative to the analytical uncertainty based on counting statistics for a single analysis. Data for this study are presented in tables as averages for each sample and for each district; standard deviations for any individual analysis based on counting statistics alone are on the order of ± 2 percent of the amount present for major elements (silicon, aluminum) and ± 10 percent of the amount present for minor elements. Standard deviations associated with the averages reported in the tables reflect the variation in the population comprised of all analyses from a given sample or location. Plots of the variation of elements across individual grains and maps of the distribution of oxides throughout individual grains are presented to illustrate the heterogeneity of the emeralds and to correlate changes in composition with changes in color.

Analytical procedures

This study was conducted by using an automated ARL-SEMQ 9-channel electron microprobe at the U.S. Geological Survey in Reston, Virginia. The instrument was equipped with fixed wavelength-dispersive spectrometers set to analyse for Si, Mg, Mn, Ca and Na, four scanning wavelength-dispersive spectrometers which were used to analyse A1, Fe, Ti, K, Cr, F, C1, V and Sc; and an energy-dispersive spectrometer, which displays the entire spectrum (atomic number 11 and above) produced by the spot that was exposed to the beam. This display of the entire spectrum permits simultaneous qualitative analysis to detect unsuspected elements. The microprobe was operated at 15 kilovolts with a beam current of 0.1 microamps. Data were collected using the $ANBA program (McGee, 1983) which incorporates a rapid data acquisition scheme with Bence-Albee empirical correction factors (Bence and Albee, 1968; Albee and Ray, 1970) for on-line data reduction. Standards consisted of natural minerals and synthetic materials, including feldspars, garnet, scandium-doped diopside, sphene, rhodonite, chromite and vanadium oxide. To monitor accuracy, a variety of working standards, including a synthetic emerald described by Hemingway and others (1986) and a natural tourmaline, were analyzed as secondary standards.

Sample preparation

Small emerald grains were mounted on glass slides with epoxy cement; larger grains were embedded in epoxy and sawed in half. All the samples were ground to a flat surface and then polished on cloth lapidary wheels with a series of diamond pastes to a final polish with 3-micron paste. The polished mounts were photographed in reflected light, and the photographs were used as "maps" of the emerald surface for locating points during microprobe analysis. The surface prepared for analysis represents only a single plane through the crystal and may not accurately reflect its true range of compositions. Similarly, the small number of emeralds available for analysis from each deposit may not adequately represent the chemical variability of the entire emerald population from a given mine. Nevertheless, some chemical signatures are apparent in the data set.

Table 6.1 Summary of microprobe analyses of emeralds and beryls by gem district.
[n.d., not   determined].

Location	Swat	Mohmand	Khaltaro	Panjsher	
No. of mines/samples	6/10	2/3	1/1	1/3	
No. of analyses[1]	151	27	13	63	
Composition	mean s.d. (range)	mean s.d. (range)	mean s.d. (range)	mean s.d. (range)	detection limit
SiO_2	62.6 1.6 (57.8 - 66.4)	62.9 0.9 (60.8 - 64.9)	64.8 0.8 (63.2 - 65.8)	64.8 1.1 (62.5 - 67.6)	0.07
Al_2O_3	13.6 0.6 (11.4 - 15.1)	14.0 0.4 (13.1 - 14.9)	17.1 0.4 (16.5 - 18.0)	16.7 0.8 (15.1 - 18.4)	0.02
FeO_2	0.60 0.36 (0.19 - 1.45)	1.27 0.28 (0.78- 1.68)	0.53 0.08 (0.32 - 0.63)	0.45 0.19 (0.14 - 0.96)	0.05
MgO	2.47 0.31 (1.89 - 3.03)	2.33 0.34 (1.93 - 3.00)	1.01 0.11 (0.69 - 1.17)	0.74 0.38 (0.22 - 1.98)	0.02
CaO	0 (0- 0.03)	0 (0 - 0.03)	0 (0 - 0.02)	0 (0 - 0.03)	0.02
Na_2O	1.90 0.35 (0.75 - 2.33)	1.88 0.38 (0.98 - 2.32)	0.92 0.05 (0.80- 0.99)	0.48 0.34 (0.12- 1.64)	0.01
K_2O	0.01 0.02 (0 - 0.07)	0.01 0.02 (0 - 0.05)	n.d.	0.01 0.02 (0 - 0.07)	0.03
TiO_2	0.01 0.02 (0 - 0.01)	0.01 0.02 (0 - 0.09)	0.01 0.01 (0 - 0.04)	0.01 0.02 (0 - 0.05)	0.03
MnO	0 (0 - 0.05)	0.01 0.01 (0 - 0.04)	0.03 0.01 (0 - 0.06)	0 (0 - 0.01)	0.04
V_2O_3	0.04 0.03 (0 - 0.12)	0.47 0.31 (0 - 0.82)	0.04 0.01 (0.02 - 0.06)	0.12 0.08 (0 - 0.31)	0.03
Cr_2O_3	1.06 0.54 (0 - 2.10)	0.08 0.10 (0 - 0.45)	0.67 0.12 (0.43 - 0.81)	0.66 0.32 (0.10 - 1.37)	0.03
Sc_2O_3	0.25 0.21 (0 - 0.76)	0.10 0.10 (0.01 - 0.32)	0.01 0.01 (0 - 0.03)	0.01 0.01 (0 - 0.03)	0.03
F	0.04 0.04 (0 -0.19)	0.04 0.03 (0 - 0.07)	n.d.	0.02 (0 - 0.12)	0.08
Cl	0.01 0.02 (0 - 0.09)	0.01 0.02 (0- 0.05)	n.d.	0.03 0.06 (0 - 0.36)	0.05
BeO(calc)[3]	13.0	13.1	13.6	13.5	
BeO[4]	9.5 - 11.6	11.0 - 11.4	11.9 - 12.0		

[1] Number of analyses that include Si, Al, Fe, Mg, Na, V, Cr determinations; fewer analyses include other elements.

[2] Total iron reported as FeO; some or all of the iron may be present as Fe_2O_3.

[3] Average of BeO values calculated for each microprobe analysis assuming an ideal 3 Be cations per 18-oxygen beryl formula unit; this represents the maximum BeO content theoretically possible for the analysis, assuming stoichiometry.

[4] BeO values reported by Snee and others (1989, chapter 5 of this volume) for analyses on samples from these localities.

MICROPROBE DATA ON EMERALDS

Emeralds are the green variety of beryl, which has the ideal formula $Be_3Al_2Si_6O_{18}$. The green color of emerald is due to small amounts of chromium. Other transition metals such as vanadium or iron, which can also substitute in the aluminum site of the ideal formula, may affect the green hue to become either a bluish or a yellowish green. Wood and Nassau (1968) used infrared spectroscopy to demonstrate that vanadium in beryl is trivalent and that vanadium can account for the green color in beryls having little or no detectable chromium. According to the comprehensive summary on emeralds and other beryls by Sinkankas (1981), the highest reported Cr_2O_3 contents for emeralds are 3.25 and 3.50 weight percent for analyses of emeralds from Muzo, Columbia reported by Vauquelin around 1800. Sinkankas points out that other investigators report only traces of chromium in Muzo emeralds and that the high values obtained in these very old analyses may be suspect.

Table 6.1 summarizes the microprobe data for the samples in terms of the average, standard deviation, and range of compositions observed in all grains of all samples from a given region. Averages for individual localities in the Swat and Mohmand districts are given in table 6.2. Data are reported in terms of oxide weight percent and the minimum detection limit for each oxide is listed in the far right column of table 6.1. Wavelength-dispersive spectrometers cannot resolve wavelengths for variable oxidation states of individual elements. Therefore, total iron is reported as FeO although all or most of the iron may be present in the emeralds as Fe_2O_3. Vanadium can also occur in multiple valence states.

GENERAL OBSERVATIONS

The following observations apply to all samples and are in general agreement with previously published analyses of emeralds from Pakistan (see table 6.3).

1. All of the samples contain FeO and MgO, and MgO/FeO generally is greater than 2.
2. CaO, K_2O, TiO_2 and MnO are virtually absent.
3. Na_2O is present in all analyses, usually in amounts greater than 0.5 percent.
4. Cr_2O_3, which is considered responsible for the green color of the emerald variety of beryl, was detected in most analyses. The highest chromium contents found in this study, up to 2 percent of the oxide, exceed the amounts found in previous studies of emeralds from Pakistan and, as far as we know, are among the highest chromium contents reported for any natural emeralds.
5. Vanadium generally occurs in amounts near or below the detection limit; however, high values for V_2O_3 are found in samples having low Cr_2O_3 contents. According to the rules of the International Confederation of Jewelry, Silverware, Diamonds, Pearls and Stones (CIBJO), use of the term "emerald" is restricted to beryl with coloration due to chromium. Therefore, the Gandao sample from the Mohmand district (table 6.2) must be called a green vanadian beryl rather than an emerald.
6. Scandium was detected in significant amounts (0.5% Sc_2O_3) in only one of the samples analyzed for scandium.
7. A few analyses for fluorine and chlorine were conducted, and most analyses fell below detection limits. These elements generally are not reported in the literature for natural beryls but could be important in the transport of beryllium in the hydrothermal environment. Chlorine contents as high as 0.3 to 0.4 weight percent are reported for some synthetic emeralds grown by hydrothermal methods (Stockton, 1984; Hanni, 1982).

Table 6.2 Summary of microprobe analyses of emeralds and beryls within given gem districts. [n.d., not determined]

Locality	Swat																Mohmand			
	emeralds																emeralds		green vanadian beryls	
Mine Sample	Mingora (2) SDA-25		Mingora (1) SEM 51-53/P3		Makhad SEM 5		Makhad ≠6		Charbagh C/5C		Gujar Kili GK21/P3						Barang SEM 4		Gandao SEM 3	
No. of grains	2		1		4		1		1		1						1		2	
No. of analyses	51		22		54		7		12		11						8		19	
Composition	mean	s.d.	mean	s.d.	mean	s.d.	mean	s.d.	mean	s.d.	mean	s.d.					mean	s.d.	mean	s.d.
SiO_2	62.8	2.2	62.7	1.7	62.2	0.9	62.5	0.3	63.3	0.2	62.3	1.3					62.1	0.7	63.2	0.7
Al_2O_3	13.1	0.5	14.2	0.5	13.68	0.66	13.80	0.34	14.1	0.3	13.5	0.4					14.1	0.2	14.0	0.5
FeO	0.91	0.21	0.52	0.05	0.67	0.40	0.44	0.04	0.23	0.02	0.25	0.02					0.94	0.22	1.40	0.18
MgO	2.50	0.32	2.46	0.11	2.37	0.38	2.68	0.07	2.67	0.04	2.57	0.05					2.77	0.24	2.14	0.14
CaO	0.01	0.01	0		0		0		0		0.01	0.01					0.02	0.01	0	
Na_2O	2.06	0.18	2.11	0.11	1.64	0.44	2.05	0.04	2.04	0.04	1.88	0.03					2.19	0.17	1.75	0.38
K_2O	0		n.d.		0.02	0.02	n.d.		n.d.		n.d.						0.02	0.03	0	
TiO_2	0.01	0.01	0.01	0.01	0.02	0.02	0.01	0.01	0		0.02	0.02					0.01	0.01	0.02	0.03
MnO	0		0		0		0				0.01	0.01					0.02	0.02	0	
V_2O_3	0.04	0.01	0.06	0.03	0.04	0.03	0.06	0.02	0.03	0.02	0.09	0.02					0.02	0.02	0.66	0.10
Cr_2O_3	1.17	0.42	0.39	0.45	1.26	0.38	0.50	0.30	0.53	0.37	1.65	0.28					0.17	0.15	0.04	0.02
F	0.02	0.03	n.d.		0.06	0.19	n.d.		n.d.		n.d.						0.05	0.03	0.04	0.03
Cl	0.01	0.01	n.d.		0.01	0.03	n.d.		n.d.		n.d.						0.02	0.03	0.01	0.02
$BeO(calc)$[1]	13.0		13.1		13.0		13.0		13.6		13.0						13.0		13.2	
BeO[2]	9.49		10.49		11.54		11.54		n.d.		11.63						10.96		11.41	

[1] Average of BeO values calculated for each microprobe analysis assuming an ideal 3 Be cations per 18-oxygen beryl formula unit; this represents the maximum BeO content theoretically possible for the analysis, assuming stoichiometry.

[2] BeO values reported by Snee and other (chapter 5 of this volume).

Table 6.3 Compilation of published chemical analyses on emeralds from Pakistan. [n.d., not determined]

Reference	Gübelin, 1982	Kazmi and others, 1986				Rafiq and Jan, 1985		Hanni, 1982	
Location	Mingora	Mingora				Bucha, Mohmand Agency		no location given	
Notes		ME7[1]		ME4, ME5[2]		light green beryl	bluish-green emerald		
		mean	s.d.	mean	s.d.				
SiO_2	n.d.	63.7	0.6	64.2	0.4	n.d.	n.d.	64.2	63.7
Al_2O_3	n.d.	12.6	0.2	12.7	0.3	n.d.	n.d.	14.8	13.7
Fe_2O_3	0.9	n.d.		n.d.		2.25[4]	2.10[4]	n.d.	n.d.
FeO	n.d.	0.21[3]	0.05	0.96[3]	0.06	n.d.	n.d.	0.4	1.3
MgO	2.6	2.48	0.23	1.93	0.09	2.18	1.91	2.7	2.4
CaO	0.0	0.02	0.02	0.02	0.02	n.d.	n.d.	n.d.	n.d.
Na_2O	2.1	n.d.		n.d.		1.87	2.14	2.1	1.9
K_2O	n.d.	n.d.		n.d.		0.27	0.08	n.d.	n.d.
TiO_2	n.d.	0.00		0.00		n.d.	n.d.	0.0	0.0
MnO	n.d.	0.03	0.02	0.02	0.01	0.10	0.10	0.0	0.1
V_2O_3	0.00	n.d.		n.d.		n.d.	n.d.	0.1	0.0
Cr_2O_3	0.66	1.12	0.37	1.61	0.15	0.27	0.83	0.8	1.4
CuO	n.d.	n.d.		n.d.		0.41	0.11	n.d.	n.d.
NiO	n.d.	n.d.		n.d.		0.04	0.02	n.d.	n.d.

[1] Average of 6 analyses on 2 chips; E. Mullen and M. Schaffer, analysts.
[2] Average of 5 analyses on 3 chips from 2 samples of euhedral emeralds from the upper level of the Mingora Mine (same location as sample SEM5 of this study); E. Mullen and M. Schaffer, analysts.
[3] Total iron reported as FeO.
[4] Total iron reported as Fe_2O_3.

CHEMICAL VARIATIONS AMONG DISTRICTS AND WITHIN DISTRICTS

Considerable overlap in composition exists among the emeralds from all four districts (table 6.1). The salient differences in chemistry may represent differences in environments of emerald formation but may also reflect sampling problems. In other words, if more analyses were acquired from specific localities, the apparent chemical distinctions among the districts might be obliterated.

The apparent chemical signatures of the districts are as follows:

1. Emeralds from the Swat and Mohmand districts have lower SiO_2 and Al_2O_3 contents than the Khaltaro and Panjsher emeralds. The Swat and Mohmand samples tend to be enriched in MgO and Na_2O compared to the other samples.
2. Swat emeralds have more Cr_2O_3 and less V_2O_3 than Mohmand emeralds. The Mohmand samples are distinct from all of the other samples due to their high iron content, low chromium content, and in the Gandao beryl sample, the high vanadium content.
3. Panjsher emeralds have the lowest average Na_2O contents. Although the average Na_2O value for the Khaltaro sample falls between the average values for the Swat and Mohmand samples (which are similar to each other) and the value for the Panjsher samples, the range in Na_2O for the Khaltaro sample shows the least spread.

Considerable overlap in composition exists among the Swat samples (table 6.2). The Charbagh and Gujarkili samples have less FeO than the other samples, and the Gujarkili sample contains significant amounts of scandium. The Mohmand emerald and green vanadian beryl samples both have lower chromium contents than emeralds from other locations; the green vanadian beryl from Gandao is slightly enriched in iron as well as vanadium and depleted in magnesium and chromium relative to the emerald from Barang. The composition of the Khaltaro emerald is not unique although it is the only sample from a pegmatitic environment: average values for SiO_2, Al_2O_3, FeO and Cr_2O_3 are similar to those for the Panjsher sample. Relatively low values for Sc_2O_3, FeO and MgO for the Khaltaro sample are consistent with the chemical signatures described by Staatz and others (1965) for beryls from pegmatites, as compared with beryls from wallrocks adjacent to pegmatites and for beryls from granites and veins. In particular, Staatz and others found virtually no scandium (generally less than 0.0005 weight percent Sc) in beryls from pegmatites whereas beryls from granites contained 0.0015 to 0.07 weight percent scandium.

ZONING

Changes in color within individual emerald crystals correlate directly with changes in composition. Figures 6.2-6.5 illustrate chemical variations across individual emerald crystals from Swat and Panjsher that exhibit several varieties of color zoning patterns. Zoning was not observed in the samples from Mohmand and Khaltaro.

The photomicrographs in photo.6.1 illustrate the variety of zoning patterns observed in emeralds from Swat. Sketch maps of these grains in figures 6.2A, 6.3A, 6.4A, and 6.5A show weight percent Cr_2O_3 for analyzed points and locations of traverses across grains; plots of the variation in selected oxides along these traverses are shown in figures 6.2B, 6.3B, 6.4B, and 6.5B.

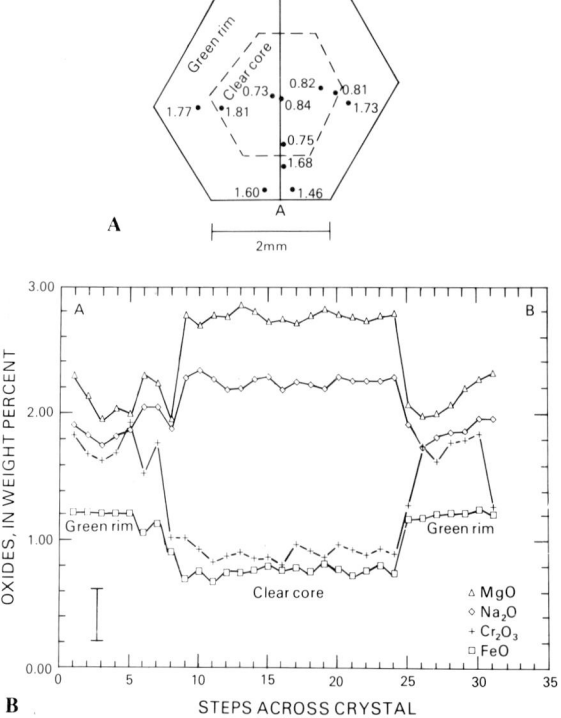

Figure 6.2. Sample SDA-25 from Mingora Mine 2. For figs. 2-5, part A of each fig. shows a sketch of the grain and has pertinent color boundaries and traverse lines marked; weight percent Cr_2O_3 values are noted next to some of the analyzed points. Part B of each fig. shows the chemical variation for selected oxides in traverses across the crystal. Traverse steps for fig. 2B are 200 microns; vertical bar in lower left corner of fig. 2B indicates analytical uncertainty.

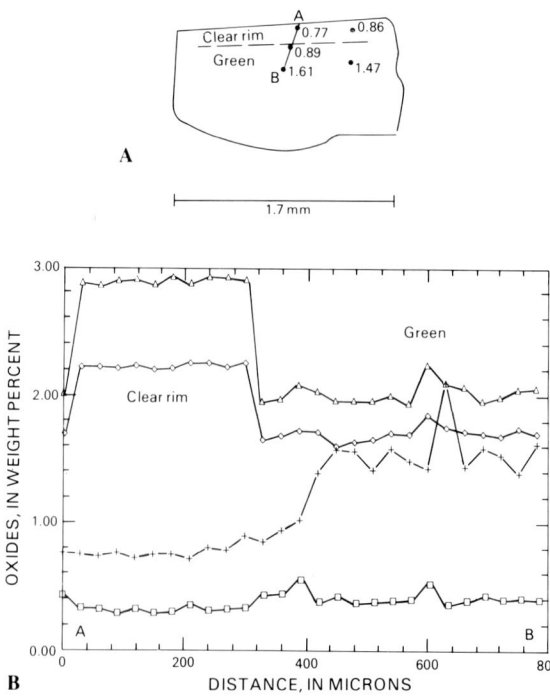

Figure 6.3. Sample SEM-5 from the Makhad Mine. See fig. 6.2. for explanation.

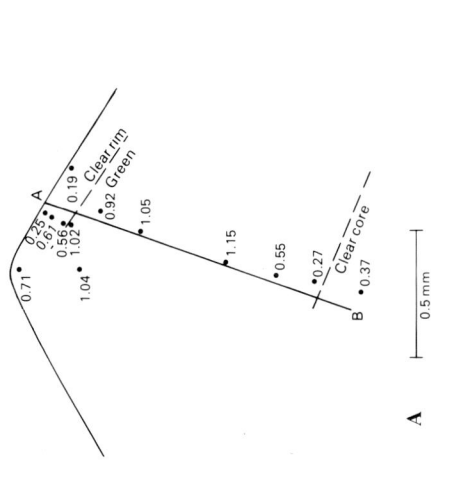

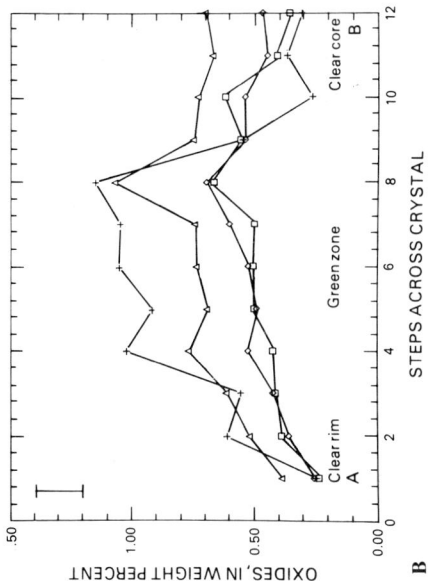

Figure 6.5. Sample SEM-1 from Panjsher, Afghanistan. See fig. 6.2 for explanation. Steps in part B are not evenly spaced; note that the Y-axis scale covers a different range from that of previous figs. Vertical bar shows analytical uncertainty.

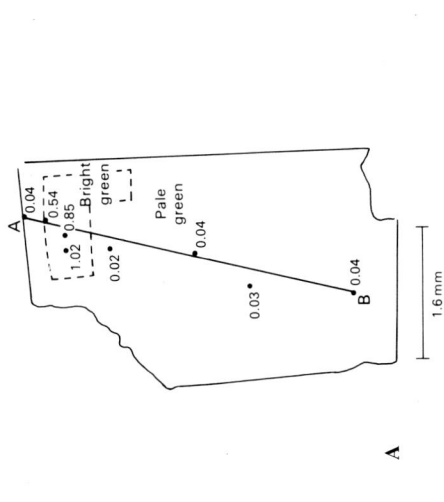

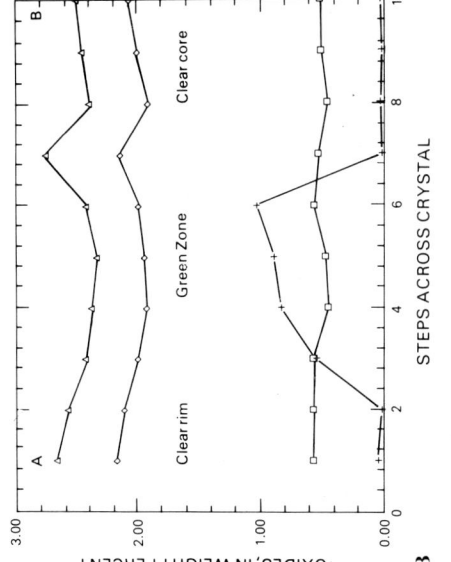

Figure 6.4. Sample SEM-51-53/P3 from Mingora Mine 1. Traverse in part B is schematic; steps are not evenly spaced. See fig. 6.2 for explanation.

In all cases, the green, brightly colored areas of the emeralds are enriched in chromium compared to the pale green or almost colorless areas of the same grain. In some cases, the apparent optical boundary correlates exactly with a change in composition. In other cases, where the color boundaries are more diffuse and are obscured by the thickness of the crystal or by dipping contacts between color zones, the change in composition occurs near the apparent color boundary. Nevertheless, the most intensely colored areas of these grains show the highest concentrations of chromium; this relationship suggests that the chromium ion largely accounts for the color.

Sinkankas (1981) described the variety of color zoning patterns observed in natural emeralds and noted that the most common pattern is a pale to colorless core surrounded by a darker colored rim. This type of zoning is observed in sample SDA-25 (fig. 6.2). In this sample, a euhedral, hexagonal cross section through the crystal reveals an almost colorless core (with Cr_2O_3 generally less than 1 percent) that is surrounded by a bright green rim (with Cr_2O_3 generally greater than 1 percent) that is marked by growth zones. Vanadium concentrations are near or below the detection limit throughout the grain. Figure 6.2B shows an inverse relationship between Cr_2O_3 and MgO or Na_2O. FeO variation generally mimics Cr_2O_3, and patterns for Na_2O and MgO are almost identical. Figure 6.3 shows the zoning pattern observed in one of several small emeralds from the Makhad mine at Swat. Although the crystal is partly broken and lacks euhedral crystal faces on all sides, a definite colorless rim has developed at one edge of the grain, in sharp contact with the rest of the grain, which is green (fig. 6.3A). A traverse in 40-micron steps across this color boundary is shown in figure6.3 (photo.6.1C). Going from the clear rim to the green zone, a sharp drop in sodium and magnesium occurs near the color boundary, and a more gradual increase in chromium and a slight increase in iron occur. As in sample SDA-25, the green part of the grain has the highest chromium content. Sample SEM 51-53/P3 is a partly broken crystal from Mine 1, the southernmost of the Mingora mines, where emeralds are concentrated around a quartz lens. The sample we studied is pale green and has an intense green L-shaped zone near one corner of the crystal (fig. 6.4A, photo. 6.1d). Throughout the pale green to colorless part of the grain, the Cr_2O_3 content is near the detection limit; however, in the intensely colored patch, Cr_2O_3 content jumps from less than 0.5 percent to values between 0.5 and 1 percent. The inverse relationship between sodium or magnesium and chromium is not as apparent in this sample as in the others (fig. 6.4B).

One Panjsher emerald, sample SEM-1, is euhedral and exhibits the most complex optical zoning pattern. The emerald has a narrow colorless rim, separated from a nearly colorless core by a 1.1-mm-wide green band. Figure 6.5 shows that, although the green middle zone of the crystal is enriched in chromium compared to the rim or core, the MgO and Na_2O contents are much lower than in the Swat emeralds and cannot be rigorously correlated with variations in Cr_2O_3.

THERMOGRAVIMETRIC ANALYSIS

Thermogravimetric analyses were conducted on two emeralds from Panjsher and two green vanadian beryls from Gandao to determine weight losses on heating, as a method to estimate water content. One small sample from each location was lightly ground and a separate large single crystal from each locality was mounted on a glass slide and sawed in half. One half was polished for microprobe analysis and the other half was gently broken under acetone in an agate mortar. Fragments from the large crystals were examined under a microscope and grains that were obviously contaminated were removed. The remaining fragments were lightly ground and

30 to 50 milligrams of the resulting powder was loaded for analysis. A small amount of the ground material was used to prepare a smear slide for x-ray diffraction. X-ray diffraction patterns of the Panjsher samples showed only beryl; however, the patterns for both Gandao samples showed the presence of a small amount of quartz. Therefore, any volatile content inferred from these samples must be considered a minimum because the starting material was impure.

Samples were heated in a Mettler thermogravimetric analyzer from room temperature to 1400 °C at a rate of 6 °C per minute in an atmosphere of flowing helium. During the runs on the large crystals, both samples were heated to 120 °C and held briefly at that temperature to remove any adsorbed water. No weight changes were detected below about 600 °C and weight loss, which apparently occurred in a single step, was completed by 930 °C; no changes occurred between 930 and 1400 °C. X-ray diffraction patterns of run products showed that both large samples retained the beryl structure up to 1400 °C and that no new phases occurred.

The Panjsher samples lost 2.28 and 2.24 percent of the total weight for runs on the small and large emeralds, respectively. Both samples retained their light green color after heating. The small Gandao sample lost 4.65 percent of its total weight on heating to 1500 °C, turned from light green to pale brown with white spots, and fused to the platinum crucible. This sample was the only one heated above 1400 °C, and no weight loss was detected at the higher temperatures. The larger crystal from Gandao lost 3.09 percent and also turned pale brown. The difference in weight loss for the two Gandao samples may reflect differences in amount of contaminants (x-ray patterns of both samples show quartz) or may reflect heterogeneity in beryls from that locality. Although release of volatiles from fluid and solid (e.g., carbonate) inclusions in the samples may contribute a small amount to the weight loss, most of the weight lost probably can be attributed to water in the beryl crystal structure. The high alkali content (average $Na_2O = 1.73$ percent) and low oxide sums for the Gandao samples compared to the Panjsher samples (average $Na_2O = 0.41$ percent), are consistent with the thermogravimetric results which show that the Gandao samples lost more water (or other volatiles) on heating. Microprobe analysis of the run product for the Panjsher sample showed that no sodium loss occurred on heating.

Microprobe data (tables 6.1 and 6.2) show that samples from both localities contain iron and chromium and (or) vanadium, all of which can account for or contribute to color in beryls (Wood and Nassau, 1968). Both iron and vanadium can occur in multiple oxidation states and although all emeralds were heated under identical conditions in an inert atmosphere, oxidation may have occurred during the runs. The Panjsher emeralds, which did not change color during heating, have an average composition of 0.73% Cr_2O_3, 0.13% V_2O_3. However, the Gandao samples, which turned from green to brown during the run, have essentially no chromium (at the microprobe detection limit) but have a high vanadium content (0.61% V_2O_3) and considerably more iron than the Panjsher sample. The color change observed for the Gandao samples is probably due to a valence change in vanadium and (or) iron or due to some interaction among transition metals induced by run conditions. Consequently, the Gandao samples are not emeralds, but are green vanadian beryls.

These results agree with previous work by Ginzburg (1955), who reported that water loss on heating occurs between 800 and 900 °C, and by Wood and Nassau (1967), who used infrared spectroscopy to identify two types of water in beryl. Wood and Nassau found that both types of water occur in the channel sites (which are empty in the ideal beryl structure) in different structural orientations. Type I water molecules are oriented parallel to the C_6 axis of the beryl crystal, whereas Type II water molecules are perpendicular to the axis. Wood and Nassau found that spectra for Type II molecular water are more pronounced in alkali-rich beryls. They concluded that the presence of a charged alkali ion (sodium or lithium) in the channel affects the orientation of the water molecules. The tightly bonded alkali ions contribute to the high thermal stability of water in beryl by effectively blocking the transport of molecules or ions through the channels. In a subsequent study, Wood and Nassau (1968) describe heating experiments on two colorless natural beryls from Brazil. The Brazilian beryls showed both water types, as well as CO_2 and Fe^{3+}. One of their samples lost 1.6 percent around 900 °C and the other still showed infrared bands for water after heating to 1200 °C, and did not give up its water and CO_2 until 1350 °C.

CRYSTAL CHEMISTRY OF THE EMERALDS

Representative microprobe analyses for samples from each of the four regions are listed in table 6.4. Mineral formulas are calculated in table 6.4 on the basis of an ideal beryl formula having 18 oxygens per formula unit. A theoretical value for BeO is computed by assuming that three beryllium cations are present. This may not be a valid assumption because, as Sinkankas points out, beryllium may substitute for or be displaced by silicon or aluminum in the beryl structure, and chemical analyses of natural beryls rarely conform to ideal beryl stoichiometry. Such a calculation does, however, provide a clue to possible maximum BeO contents for comparison with beryllium determinations by other analytical techniques and the calculation provides a check on oxide sums. The theoretical BeO contents calculated for the analyses in table 6.4 are generally only a percent or two higher than the BeO values that Snee and others report in chapter 5 of this volume for other samples from some of the same localities. The oxide sums, including calculated BeO contents, are about 96 percent for the Charbagh and Gandao samples whereas the oxide sums for the Khaltaro and Panjsher samples average around 98 percent. Although the "missing" 2 to 4 percent may include elements such as lithium, the high sodium contents and the weight losses on heating suggest that considerable water may be present. Data reported by Snee and others in chapter 5 of this volume show that cesium is not an important constituent of these emeralds. Most complete beryl analyses in the literature report water contents in the range of 1 to about 4 percent H_2O (Sinkankas, 1981). Hawthorne and Černý (1977) examined a large number of published beryl analyses in conjunction with their study of the structure of a cesium-lithium beryl. They showed that the sodium content of beryl may determine the minimum amount of water present because sodium is bonded to one or two H_2O molecules in the channel sites. Their study showed that for most alkali beryl analyses there are at least twice as many water molecules as sodium cations per formula unit.

If the average amount of water (or other volatiles) obtained from thermogravimetry on the Panjsher samples, 2.26 percent, is added to the oxide sums in table 6.4, the totals approach 100 percent. In terms of water, this translates to about 0.7 H_2O molecule per 18-oxygen formula unit, considerably more than twice the number of sodium cations in the formula. These numbers are regarded as crude approximations at best, since calculation of the formula relies on assumptions about BeO contents, valence states, and lack of other elements or vacancies. This calculation ascribes weight loss solely to water content.

Similarly, addition of 3 percent or more to the Gandao oxide sums, which average around 96 percent including calculated BeO contents, seems to account for all the major constituents of the beryl from that locality. A water content of 3.09 percent (from thermogravimetry on the smaller of the two Gandao beryls) would contribute 0.98 water molecules to the formula whereas a water content of 4.65 percent would contribute nearly 1.5 water molecules. Estimates of the number of water molecules exceed the maximum number of sodium cations (0.40, observed for Gandao analyses in table 6.4) by a factor of 2 or more and agree with the conclusions of the Hawthorne and Černý study.

Most of the cations found as substituents in beryl are too large to fit in the sites normally occupied by Be^{2+} or Si^{4+}, which have ionic radii of 0.35 and 0.42 angstrom, respectively. The octahedrally coordinated Al^{3+} site however, can accomodate the larger cations, such as Fe^{2+}, Cr^{3+}, Sc^{3+} and V^{3+}. The exact nature of the chemical substitutions in the emeralds cannot be rigorously determined from microprobe data alone because variations in beryllium content as a function of variations in other elements are not known. The apparent inverse zonation between Mg (+2 charge) and Na (+1 charge) versus Cr (+3 charge) and Fe (+2 or +3 charge) however, suggests that sodium may be involved in coupled substitutions with magnesium to maintain charge balance. Deer and others (1962) reviewed various proposed formulas for alkali-bearing beryls. These formulas involve substitution of alkalis for beryllium with hydration (e.g., $Be_{3-n/2}$ $(Na,Li,Cs)_n Al_2 Si_6 O_{18} \cdot pH_2O$, where n=0-1 and p=0.2-0.8 (Ginzburg, 1955).

Table 6.4 Representative microprobe analyses of emeralds and green vanadian beryls. [n.d., not determined]

Location	Charbagh, Swat sample C/5C					Khaltaro (pegmatite)				
Analysis	1	2	3	4	5	1	2	3	4	5
SiO_2	63.18	63.36	63.16	62.91	63.49	65.79	65.31	65.19	64.90	63.18
Al_2O_3	14.08	14.29	13.86	14.01	14.60	17.10	17.24	17.09	17.95	16.45
FeO	0.24	0.25	0.19	0.21	0.23	0.58	0.53	0.53	0.32	0.63
MgO	2.63	2.70	2.67	2.68	2.62	1.04	1.04	1.00	0.69	1.04
CaO	0.01	0.02	0.01	0.01	0.00	0.00	0.01	0.01	0.2	0.00
Na_2O	2.00	2.04	2.07	2.06	1.98	0.90	0.93	0.91	0.80	0.95
K_2O	n.d.	n.d.	n.d.	n.d.	n.d.	n.d.	n.d.	n.d.	n.d.	n.d.
TiO_2	0.02	0.00	0.00	0.00	0.00	0.00	0.00	0.01	0.02	0.04
MnO	0.00	0.00	0.00	0.00	0.00	0.02	0.02	0.03	0.04	0.06
V_2O_3	0.01	0.04	0.00	0.04	0.05	0.03	0.03	0.03	0.02	0.03
Cr_2O_3	0.40	0.27	0.93	1.07	0.21	0.76	0.58	0.72	0.43	0.75
Sc_2O_3	n.d.	n.d.	n.d.	n.d.	n.d.	0.02	0.01	0.01	0.01	0.03
BeO^1	13.14	13.20	13.16	13.16	13.25	13.79	13.71	13.68	13.67	13.28
Sum	95.92	96.41	96.05	96.35	96.60	100.03	99.41	99.21	98.87	96.44

Cations based on 18 oxygens, 3 Be atoms

Si	6.00	5.99	5.99	5.97	5.98	5.96	5.95	5.95	5.93	5.94
Al	1.58	1.59	1.55	1.57	1.62	1.83	1.85	1.84	1.93	1.82
Fe^{2+}	0.02	0.02	0.02	0.02	0.02	0.04	0.04	0.04	0.02	0.05
Mg	0.37	0.38	0.38	0.38	0.37	0.14	0.14	0.14	0.09	0.15
Ca	0	0	0	0	0	0	0	0	0	0
Na	0.37	0.37	0.38	0.38	0.36	0.16	0.16	0.16	0.14	0.17
Ti	0	0	0	0	0	0	0	0	0	0
Mn	0	0	0	0	0	0	0	0	0	0
V	0	0	0	0	0	0	0	0	0	0
Cr	0.03	0.02	0.07	0.08	0.02	0.05	0.04	0.05	0.03	0.06
Be	3.00	3.00	3.00	3.00	3.00	3.00	3.00	3.00	3.00	3.00

Table 6.4 (Continued)

Location	Makhad Mine, Swat district			Gandao, Mohmand district[3]			Panjsher, Afghanistan		
Analysis	1	2	3	1	2	3	1	2	3
SiO_2	62.8	62.4	62.9	63.3	63.4	64.2	66.0	67.1	65.1
Al_2O_3	14.2	13.9	13.5	13.8	13.9	13.7	18.2	18.2	17.1
FeO	0.47	0.49	0.47	1.68	1.36	1.50	0.27	0.27	0.46
MgO	2.69	2.72	2.64	1.95	2.33	2.15	0.22	0.31	0.75
Na_2O	2.05	2.09	2.03	1.99	2.11	2.11	0.21	0.30	0.70
V_2O_3	0.05	0.06	0.06	0.69	0.53	0.73	0.08	0.07	0.03
Cr_2O_3	0.23	0.35	0.33	0.02	0.04	0.05	0.19	0.23	0.10
Sc_2O_3	0.17	0.16	0.19	n.d.	n.d.	n.d.	n.d.	n.d.	n.d.
BeO^1	13.1	13.0	13.0	13.2	13.2	13.3	13.8	14.0	13.5
Sum	95.8	95.2	95.1	96.6	96.9	97.7	98.9	100.5	97.7
Weight loss[2]		n.d.			3.1%			2.2%	

Cations based on 18 oxygens, 3 Be atoms

Si	5.98	5.99	6.03	6.01	5.99	6.01	5.99	6.00	6.00
Al	1.60	1.57	1.52	1.54	1.55	1.52	1.95	1.92	1.85
Fe^{2+}	0.04	0.04	0.04	0.13	0.11	0.12	0.02	0.02	0.04
Mg	0.38	0.39	0.38	0.28	0.33	0.30	0.03	0.04	0.10
Na	0.38	0.39	0.38	0.37	0.39	0.38	0.04	0.05	0.13
V	0.00	0.00	0.00	0.05	0.04	0.05	0.01	0.01	0.00
Cr	0.02	0.03	0.03	0.00	0.00	0.00	0.01	0.02	0.01
Sc	0.01	0.01	0.02	—	—	—	—	—	—
Be	3.00	3.00	3.00	3.00	3.00	3.00	3.00	3.00	3.00

Cations based on 18 oxygens, BeO from Snee and others (this volume).

Si	6.10	6.10	6.15	6.14	6.13	6.16	6.12	6.14	6.11
Al	1.63	1.60	1.56	1.58	1.58	1.55	1.99	1.96	1.89
Fe^{2+}	0.04	0.04	0.04	0.14	0.11	0.12	0.02	0.02	0.04
Mg	0.39	0.40	0.38	0.28	0.34	0.31	0.03	0.04	0.10
Na	0.39	0.40	0.38	0.37	0.40	0.39	0.04	0.05	0.13
V	0.00	0.00	0.00	0.05	0.04	0.06	0.01	0.01	0.00
Cr	0.02	0.03	0.03	0.00	0.00	0.00	0.01	0.02	0.01
Sc	0.01	0.01	0.02	—	—	—	—	—	—
Be	2.70	2.71	2.71	2.66	2.65	2.63	2.68	2.65	2.72

[1] Theoretical amount of BeO computed assuming 3.00 Be cations per formula unit; since Al or Si can substitute in the Be site in the beryl structure, this assumption may not be valid.

[2] Weight loss from heating (other halves of emerald crystals used to obtain these microprobe analyses) from room temperature to 1400 °C in a thermogravimetric analyzer.

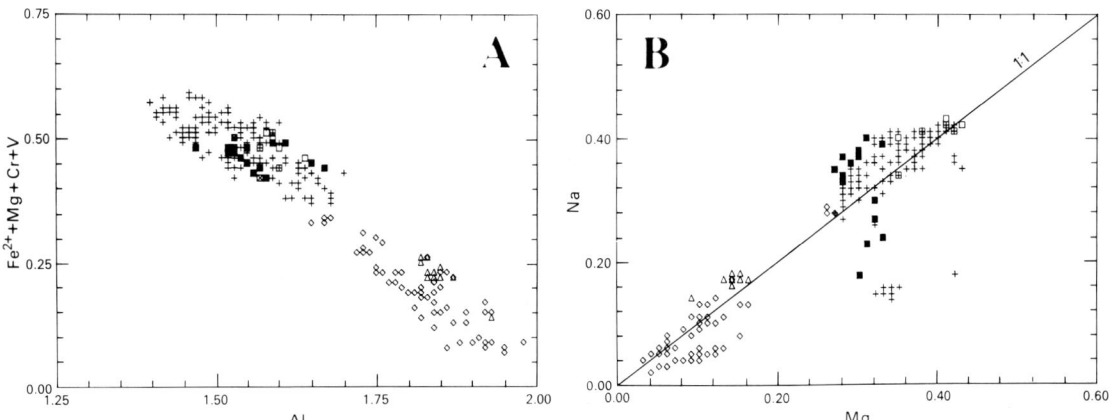

Figure 6.6. Plots of cations for all the emeralds and green vanadian beryls in this study, in terms of cations per 18 oxygen formula unit computed with the assumption of ideal beryl stoichiometry having 3Be cations per formula.
(A) Al vs others octahedral cations, $Fe^{2+}+Mg+Cr+V$. **(B)** Na vs Mg; 1:1 line shown for reference. Symbols are as follows; crosses, Swat samples; open squares, Mohmand (Barang); filled squares, green vanadian beryls from Mohmand (Gandao); triangles, Khaltaro pegmatite; diamonds, Afghanistan (Panjsher).

Calculation of formulas using the analyzed BeO contents from Snee and others (table 5.1, chapter 5 of this volume) do not substantiate an inverse relationship between beryllium and sodium for the Pakistan emeralds. Figure 6.6A shows an inverse relationship between the number of aluminum cations (Al) and the sum of other cations (Cr+Fe+Mg+V) in the octahedral site. Figure 6.6B shows that the correlation between the number of magnesium cations and the number of sodium cations per formula unit is nearly 1:1. Since magnesium is the most abundant of the cations likely to substitute for aluminum in the beryl structure, these relationships support a substitution of magnesium for aluminum in the octahedral site coupled (for charge balance) with a substitution of sodium for a vacancy in the channel site. This type of substitution suggests solid solution towards the "FeMag" beryl end member proposed by Schaller and others (1962): $(Na, K, Cs) Be_3R^{3+}R^{2+}Si_6O_{18}$, where $R^{3+} = Al, Fe^{3+}, Cr, Sc$ and $R^{2+} = Fe^{2+}, Mn, Mg$. Schaller and others based their study on an unusual, iron- and cesium-rich blue beryl having a high refractive index from a pegmatite dike in schist and gneiss from Mohave County, Arizona.

OTHER MICROPROBE DATA

Microprobe data were obtained for several minerals in a thin section of the carbonate rock that hosts the emerald mineralization at Mingora Mine 3 in the Swat area. No emeralds occur in the section; however, small, green euhedral crystals of chromium-rich tourmaline (photo.6.2) occur in thin, wispy veinlets of green mica (fuchsite) throughout the carbonate. These tourmalines are up to 0.1 mm in diameter, have hexagonal cross sections, and mimic emeralds in habit and color. Microprobe analyses for these tourmalines and an analysis of a very similar tourmaline from calcareous rocks at Alpurai, near Mingora (described by Jan and others, 1972), are given in table 6.5. At Alpurai, tourmaline occurs in calcite or between calcite and quartz in fuchsite-bearing calcareous rocks that are in contact with serpentinite. All of the tourmaline in

Mineral Chemistry of Emeralds: an Electron Microprobe Study

Table 6.5 Tourmaline analyses [n.d., not determined]

Location	Alpurai, Swat[1]	Mingora Mine 3			
SiO_2	36.40	35.89	36.17	36.17	36.31
Al_2O_3	29.00	28.86	27.97	27.72	29.40
FeO	1.20	1.02	1.00	1.26	1.14
MgO	8.20	9.11	9.14	9.65	9.50
CaO	0.03	0.10	0.07	0.04	0.05
Na_2O	2.80	2.59	2.64	2.13	2.23
K_2O	0.04	0.00	0.00	0.00	0.00
TiO_2	0.09	0.14	0.11	0.01	0.05
V_2O_3	0.25	0.07	0.04	0.06	0.09
MnO	n.d.	0.02	0.03	0.02	0.01
Cr_2O_3	8.50	8.18	8.78	7.90	6.52
F	n.d.	0.41	0.37	0.00	0.27
Cl	n.d.	0.00	0.00	0.00	0.00
B_2O_3*	10.69	10.63	10.59	10.53	10.63
Sum[3]	97.20	96.85	96.57	95.49	96.09

Cations based on 29 oxygens

Si	5.92	5.87	5.94	5.97	5.94
Al	5.56	5.56	5.38	5.40	5.67
Fe^{2+}	0.16	0.14	0.14	0.17	0.16
Mg	1.99	2.22	2.24	2.37	2.31
Ca	0.01	0.02	0.01	0.01	0.01
Na	0.88	0.82	0.84	0.68	0.71
K	0.01	0	0	0	0
Ti	0.01	0.02	0.01	0	0.01
Mn	—	0	0	0	0
Cr	1.09	1.06	1.14	1.03	0.84
F	—	0.21	0.19	0	0.14
Cl	—	0	0	0	0
B	3.00	3.00	3.00	3.00	3.00

[1] Analysis from Jan and others, 1972.
[2] Sum corrected for oxygen equivalents of fluorine and chlorine.
* Calculated assuming tourmaline stoichiometry (see Henry and Guidotti, 1985)

the Mingora section occurs within the fuchsite veinlets, and the host rock is predominantly composed of ferroan magnesite and quartz. Analyses of fuchsite, which contains over 7% Cr_2O_3 and 0.5% NiO, and partial analyses for several carbonate grains are given in table 6.6.

Table 6.6 Other microprobe data for minerals in the carbonate host rock for gem mineralization at Mingora Mine 3.[n.d., not determined]

Mineral	mica fuchsite		carbonates[1] magnesite[2]		dolomite[3]	talc
Grain no. (No. of analyses)	1(2)	2(2)	1(2)	2(2)	1(1)	1(1)
SiO_2	45.89	46.79	0.08	0.15	0.00	58.81
Al_2O_3	24.99	25.47	0.00	0.03	0.00	0.79
FeO	1.56	1.73	5.12	4.55	1.46	2.13
MgO	3.95	4.08	39.37	39.05	20.13	28.01
CaO	0.05	0.04	n.d.[4]	0.12	27.32	0.00
Na_2O	0.29	0.28	0.01	0.00	0.00	0.02
K_2O	10.18	9.77	n.d.	0.00	0.00	0.00
TiO_2	0.20	0.19	n.d.	0.00	0.05	n.d.
MnO	0.00	0.02	n.d.	0.01	0.06	0.00
V_2O_3	n.d.	n.d.	0.04	0.00	0.00	0.03
NiO	0.43	0.88	n.d.	n.d.	n.d.	n.d.
Cr_2O_3	7.33	6.57	0.12	0.01	0.00	0.16
F	0.21	0.19	n.d.	0.00	0.00	0.05
Cl	0.01	0.01	n.d.	0.00	0.00	0.00
Sum	95.11	96.01	44.75	43.93	49.02	90.00

[1] Carbonate analyses are considered semi-quantitative because carbonate standards were not used for these analyses; 0.00 shows that an element was sought but not detected.

[2] Magnesite, $MgCO_3$, a common alteration product of serpentine, exhibits solid solution towards siderite, $FeCO_3$; most magnesite contains 50 to 52% CO_2.

[3] Dolomite, $CaMgCO_3$, typically contains 43 to 48% CO_2.

[4] Simultaneous examination of the spectrum for this grain from an energy dispersive spectrometer showed no Ca present.

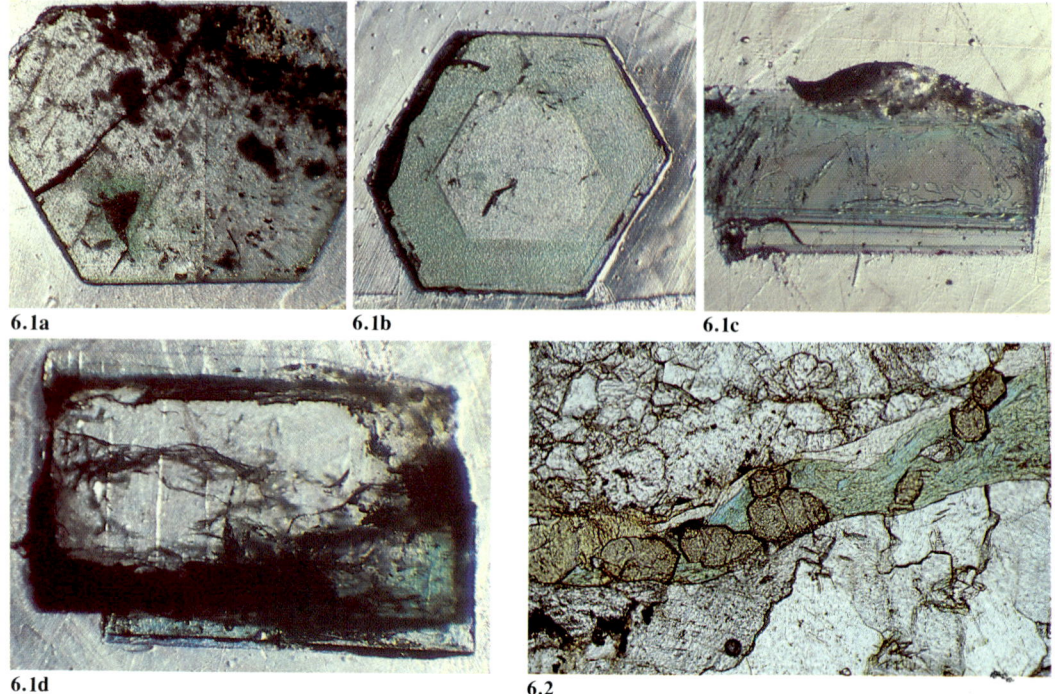

6.1a   **6.1b**   **6.1c**

6.1d   **6.2**

Photo 6.1. Photomicrographs of emeralds from the Swat district, Pakistan:
a) Euhedral emerald from Makhad Mine, sample Makhad No. 6. showing concentration of green color around inclusion: scale represents 5.4 mm. b) Euhedral emerald from Mine 2 at Mingora, sample SDA-25, which has a nearly colorless core surrounded by a green rim; scale represents 2.0 mm. c) Subhedral emerald lath from Makhad Mine, sample SEM-5, which has a colorless rim along one edge of the grain; the rest of the grain is green; scale represents 1.7 mm. d) Part of an emerald lath from Mine 1 at Mingora, sample SEM 51-53/P3, which has an L-shaped zone of intense green color; scale represents 2.4 mm. (photo *J. M. Hammarstrom*).

Photo 6.2. Photomicrograph of a chromium-rich tourmaline in a fuchsite vein in a thin section of carbonate host rock at Mingora mine 3. (photo *A. H. Kazmi*).

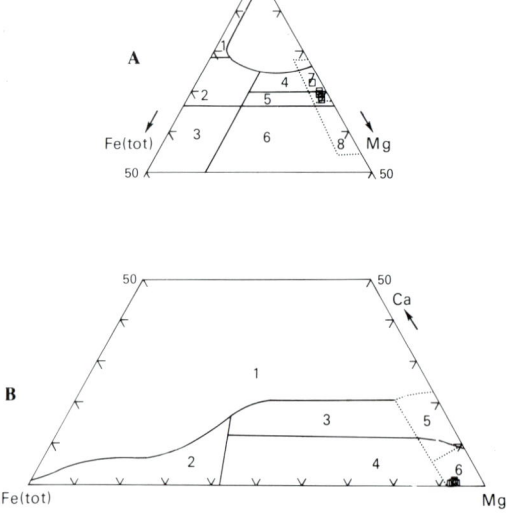

Figure 6.7. Ternary diagrams for Swat tourmalines (from Henry and Guidotti, 1985, figs. 1 and 2). **(A)** Cation proportions of Al, total Fe and Mg from table 6.5. Fields designate ranges of compositions for tourmalines from different rock types, as follows: (1) Li-rich granitoid pegmatites and aplites, (2) Li-poor granitoids and associated pegmatites and aplites, (3) Fe^{3+}-rich quartz-tourmaline rocks, (4) metapelites and metapsammites associated with an Al-saturating phase, (5) metapelites and metapsammites not associated with an Al-saturating phase, (6) Fe^{3+}-rich quartz-tourmaline rocks, calc-silicate rocks and metapelites, (7) low-Ca metaultramafics and Cr, V-rich metasediments, and (8) metacarbonates and meta-pyroxenites. **(B)** Cation proportions of Ca, total Fe and Mg for Swat tourmalines. Fields 1 and 2 are the same as in part A. Other fields: (3) Ca-rich metapelites, metapsammites and calc-silicate rocks, (4) Ca-poor metapelites, metapsammites and quartz-tourmalines rocks. (5) metacarbonates, and (6) metaultramafics.

Henry and Guidotti (1985) described methods for calculating mineral formulas from microprobe analyses of tourmalines and demonstrated the use of tourmaline composition as a petrogenetic indicator. Formulas for the Swat tourmalines in table 6.5 are computed by this technique and are plotted on diagrams used by Henry and Guidotti (1985) to show compositional ranges of tourmalines from different rock types in figure 6.7. The Swat tourmalines plot entirely within the fields for tourmalines from low-calcium metaultramafic rocks and chromium-, vanadium-rich metasediments. This correlation is consistent with the conclusion by Kazmi and others (1986) that the talc-dolomite schist and emerald mineralization in the Mingora ophiolitic melange represent alteration of a serpentinite melange.

DISCUSSION

Most previous studies of chemical variations in emeralds, e.g., Hanni (1982), Schrader (1983), and Stockton (1984), have been directed toward development of chemical criteria to distinguish between natural and synthetic gemstones and to fingerprint particular localities. Gübelin (1982) compared a partial analysis of an emerald from Mingora, Pakistan, with emeralds from other deposits and demonstrated that the Mingora stones are unusually rich in chromium, magnesium and sodium. This study confirms earlier work on the general chemical characteristics of emeralds from Pakistan but demonstrates that considerable variation can occur within individual crystals and among different mining districts from the same region. For example, within the Mohmand district, the emerald from Barang contains a small amount of chromium and almost no vanadium whereas the samples from Gandao are green vanadian beryls containing almost no chromium. The variations in compositions encountered in the emeralds and green vanadian beryls in this study are summarized on ternary diagrams in figure 6.8 in terms of the oxides of the cations that substitute for aluminum in the beryl structure and that account for the green color. Different symbols represent different mining districts. Overlaps in composition among districts occur, but the Panjsher and Khaltaro samples are clearly distinct from the Swat and Mohmand samples in figure 6.8A, and the contrast in vanadium content between the two Mohmand samples is evident in figure 6.8B. Although the Khaltaro sample is from a pegmatite and all the other emeralds are associated with carbonate schists or quartz and carbonate veins, no distinctive chemical signature is apparent in figure 6.8 to suggest a different paragenesis.

Analyses of emeralds from other areas cited by Sinkankas (1981) are listed in table 6.7. These analyses are plotted on ternary diagrams in figure 6.9, which includes microprobe data for natural emeralds reported by Hanni (1982). With the exception of the unusual Colombian emerald occurrences in shales, all of the deposits are in biotite schists, pegmatites, or schists associated with pegmatites. Comparison of figures 6.8 and 6.9 shows that some of the Pakistan emeralds are enriched in chromium compared to most other occurrences. Only the Colombian emeralds are distinct because of the absence of iron.

The formation of emerald deposits requires the availability of beryllium and the transition metal chromium, as well as pressure and temperature conditions necessary for beryl crystallization. For the Pakistan emerald deposits, the metaultramafic talc-carbonate schists provide the chromium, but the source of the beryllium is not readily apparent. At Gravelotte, in the northeastern Transvaal region of Africa, emeralds occur in biotite schists which are intruded by beryl-bearing granitic pegmatites (Sinkankas, 1981). These pegmatites are considered the source of the beryllium for emerald crystallization in the schist. Examination of a sample of emerald-bearing schist from Gravelotte (U.S. National Museum sample 120556) showed that

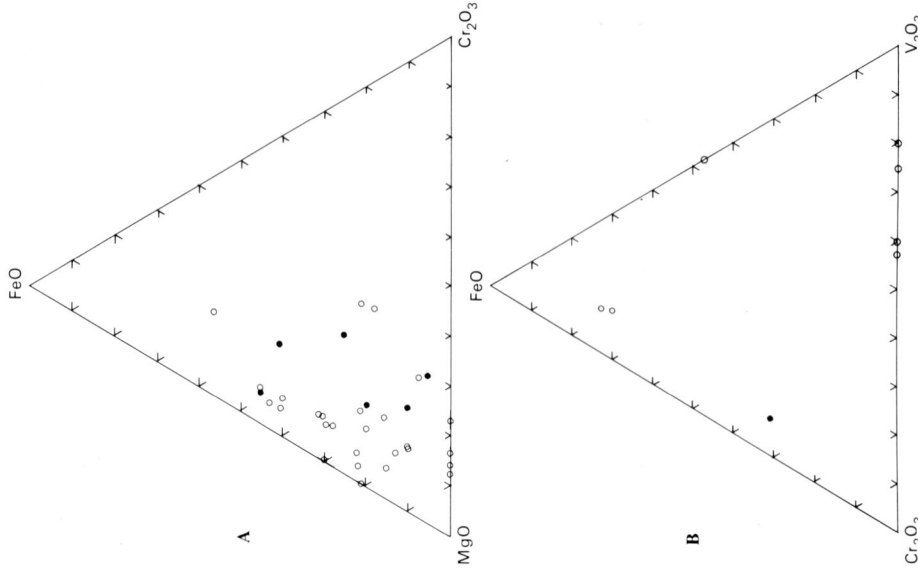

Figure 6.9. Ternary diagrams as shown in figure 6.8 for other emerald analyses, taken from the literature. Filled symbols represent analyses for emeralds and green vanadian beryls from Pakistan (table-6.3). Sources of data are the analyses in table-6.7 compiled from Sinkankas (1981) and the microprobe data for natural emeralds from Hanni (1978).

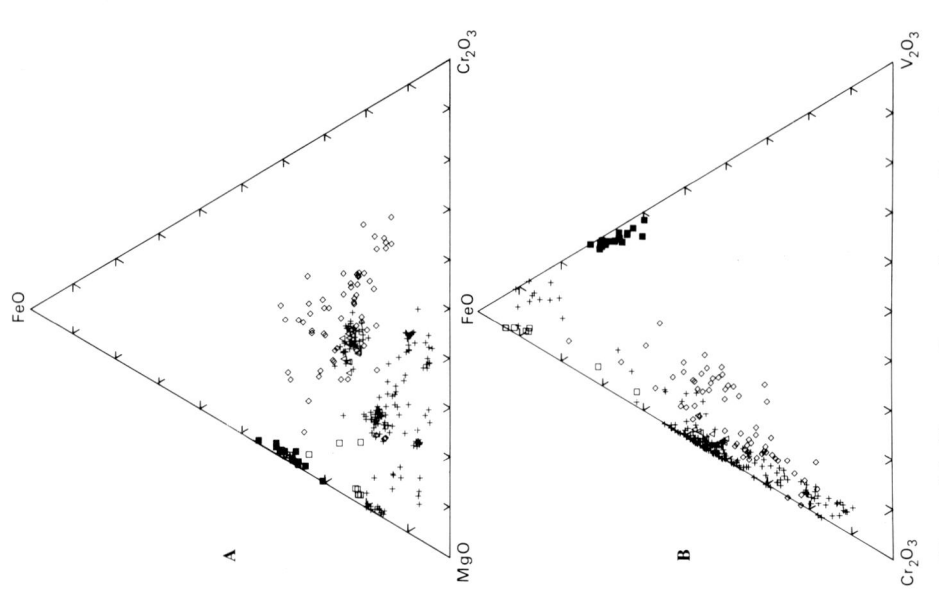

Figure 6.8. Ternary diagrams for emeralds and green vanadian beryls in this study, in terms of oxides percents: (A) FeO, MgO, Cr_2O_3; B) FeO, Cr_2O_3, V_2O_3. See figure 6.6 for explanation of symbols.

Table 6.7 Selected emerald analyses from the literature.
[Analyses are from Sinkankas, 1981; Graziani and others 1983; and Hanni and Klein, 1982]

Location	Urals, USSR	Miku, Zambia	Kitwe, Zambia	Sandanawa, Zimbabwe	Chingachura, Zimbabwe	Ankadilalana Mine, Madagascar	Salzburg, Austria	Pila, Bulgaria	Muzo, Colombia	Muzo, Colombia	Transvaal, S. Africa
Occurrence	biotite schist	biotite schist	biotite schist/ pegmatite	tremolite schists/ pegmatites	tremolite schists/ pegmatites	mica schist	biotite schist	pegmatite	calcite veins	calcite veins	biotite schist/ pegmatites
SiO_2	64.69	62.23	63.91	65.0	63.84	63.25	63.54	63.63	67.85	65.25	61.20
Al_2O_3	15.16	15.41	18.88	14.2	19.00	14.23	17.25	15.95	17.95	17.62	17.20
Fe_2O_3	0.35	0.04	—	0.5	—	1.46	0.71	0.48	—	1.00	1.80
FeO	—	0.07	0.69	—	0.30	1.96	0.84	—	—	—	—
MgO	1.89	0.76	0.83	3.0	0.75	0.00	0.78	1.00	0.9	0.43	1.70
CaO	0.80	0.31	0.96	—	—	1.56	1.42	1.53	—	—	0.17
Na_2O	1.80	2.63	1.45	2.0	2.03	0.11	0.14	0.90	0.7	—	0.85
K_2O	tr	2.89	0.01	—	0.05	0.19	0.12	0.01	—	—	—
Cr_2O_3	0.25	0.33	0.08	0.5	0.6	—	—	0.004	tr	0.26	0.25
H_2O	1.29	2.59	1.45	—	1.07	—	2.97	1.16	—	1.55	1.85
V_2O_3	—	—	—	—	—	0.01	—	—	—	—	—
BeO	13.37	11.90	13.70	13.6	13.28	—	13.07	13.10	12.4	13.89	14.60

[1] Total iron as FeO.

the schist is essentially comprised of chromium-bearing phlogopite (42.7% SiO_2, 12.8% Al_2O_3, 20.6% MgO, 5.4% FeO, 0.2% Cr_2O_3, 0.1% TiO_2, 0.1% Na_2O, 8.0% K_2O and 3.2% F). Thus the schist itself appears to provide the chromium for the emeralds. The African emeralds are similar to the Pakistan emeralds in that they are relatively rich in chromium, magnesium and sodium and are poor in calcium (table 6.7).

CONCLUSIONS

1. Emeralds from four gem mining regions in Pakistan and Afghanistan can be chemically characterized by high contents of MgO (0.3% to 3%), FeO (0.1% to 1.5%) and Na_2O (0.1% to 2.3%).

2. Chromium is responsible for the green coloration of the emeralds. Samples from the Gandao mine in the Mohmand district are unique because they lack chromium but contain about 0.6% V_2O_3; the Gandao samples are green vanadian beryls rather than true emeralds according to currently accepted industry rules for naming gemstones.

3. None of the emeralds is homogeneous; many display color zoning that correlates directly with variations in chromium content. Iron generally mimics the zoning pattern for chromium, whereas the zoning patterns for sodium and magnesium are commonly inverse to those for iron and chromium.

4. Emeralds from a pegmatite locality (Khaltaro) and a carbonate vein/shear zone locality (Panjsher) are chemically distinct from emeralds in talc-carbonate schists in the Swat and Mohmand areas.

5. Emeralds from Pakistan are among the most chromium-rich reported, but are otherwise similar to emeralds from other schist-type deposits. The magnesium-rich nature of the emeralds and limited data on associated minerals suggest that the emerald mineralization may be confined to areas where ultramafic rocks rich in magnesium, chromium and nickel interacted with beryllium-bearing fluids.

SUGGESTIONS FOR FUTURE RESEARCH

Although emeralds have received much attention in the gemological literature, surprisingly little recent work has been done on the detailed petrology of emerald deposits. Several recent studies (e.g., Barton, 1986; Cemic and others, 1986) have addressed the phase relations of beryllium systems and models for the hydration of beryl, but the effects of substitutions on beryl stability, such as those present in the Pakistan emeralds, have not been studied. Detailed study of the Pakistan emeralds and the petrology of their host rocks, in light of recent experimental work, would undoubtedly further the understanding of the petrogeneses of these types of deposits. Furthermore, the unusually high concentrations of magnesium, chromium, and sodium in these emeralds pose interesting crystal-chemical problems. One such problem is the apparent inverse zoning relationship between chromium and magnesium, which substitute for aluminum in octahedral coordination, and sodium, which is thought to occur bonded to water in the channel sites of the beryl structure. Spectroscopic studies would help determine the nature of the water and (or) other channel constituents, such as helium or carbon dioxide, in these emeralds. Investigation of the effects of varying the amounts of subordinate pigmenting transition metal ions (Fe and V) on the color tones (bluish green, blue green, pure green, yellowish green, and yellow green) of the true emeralds would also be highly elucidating.

ACKNOWLEDGMENTS

This paper benefited greatly from critical reviews by Larry Snee, Priestley Toulmin, A.H. Kazmi, and Dr. Edward Gübelin. The author wishes to thank Priestley Toulmin for his knowledgeable assistance with thermogravimetric analysis and for his helpful discussions on emerald chemistry. I also thank John White of the U.S. National Museum for providing access to the museum's emerald collection and for providing the Gravelotte sample.

This work was done as part of a cooperative study among the Geological Survey of Pakistan, Oregon State University, and the U.S. Geological Survey. A.H. Kazmi and Larry Snee provided all samples and background information. I am especially indebted to them for introducing me to the study of Pakistan emeralds and for their enthusiastic guidance throughout the project.

REFERENCES

Albee, A.L., and Ray, L., 1970, Correction factors for electron probe microanalysis of silicates, oxides, carbonates, phosphates, and sulfates: Analytical Chemistry, v. 42, p. 1408-1414.

Barton, M.D., 1986, Phase equilibria and thermodynamic properties of minerals in the BeO—Al_2O_3—SiO_2—H_2O (BASH) system, with petrologic applications: American Mineralogist, v. 71, p. 277-300.

Bence, A.E., and Albee, A.L., 1968, Empirical correction factors for the electron microanalysis of silicates and oxides: Journal of Geology, v. 76, p. 382-403.

Cassedanne, J.P., and Sauer, D.A., 1984, The Santa Terezinha de Goias emerald deposit: Gems and Gemology, v. 20, p. 4-13.

Cemic, L., Langer, K., and Franz, G., 1986, Experimental determination of melting relationships of beryl in the system BeO—Al_2O_3—SiO_2—H_2O between 10 and 25 k bar: Mineralogical Magazine, v. 50, p. 55-61.

Deer, W.A., Howie, R.A., and Zussman, J., 1962, Beryl, in Rock-forming Minerals, Volume 1, *Ortho— and ring silicates:* William Clowes and Sons, Limited, London, p. 256-267.

Dunn, P.J., 1977, The use of the electron microprobe in gemmology: Journal of Gemmology, v. 15, p. 248-258.

Ginzburg, A.I., 1955, On the question of the composition of beryl: Trudy Mineralogischeskogo Muzeya, Akad. Nauk SSSR, Leningrad, v. 7, p. 56-69.

Graziani, G., Gübelin. E.J., and Lucchesi, S., 1983. The genesis of an emerald from the Kitwe district, Zambia: Neues Jahrbuch für Mineralogie, Monatschefte, v. 4, p. 175-186.

Gübelin, E.J., 1969, On the nature of mineral inclusions in gemstones: Journal of Gemmology, v. 11, p. 149-192.

Gübelin, E.J., 1974, *Internal world of gemstones:* ABC Zürich, Switzerland, 233 p.

Gübelin, E.J., 1982, Gemstones of Pakistan: emerald, ruby, and spinel: Gems and Gemology, v. 18, p. 123-139.

Gübelin, E.J. 1989, Gemological characteristics of Pakistani emeralds: *in* A.H. Kazmi and L.W. Snee, editors, *Emeralds of Pakistan: Geology, Gemology and Genesis,* Von Nostrand Reinhold, New York, p. 75-92.

Gübelin, E.J. and Koivula, J.I., 1986, *Photoatlas of inclusions in gemstones:* ABC Zürich, Switzerland, 532 p.

Hanni, H.A., 1982, A contribution to the separability of natural and synthetic emeralds: Journal of Gemmology, v. 18, no. 2, p. 138-144.

Hanni, H.A. and Klein, H.H., 1982, Ein smaragdvorkommen in Madagaskar: Deutschen Gemmologischen Gesellschaft, Zeitschrift, v. 21, p. 71-77.

Hawthorne, F.C. and Černý, P., 1977, The alkali-metal positions in Cs-Li beryl: Canadian Mineralogist, v. 15, p. 414-421.

Hemingway, B.S., Barton, M.D., Robie, R.A., and Haselton, H.T., Jr. 1986, Heat capacities and thermodynamic functions for beryl $Be_3Al_2Si_6O_{18}$, phenakite, Be_2SiO_4, euclase, $BeAlSiO_4$ (OH), bertrandite, $Be_4Si_2O_7(OH)_2$, and chrysoberyl, $BeAl_2O_4$: American Mineralogist, v. 71, p. 557-568.

Henry, D.J., and Guidotti, C.V., 1985, Tourmaline as a petrogenetic indicator mineral: An example from the staurolite-grade metapelites of NW Maine: American Mineralogist, v. 70, p. 1-15.

Jan, M. Qasim, Kempe, D.R.C., and Symes, R.F., 1972, A chromian tourmaline from Swat, West Pakistan: Mineralogical Magazine, v. 38, p. 756-759.

Kazmi, A.H., Anwar, J., Hussain., S., Khan., T., and Dawood, H., 1989, Emerald deposits of Pakistan: *in* A.H. Kazmi and L.W. Snee, editors, *Emeralds of Pakistan: Geology, Gemology and Genesis,* Van Nostrand Reinhold, New York p. 39-74.

Kazmi, A.H., Lawrence, R.D., Anwar, J., Snee, L.W., and Hussain, S., 1986, Mingora emerald deposits (Pakistan): suture associated gem mineralization: Economic Geology, v. 81, p. 2022-2028.

Kazmi, A.H., Lawrence, R.D., Dawood, H., Snee, L.W., and Hussain, S., 1984, Geology of the Indus suture zone in the Mingora-Shangla area of Swat, N. Pakistan: University of Peshawar, Geological Bulletin, v. 17, p. 127-144.

Kazmi, A.H., Peters, J.J. and Obodda, H.P., 1985, Gem pegmatites of the Shingus-Dusso area, Gilgit, Pakistan: The Mineralogical Record, v. 16, p. 393-411.

McGee, J.J., 1983, $ANBA—a rapid, combined data acquisition and correction program for the SEMQ electron microprobe: U.S. Geological Survey Open-File Report 83-817, 47 p.

Rafiq, M. and Jan, M.Q., 1985, Emerald and green beryl from Bucha, Mohmand Agency, NW Pakistan: Journal of Gemmology, v. 19, no. 5, p. 404-411.

Schaller, W.T., Stevens, R.E., and Jahns, R.H., 1962, An unusual beryl from Arizona: American Mineralogist, v. 47, p. 672-699.

Schrader, H.W., 1983, Contributions to the study of the distinction of natural and synthetic emeralds: Journal of Gemmology, v, 18, no. 6, p. 530-543.

Sinkankas, J. 1981, *Emeralds and other beryls:* Chilton Book Company, Radnor, Pennsylvania, USA, 665 p.

Smith, D.G.W., ed., 1976, *Short course in microbeam techniques:* Mineralogical Association of Canada, Short Course Handbook, Vol. 1, May 1976, Co-op Press, Edmonton, Canada, 186 p.

Snee, L.W., Foord, E.E., Hill, B., and Carter, S.J., 1989. Regional chemical differences among emeralds and host rocks of Pakistan and Afghanistan: implications for the origin of emerald: *in* A.H. Kazmi and L.W. Snee, editors, *Emeralds of Pakistan: Geology, Gemology and Genesis,* Van Nostrand Reinhold, New York, p. 93-124.

Staatz, M.H., Griffiths, W.R., and Barnett, P.R., 1965, Differences in the minor element composition of beryl in various environments: American Mineralogist, v. 50, 1783-1795.

Stockton, C. M., 1984, The chemical distinction of natural from synthetic emeralds: Gems and Gemology, v. 20, no. 3, p. 141-145.

Wood, D.L., and Nassau, K., 1967, Infrared spectra of foreign molecules in beryl: Journal of Chemical Physics, v. 47, no. 7, p. 2220-2228.

Wood, D.L., and Nassau, K., 1968, The characterization of beryl and emerald by visible and infrared absorption spectroscopy: American·Mineralogist, v. 53, p. 777-800.

A Reconnaissance Study of the Fluid Inclusion Geochemistry of the Emerald Deposits of Pakistan and Afghanistan

Robert R. Seal, II

INTRODUCTION

Emerald and other gem deposits have not been the focus of the same intensity of scientific examination as other economically-exploitable mineral deposits. However, geochemical investigations, such as fluid inclusion studies, hold equal potential for increasing our understanding of the conditions accompanying emerald deposition. To this end, a reconnaissance fluid inclusion study was undertaken to document the geochemical characteristics of the hydrothermal fluids responsible for emerald mineralization in the Swat district of Pakistan. A single sample from the Panjsher district of Afghanistan was also examined for comparison purposes.

Despite the excellent geological framework for the district (Kazmi and others, 1986; Jan and others, 1981), the implications of this study are limited by the scarcity of experimental studies of beryl stability (Barton, 1986) and solubility (Renders and Anderson, 1987) under a wide range of geological conditions. Furthermore, comparable fluid inclusion studies of other emerald deposits are also scarce (e.g., Roedder, 1963; Ottaway and others, 1986; Kozlowski and others, 1988).

ANALYTICAL METHODS

A variety of analytical techniques was employed to document the physical and chemical characteristics of the inclusion fluids. These techniques included standard petrographic examination, scanning electron microscopic examination and microthermometric analysis.

Transmitted-light, petrographic studies were conducted to: 1) assess the relative abundances of primary, pseudosecondary, and secondary fluid inclusions; 2) evaluate the degree of post-entrapment modification (necking) of inclusions; 3) identify daughter minerals in inclusions; and 4) provide a preliminary assessment of the presence or absence of boiling or fluid immiscibility during emerald deposition. Scanning electron microscopy (SEM) was employed to aid in the identification of daughter minerals (Metzger and others, 1977) hosted by the fluid inclusions.

Microthermometric data were obtained from thin (0.25 to 0.50 mm thick) doubly-polished wafers of emerald or quartz, using a modified USGS gas-flow heating/cooling stage (photo. 7.1). Low temperature analyses were conducted using liquid nitrogen-cooled nitrogen gas. The warming rate was controlled by a heating coil mounted in the stage housing. High temperature analyses were conducted using heated compressed air. Calibration of the stage was accomplished using commercially-available synthetic fluid inclusions (I.V.T.S. Hanel, pers. comm., 1986). The precision of the stage is estimated at $\pm 0.2°C$ for the temperature range in question (Haynes, 1985). The reproducibility of preliminary observations from inclusions in both emerald and quartz indicated that replicate observations were unnecessary. Furthermore, the common decrepitation of inclusions during homogenization runs also precluded replicate observations.

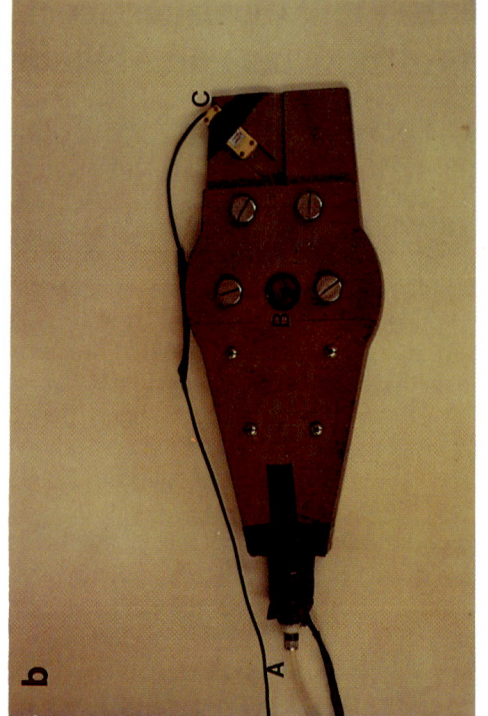

Photo 7.1. Fluid Inclusion Microthermometric Laboratory

a) Laboratory set-up for cooling experiments, a sample is placed in the USGS gas-flow stage (E; see also Fig. 1b), which is mounted on a standard petrographic microscope (D). Cooling experiments are conducted using nitrogen gas (A), which has been cooled by passing it through a copper coil, submersed in a dewer filled with liquid-nitrogen (C), which is then passed over the sample located in the stage (E). Cooling and heating rates are controlled at the control panel (B). Sample temperatures are obtained from the digital display at the top of the control panel. For heating experiments, the gas-flow is provided by heated, purified compressed air which has not passed through the cooling coil, submersed in the dewer.

b) Close-up of modified USGS gas-flow stage.

    (A) Air Input. In this photograph, the air supply is purified compressed air. Compare with figure 1a. The air is heated by a coil mounted in the stage housing between the air input (A) and the sample chamber (B).

    (B) Sample Chamber. Note thermocouple in contact with the sample.

    (C) Thermocouple.

PETROGRAPHY

Swat emeralds contain abundant primary, pseudosecondary, and secondary fluid inclusions as noted by Gübelin (1982). Fluid inclusions hosted by unfractured emeralds are generally scarce. They are localized along growth zones and elongated parallel to the *c*-axis, thus suggesting a primary origin (photo. 7.2a, b). The unfractured emeralds also contain flat inclusions oriented along the (0001) cleavage suggesting primary or pseudosecondary origins (photo. 7.2c). In addition to these two types of inclusions, fractured emeralds also contain abundant secondary inclusions filling oblique fractures. Generally, all of the inclusions hosted by emerald are less than 50 μm in maximum dimension, but inclusions up to 200 μm are not uncommon. Fluid inclusions hosted by quartz occur either as large (up to 50 μm) isolated inclusions, locally forming negative crystals, suggesting of a primary origin, or as smaller ($<15\mu$m) fracture-filling secondary inclusions.

Two-phase (liquid + vapor) fluid inclusions are the most abundant inclusion type in both hosts at room temperature, regardless of origin. The cooling of these inclusions below room temperature, however, results in the condensation of a thin film of a second immiscible liquid phase prior to freezing, identified as CO_2-liquid (see below). In all cases, the liquid/vapor ratio in cogenetic inclusions is constant, indicating a homogeneous, one-phase fluid at the time of entrapment (i.e., *no* boiling or CO_2 effervescence). Slight variations in liquid/vapor ratios are only apparent in inclusions which have obviously undergone necking.

Daughter minerals were only locally observed in emerald-hosted fluid inclusions from the Swat district. Sample C/5C (Charbagh) contains hematite daughters, verified by SEM and energy dispersive X-ray analysis, in primary or pseudosecondary inclusions forming negative crystals perpendicular to the *c*-axis (photo. 7.3). Several samples from elsewhere in the district contain apparently isotropic solids, which SEM examination proved to be epitaxial overgrowths of emerald, giving relief to the interior of inclusions (photo. 7.3).

One emerald sample (SEM-1) from the Panjsher area of Afghanistan is distinctly different from those of the Swat district. It contains numerous daughter minerals in inclusions oriented parallel to the *c*-axis (photo. 7.2d). Up to possibly five distinct solid phases, in addition to liquid and vapor, have been observed in individual inclusions: three isotropic (?), one anisotropic and one opaque. The cubic habit of one of the isotropic phases is suggestive of halite. The heating and dissolution behavior of a second solid is consistent with sylvite. The identities of the remaining phases are uncertain. The possibility exists that some of the bodies identified as isotropic daughters may in fact be epitaxial overgrowths of emerald, as described above. The small size of the available sample and the elongated nature of these inclusions, precluded detailed SEM analysis of these inclusions and their daughter minerals.

RESULTS

The results of the microthermometric analyses are presented in table 7.1 for fluid inclusions from the Swat district. These data are discussed below under the categories of 'Cooling Phenomena', 'Freezing Phenomena', and 'Heating Phenomena'. The category 'Cooling Phenomena' consists of semi-quantitative observations on inclusions during cooling to -130°C; 'Freezing Phenomena' consists of observations during warming from -130°C to the temperature of the last disappearance of solid; and 'Heating Phenomena' comprises observations during heating above room temperature.

Table 7.1 Microthermometric data from Swat District.

SAMPLE	HOST	TYPE	PHASES	T_{hCO_2}	T_{mCO_2}	T_m	T_h	T_d
C/5C—1	EM	P,PS	2			7.9	326.4	
C/5C—2	EM	P,PS	2			6.2	323.8	
C/5C—3	EM	P,PS	2			7.2	322.1	
C/5C—4	EM	P,PS	2			7.7	329.3	
C/5C—5	EM	P,PS	2			7.0	330.6	
C/5C—6	EM	P,PS	2			3.5		275.0
C/5C—7	EM	P,PS	2			3.5		275.0
C/5C—8	EM	P,PS	2			2.7	284.6	
C/5C—9	EM	P,PS	2			3.5	290.8	
C/5C—10	EM	P,PS	2			4.1	304.9	
ES—21—1	QTZ	PS,S	3	0.0	—58.7	11.1		213.0
ES—21—2	QTZ	PS,S	3	0.0	—58.0	11.0		150.0
ES—21—3	QTZ	PS,S	3	0.0	—58.4	11.4		151.0
ES—21—4	QTZ	PS,S	3	0.0	—57.9	12.8		160.0
ES—21—5	QTZ	PS,S	3	0.0	—57.7	12.6		160.0
ES—21—6	QTZ	P	3	0.0	—58.0	10.4		167.0
ES—21—7	QTZ	P	3	0.0	—57.8	8.5		254.0
ES—21—8	QTZ	P,PS	3	25.0	—58.8	9.1		247.0
ES—21—9	QTZ	P,PS	2					245.0
ES—21—10	QTZ	S	2					220.0
ES—21—11	QTZ	S	2					250.0
ES—21—12	QTZ	S	2					251.0
ES—21—13	QTZ	P,PS	2	0.0				243.0
ES—21—14	QTZ	P,PS	2	0.0	—57.2			236.0
ES—21—15	QTZ	P,PS	2					270.0
ES—21—16	QTZ	P,PS	2	0.0				230.0
ES—21—17	QTZ	PS,S	2					191.0
ES—21—18	QTZ	PS,S	2					166.0
M—2—1	EM	PS,S	3	0.0	—57.4	9.5		300.0
M—2—2	EM	PS,S	3	0.0	—57.3	9.8		295.0
M—2—3	EM	PS,S	3	0.0	—57.3	10.1		297.0
M—2—4	EM	PS,S	3	0.0	—56.6	10.9		267.0
M—2—5	EM	PS,S	3	0.0	—56.6	10.3		267.0
M—2—6	EM	PS,S	3	0.0	—56.8	11.2		267.0
SDA/25A—1	EM	PS,S	3	0.0	—58.2	11.1		278.0
SDA/25A—2	EM	PS,S	3	0.0	—58.3	11.0		250.0
SDA/25A—3	EM	PS,S	3	0.0	—58.2	11.0		230.0
SDA/25A—4	EM	P	3	0.0	—58.6	11.4		251.0
SDA/25A—5	EM	P	3	0.0	—58.4	11.6		250.0
SDA/25A—6	EM	P	3	0.0	—58.6	11.4		251.0
SDA—25C—1	EM	P	3	5.0	—56.6	8.9		277.0
SDA—25C—2	EM	P	3	5.0	—56.6	9.2		196.0
SDA—25C—3	EM	P	3	5.0	—56.8	9.1		281.0
SDA—25C—4	EM	P,PS	3	0.0	—58.1	11.8		284.0
SDA—25C—5	EM	P,PS	3	0.0	—58.1	11.3		284.0
SDA—25C—6	EM	P,PS	3	0.0	—58.1	11.9		284.0
SDA—25C—7	EM	P,PS	3	0.0	—58.1	12.0		284.0
SEM—M1—1	EM	P,PS	2			—1.2		262.0
SEM—M1—2	EM	P,PS	2			—1.0		264.0
SEM—M1—3	EM	P,PS	2			—1.2	278.0	
SEM—M1—4	EM	P,PS	2			4.5	278.0	
SEM—M1—5	EM	P,PS	2			4.6		307.0

Table 7.1 (Continued)

SAMPLE	HOST	TYPE	PHASES	T_hCO_2	T_mCO_2	T_m	T_h	T_d
SEM—M2—1	EM	PS,P	3	—15.0	—58.9	11.1	324.0	
SEM—M2—2	EM	PS,P	3	—15.0	—59.2	11.3	318.0	
SEM—M2—3	EM	PS,P	3	—15.0	—59.4	10.8	313.0	
SEM—M2—4	EM	PS,P	3	—15.0	—58.7	11.1	322.0	
SEM—M2—5	EM	PS,P	3	—15.0	—58.5	10.9	324.0	
SEM—M2—6	EM	PS,S	3	—10.0	—59.7	10.9	305.0	
SEM—M2—7	EM	PS,S	3	—10.0	—59.5	11.9		275.0
SEM—M2—8	EM	PS,S	3	—10.0	—59.2	11.0		215.0
SEM—M2—9	EM	PS,S	3	—10.0	—59.3	11.7		309.0
SEM—M2—10	EM	PS,S	3	—10.0	—59.1	11.4		309.0
SEM—M2—11	EM	PS,S	3	—15.0	—59.4	11.7	317.0	
SEM—M2—12	EM	PS,S	3	—15.0	—59.4	11.5	313.0	
SEM—M2—13	EM	PS,S	3	—15.0	—59.4	11.4	314.0	
SEM—M2—14	EM	PS,S	3	—15.0	—59.3	11.3	316.0	
SEM—M2—15	EM	PS,S	3	—15.0	—59.0	10.6	328.0	
SEM—M2—16	EM	PS,S	3	—15.0	—59.0	10.5	328.0	
SEM—M2—17	EM	P,PS	3	—10.0	—59.3	11.3		250.0
SEM—M2—18	EM	P,PS	3	—10.0	—59.3	11.3		281.0
SEM—M2—19	EM	P,PS	3	—10.0	—59.4	11.4		281.0
SEM—M2—20	EM	P,PS	3	—10.0	—59.5	11.4		200.0
SEM—M2—21	EM	P,PS	3	—10.0	—59.4	11.2		200.0
SEM—4—1	EM	P,PS	3	10.0	—57.3	7.5		280.0
SEM—4—2	EM	P,PS	3	10.0	—57.2	7.2		271.0
SEM—4—3	EM	P,PS	3	10.0	—57.2	7.1		250.0
SEM—4—4	EM	P,PS	3	10.0	—57.5	7.4		190.0
SEM—4—5	EM	P,PS	3	10.0	—57.5	7.3		218.0
SEM—51/P2—1	EM	P,PS	2			6.0		298.0
SEM—51/P2—2	EM	P,PS	2			6.0		298.0
SEM—51/P2—3	EM	P,PS	2			5.7		298.0
SEM—51/P2—4	EM	P,PS	2			6.1	349.0	
SEM—51/P2—5	EM	P,PS	2			5.5		310.0
SEM—51/P2—6	EM	S	3	—20.0	—60.5	12.3		253.0
SEM—51/P2—7	EM	S	3	—20.0	—60.5	11.7		284.0
SEM—51/P2—8	EM	S	3	—20.0	—60.3	12.1		284.0
SEM—51/P2—9	EM	S	3	—20.0	—60.4	12.2		277.0
SEM—51/P2—10	EM	S	3	—20.0	—60.6	12.5		288.0
SEM—51/P2—11	EM	S	3	—20.0	—60.6	12.2		255.0

EM: Emerald; QTZ: Quartz

P: Primary; PS: Pseudosecondary; S: Secondary

Temperatures in Degrees Celsius

Mine 1, Mingora: SEM-M1

Mine 2, Mingora: SEM-M2, M-2, SDA/25, SDA-25, SEM-51/P2

Charbagh: C/5C

Barang: SEM-4

T_hCO_2: Homogenization Temperature of CO_2; T_mCO_2: Melting Temperature of CO_2; T_m: Final Melting Temperature of Ice or Clathrate;

T_h: Homogenization Temperature; T_d: Decrepitation Temperature

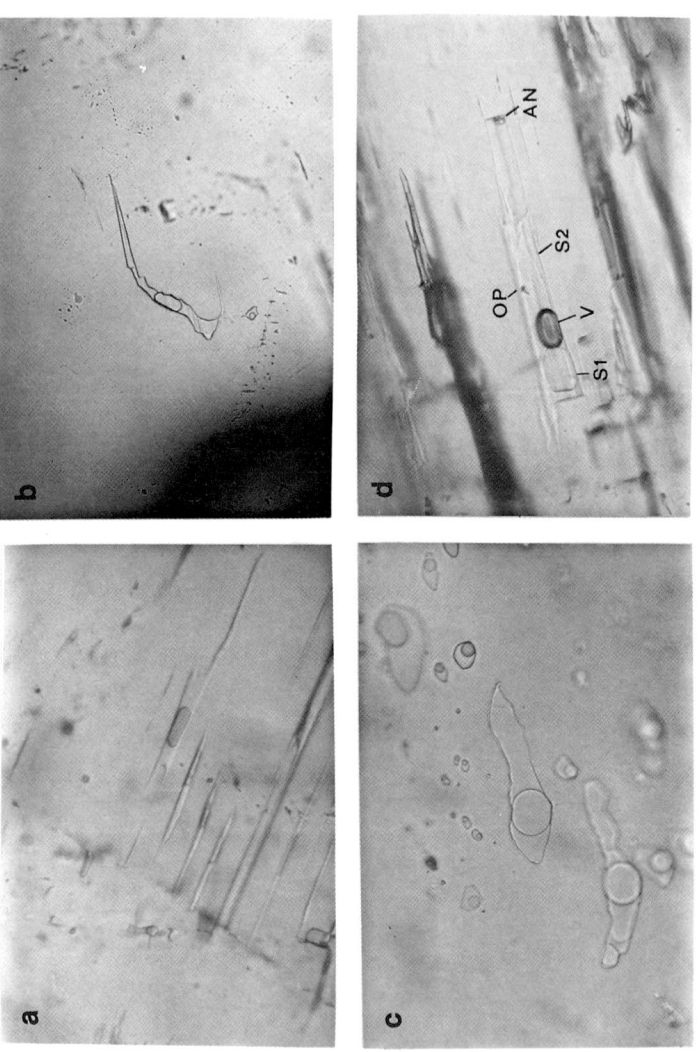

Photo 7.2. Photomicrographs of Emerald-hosted Fluid Inclusions. Plane-polarized transmitted light.

a) Primary, two-phase (H_2O+vapor) fluid inclusions, oriented parallel to the c-axis of the host emerald. Note that CO_2—liquid condensed from the vapor phase on cooling (see text). Sample SEM-M2 from the Swat district. Field of View: 250μm.

b) Primary, two-phase (H_2O+ vapor) fluid inclusion, elongated parallel to the c-axis of the host emerald. Note that CO_2—liquid condensed from the vapor phase on cooling (see text). Sample SDA-25 from the Swat district. Field of View: 400μm.

c) Primary or pseudosecondary, two-phase (H_2O+vapor) fluid inclusions, oriented parallel to the (0001) cleavage of the host emerald. Note that CO_2-liquid condensed from the vapor phase on cooling (see text). Sample SDA/25 from the Swat district. Field of View: 200 μm.

d) Primary, multiphase fluid inclusion, oriented parallel to the c-axis of the host emerald. Note that CO_2-liquid condensed from the vapor phase on cooling (see text). S1: Halite; S2: Sylvite?; OP: Unknown opaque daughter mineral; AN: Unknown anisotropic daughter mineral; V: Vapor phase. Sample SEM-1 from the Panjsher area, Afghanistan. Field of View: 150 μm.

Photo. 7.3: SEM photomicrograph of an opened, daughter mineral-bearing fluid inclusion. The inclusion is oriented parallel to teh (0001) cleavage of the host emerald. Note the negative crystal (hexagonal) form of the inclusion. Also note the hematite daughter (mottled gray to white, upper right portion of inclusion) and the "stepped" inner surface of the inclusion (upper left portion of inclusion). See text. Sample C/5C from the Charbagh are, Swat district. Field of View: 250 ^{7}m.

Cooling Phenomena

The cooling of two-phase (liquid+ vapor) aqueous fluid inclusions resulted in the condensation of a thin film of a second immiscible liquid phase prior to freezing. Lower temperature observations, described below, confirmed this phase as a CO_2-rich liquid. Because the majority of these observations occur below approximately 5°C, this condensation of CO_2 liquid occurs along the metastable extension of the CO_2 liquid-vapor curve in the clathrate field in a water-dominated CO_2-H_2O-NaCl system (Hedenquist and Henley, 1985). Ice and clathrate were only able to nucleate after significant supercooling below these temperatures. Thus, any T_h observation on the carbonic phase during warming, an observation critical to accurately determining the purity of the CO_2, may be in error due to clathrate formation.

Freezing Phenomena

Pure CO_2 solid, in the presence of vapor, melts to liquid at -56.6°C. Observations for two-phase inclusions from both emerald and quartz range from -56.6 to -60.6°C, suggesting that another component, such as methane (CH_4), is present in some of the inclusions (fig. 7.1). Triple point observations, combined with homogenization temperatures for the carbonic phase, can be used to define X_{CH_4}(Burruss,1981). Assuming an average T_h for the CO_2-bearing phase of 0°C, consistent with the T_hCO_2 observations, a lower most T_mCO_2 the melting temperature of CO_2 solid in the presence of CO_2 liquid and vapor, of -60.6°C places an upper limit on X_{CH_4} of approximately 22 mole % of the CO_2-rich phase (Burruss, 1981). A comparison of triple point measurements for emerald-hosted and quartz-hosted inclusions, and primary, pseudosecondary and secondary inclusions from both hosts reveals no significant differences in T_mCO_2 (fig. 7.1).

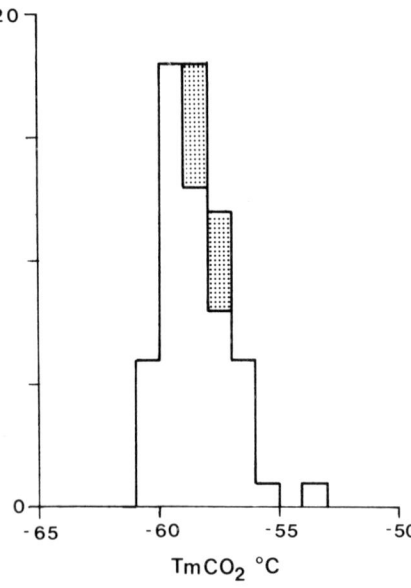

Figure 7.1: T_mCO_2 histogram for all fluid inclusions. Note that the two observations above the CO_2 triple point, -56.6°C, are from the Panjsher emerald (see text). Stippled: Quartz-hosted; Open: Emerald-hosted.

Apparent eutectic temperatures (T_e) for inclusions that are undersaturated with respect to NaCl ranged from -21 to -36°C. These observations bear a high degree of inaccuracy, because even under optimum conditions, the first fraction of solid to melt is difficult to observe. Furthermore, experimental data are lacking to permit accurate interpretation of these data in multi-component systems, especially those containing CO_2 and CH_4 (Roedder, 1984). Thus, these data may only be broadly interpreted as indicating that these inclusions contain dissolved salts ranging from nearly pure NaCl±KCl to mixed, multi-component salts (Crawford, 1981).

Clathrate final melting temperatures for quartz-hosted inclusions ranged from 9.1 to 12.8°C, with no significant differences between primary, pseudosecondary, or secondary inclusions (fig. 7.2a). Clathrate (and ice) final melting temperatures for two-phase emerald-hosted inclusions are more variable that those in quartz; emerald-hosted secondary and pseudosecondary inclusions range from 9.5 to 12.5°C, whereas primary and pseudosecondary inclusions vary from -1.2 to 11.6°C (figs. 7.2b, c). Due to the chemical complexity of the system, accurate salinities are difficult to estimate from these inclusions. Collins (1979) investigated ice/clathrate melting relations in the system H_2O-CO_2-NaCl; CO_2 clathrate melts at 10°C in a 0% NaCl solution, whereas clathrate eutectic melting occurs at -10°C and 24.2 weight % NaCl. Triple point and homogenization data from the present study, however, indicate that CH_4 is present in these inclusions in amounts up to 22 mole %. Methane is qualitatively known to counteract the effect of NaCl on clathrate melting, raising melting temperatures to above that for the pure H_2O-CO_2-NaCl system (Collins, 1979). Although no experimental data are available for the system H_2O-CO_2-NaCl, topological similarities with the systems H_2O-NaCl and H_2O-CO_2-NaCl (Collins, 1979; Bozzo and others, 1975) suggest that a 12.8°C range in T_m (11.6 to -1.2°C) for emerald-hosted, two-phase primary inclusions is consistent with a range of approximately 20 equivalent weight % NaCl.

Heating Phenomena

Homogenization data for fluid inclusions hosted by both quartz and emerald are scarce due

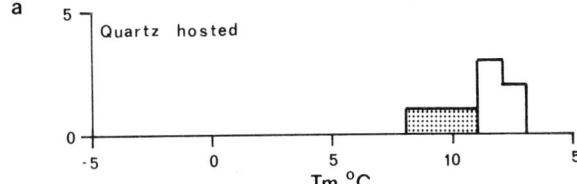

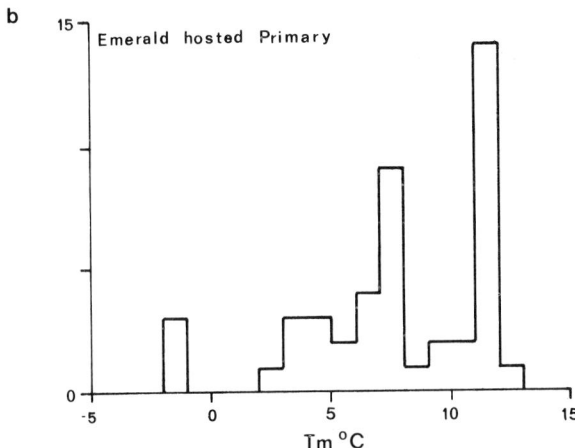

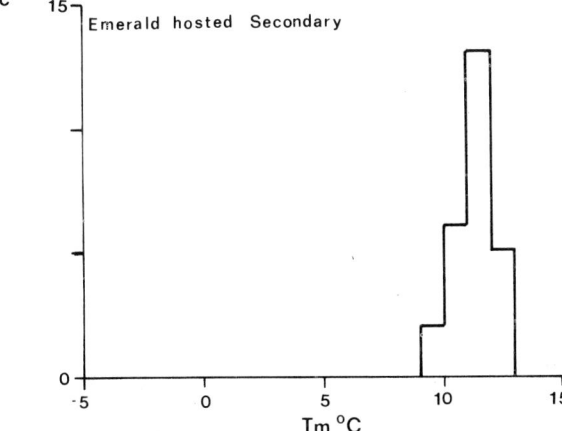

Figure 7.2. T_m histogram for all fluid inclusions (see text).

    a) Quartz-hosted fluid inclusions. Stippled: Primary inclusions; Open: Secondary inclusions.

    b) Emerald-hosted primary (and pseudosecondary) inclusions. Note the variability in the observations.

    c) Emerald-hosted secondary (and pseudosecondary inclusions).

to the common decrepitation of inclusions prior to homogenization (fig. 7.3). The close correspondence of decrepitation data with the limited amount of homogenization data, however, suggests that the former may be a representative lower limit for the latter (fig. 7.3). Decrepitation temperatures for quartz-hosted inclusions ranged from 150 to 270°C; no quartz-hosted inclusions homogenized prior to decrepitation (fig. 7.3). Decrepitation temperatures for emerald-hosted inclusions ranged from 190 to 310°C, whereas homogenization temperatures for two-phase inclusions ranged from 278 to 349°C (fig. 7.3).

The lack of evidence for boiling or fluid immiscibility at the time of entrapment for all inclusions indicates that these homogenization (and decrepitation) data require a correction for confining pressure to be representative of true entrapment temperatures. A burial depth of 1 to 3 km (L.W. Snee, pers. comm., 1986) is equivalent to confining pressures ranging from 100 to 300 bars for hydrostatic pressure and 300 to 900 bars for lithostatic pressure. The ubiquitous decrepitation of quartz-hosted inclusions indicates that the pressures within the inclusions sufficiently exceeded the external confining pressures on the sample in which decrepitation occurred. For moderate-sized ($>10\mu$m) quartz-hosted inclusions, this requires internal pressures in excess of 850 bars (Roedder, 1984). This qualitatively implies that the confining pressure at the time of quartz deposition was also in excess of 850 bars.

Approximating the fluids as a pure NaCl-H_2O solution with 5 weight % NaCl, pressure corrections would range from essentially nil at 100 bars to approximately 100°C at 900 bars confining pressure (Potter, 1977).

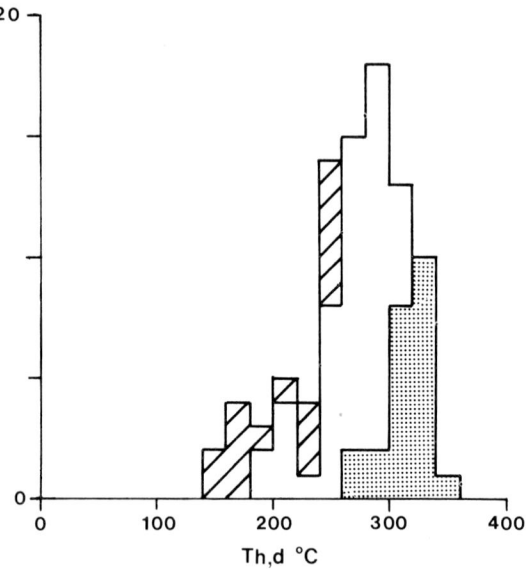

Figure 7.3. $T_{h(d)}$ histogram for all fluid inclusions. Stippled: Emerald-hosted homogenization; Open: Emerald-hosted decrepitation; Slashed: Quartz-hosted decrepitation. See text.

Panjsher Emeralds

Microthermometric data for the emerald sample from the Panjsher area of Afghanistan is presented in table 7.2. CO_2 triple point measurements for two daughter-bearing inclusions in sample SEM-4 of -55.8 and -53.7°C (i.e., above the CO_2 triple point) are anomalous, and may represent interaction between the CO_2-rich solid and the high salinity brine found in these inclusions. Unfortunately, the experimental data to evaluate this hypothesis are lacking.

Bulk salinity for those inclusions saturated with respect to NaCl are also difficult to estimate because of the unknown identities of the other daughter solids, and the inability to dissolve these solids during heating. However, if it is assumed that some of the apparently isotropic daughters are actually epitaxial growths of emerald inside the inclusions, salinity can be estimated using the dissolution temperatures of halite (98.5 and 106.6°C, respectively, in two inclusions) and the dissolution temperature of probable sylvite (42.7°C in one inclusion). These data indicate a salinity of approximately 21 weight % NaCl and 16 weight % KCl (Roedder, 1984).

Three daughter mineral-bearing inclusions in sample SEM-1 homogenized to liquid at 147, 158 and 201°C, respectively, without prior dissolution of all of their daughter minerals (table

Table 7.2. Microthermometric data from Panjsher emerald

SAMPLE	HOST	TYPE	PHASES	T_{hCO_2}	T_{mS_1}	T_{mS_2}	T_{mS_3}	T_{mS_4}	T_{mS_5}	$T_{h(v)}$	T_d
SEM—1—1	EM	P	7	—53.7	98.5	na	na	na	na	201.0	
SEM—1—2	EM	P	4		na	na				158.8	
SEM—1—3	EM	P	4		na	na				147.5	
SEM—1—4	EM	P	7	—55.8	106.3	42.7	na	na	na		225.0

P: Primary Inclusion, T_{mCO_2}: Melting Temperature of CO_2, T_{mS_1}: Dissolution Temperature of Halite, T_{mS_2}: Dissolution emperature of Sylvite (?), $T_{mS_{3-5}}$: Dissolution Temperatures of Unidentified Solids, $T_{h(v)}$: Liquid to vapor homogenization, T_d: Decrepitation Temperature, na: Data not observed.

7.2); thus, interpretation of these homogenization data are problematic. The lower homogenization temperatures, however, suggest either that these emeralds formed at lower temperatures than those from the Swat district or that the confining pressure accompanying emerald mineralization in the Panjsher area was significantly higher than that in the Swat district.

DISCUSSION

Nature of Fluids

On the basis of homogenization and decrepitation, and melting data, the fluids associated with emerald and quartz deposition can qualitatively be divided into at least two populations. The limited amount of data prohibits a rigorous statistical treatment of these populations. The primary fluid inclusions hosted by emeralds tend to have lower clathrate melting temperatures than either the emerald-hosted secondary inclusions or the quartz-hosted primary and secondary inclusions (fig. 7.2). Emerald-hosted secondary inclusions cluster at higher T_m values, similar to those of quartz-hosted inclusions, but at slightly higher homogenization(decrepitation) temperatures. Primary inclusions in quartz tend to have slightly higher homogenization (decrepitation) temperatures and lower clathrate melting temperatures than quartz-hosted secondary inclusions. Therefore, the primary inclusions in emerald appear to be distinctive from the secondary inclusions in emerald and both primary and secondary inclusions in quartz. This suggests that they represent a distinctly different fluid. Thus, the fluid inclusion data from the present study suggest the presence of at least two fluids associated with emerald and quartz deposition. The apparent range of 20 equivalent weight % NaCl exhibited by inclusions hosted by emerald and quartz correlates with a slight decrease in temperature (table 7.1). It should be pointed out that this trend however, could merely be an artifact of the limited data set. Regardless of origin, however, fluid inclusion $T_{h(d)}$ observations range from 150 to 349°C for two-phase inclusions. Thus, the 20 weight % decrease in salinity, accompanied by a decrease in temperature could be interpreted as the mixing of a high temperature, high salinity fluid with a lower temperature, lower salinity fluid.

On a district-scale, there appears to be no significant differences in homogenization (decrepitation) temperatures between the Charbagh, Mingora (Mines 1 and 2), and Barang areas (tables 7.1 and 7.2). However, on the basis of final melting temperatures (T_m), the Mine 2 samples exhibit higher melting temperatures than the Mine 1, Charbagh, and Barang samples, suggesting lower salinities for the Mine 2 fluid inclusions.

Implications for Emerald (Beryl) Deposition

The strength of the conclusions drawn herein with regards to controls on emerald deposits in the Swat district is greatly limited by the reconnaissance nature of the sampling and the limited data pertaining to beryl solubility and Be-complexing. Future fluid inclusion work should involve detailed sampling of mineralized and unmineralized zones on the deposit-scale, accompanied by petrographic work to constrain equilibrium phase assemblages. Such data would enable those factors which control emerald deposition to be more accurately identified, particularly in light of the recent work of Barton (1986). Controls on the availability of Cr and V are best evaluated by more conventional geological means other than fluid inclusion analysis.

Homogenization temperatures for primary fluid inclusions hosted by emerald range from 278 to 349°C, which is approximately 40°C above the beryl hydration reaction at low pressures, whereas at 900 bars, corrected entrapment temperatures (378 to 449°C) are approximately 80°C above this reaction (Barton, 1986). Therefore, the emerald-depositing fluids were well within the pressure-temperature limits of beryl stability. Decrepitation temperatures for some of the quartz- and emerald-hosted inclusions, however, are below this reaction suggesting that temperature may have been important in determining beryl distributions.

Renders and Anderson (1987) conclude that in slightly acidic waters, where beryllium is transported as hydroxy-complexes, a shift in pH, either through wall-rock alteration or CO_2 loss, would promote beryl deposition. They also conclude that decreasing temperature would be unimportant in destabilizing Be-hydroxy complexes. For the Swat emeralds, fluid inclusion data indicate that CO_2 loss (effervescence) was not active during emerald deposition.

The high salinity of some of the fluid inclusions suggests that Be-chloride complexes may have also been important. The fluid mixing implied by the temperature-salinity relationships of the fluid inclusion data may have promoted emerald deposition by destabilizing chloride complexes through cooling and by decreasing total chloride concentration through dilution. Fluid mixing may also have been important as a means of combining Be, Al, and Si from different sources to exceed the solubility of beryl.

Broad constraints can be placed on fluid sources with salinities and homogenization temperatures. The moderate to high salinities and temperatures of primary emerald-hosted inclusions are consistent with either magmatic, metamorphic, or sedimentary basin-derived sources, whereas the low salinity fluids are consistent with heated meteoric or metamorphic sources. Further distinction of fluid sources must rely on stable isotope data.

Comparison with Other Studies

Other fluid inclusion studies of emerald deposits are scarce in the literature. Roedder (1963) reports a limited amount of data on the petrography and freezing behavior of fluid inclusions hosted by Colombian emeralds. His findings were similar to those from the present study; primary fluid inclusions contain minor amounts of CO_2, near CO_2-liquid saturation. He also reports NaCl daughters, as in the Panjsher emerald (photo. 7.2d). In a more detailed study of the Colombian Muzo emeralds, Ottaway and others (1986) report salinities of approximately 40 weight % NaCl±KCl, similar to those observed in the Panjsher emeralds, and homogenization temperatures of 324±10°C, similar to those for the Swat emeralds. Kozlowski and others (1988) report salinities of approximately 40 weight % from emeralds from the Somondoco district, Colombia, with homogenization temperatures in excess of 470°C.

CONCLUSIONS

The characteristics of the fluids associated with emerald and quartz deposition in the emerald deposits of the Swat district can be summarized as follows:

1. salinities ranged from essentially nil to approximately 20 weight % dissolved salts;
2. the inclusions contain up to 3 mole % of a CO_2-rich component, with up to 0.22 X_{CH_4};
3. homogenization and decrepitation temperatures ranged from 150 to 349°C, which is equivalent to entrapment temperatures of 250 to 449°C at a confining pressure of 900 bars; and
4. boiling and fluid immiscibility were absent in all stages of hydrothermal activity recorded by these fluid inclusions.

On the district-scale, the T_m and $T_{h(d)}$ data from the present study, particularly those from emerald-hosted primary fluid inclusions, suggest that mixing of a high temperature, high salinity fluid with a lower temperature, less saline fluid may have been important in mineral deposition. The only anomalous features of the fluid inclusions, observed in this reconnaissance study, are that the fluid inclusions from Mine 2 are less saline than those from Mine 1, Charbagh, and Barang, and that the fluid inclusions from Charbagh locally contain hematite daughters. All other fluid inclusion characteristics from the Swat district are indistinguishable from one mine to another.

ACKNOWLEDGMENTS

Discussions with W.C. Kelly and L.W. Snee were greatly appreciated. This research was supported by NSF Grant EAR-8025263 to W.C. Kelly. A critical review by John Chesley was appreciated.

REFERENCES

Barton, M.D., 1986, Phase equilibria and thermodynamic properties of minerals in the BeO-Al$_2$O$_3$-SiO$_2$-H$_2$O (BASH) system with petrologic applications: American Mineralogist, v. 71, p. 277-300.

Bozzo, A.T., Chen, H.S., Kass, J.R., and Barduhn, A.J., 1975, The properties of the hydrates of chlorine and carbon dioxide: Desalination, v. 16, p. 303-320.

Burruss, R.C., 1981, Analysis of phase equilibria in C-O-H-S fluid inclusions: Mineralogical Association of Canada Short Course Handbook 6, p. 39-74.

Collins, P.L.F., 1979, Gas hydrates in CO$_2$-bearing fluid inclusions and the use of freezing data for estimation of salinity: Economic Geology, v. 74, p. 1435-1444.

Crawford, M.L., 1981, Phase equilibria in aqueous fluid inclusions: Mineralogical Association of Canada Short Course Handbook 6, p. 74-100.

Gübelin, E.J., 1982, Gemstones of Pakistan: emerald, ruby, and spinel: Gems and Gemology, v. 18, p. 123-139.

Haynes, F.M., 1985, Determination of fluid inclusion compositions by sequential freezing: Economic Geology, v. 80, p. 1436-1439.

Hedenquist, J.W., and Henley, R.W., 1985, The importance of CO$_2$ on freezing point measurements of fluid inclusions: evidence from active geothermal systems and implications for epithermal ore deposition: Economic Geology, v. 80, p. 1379-1406.

Jan, M.Q., Kamal, M., and Khan, M.I., 1981, Tectonic control over emerald mineralization in Swat: University of Peshawar, Geological Bulletin, v. 14, p. 101-109.

Kazmi, A.H., Lawrence, R.D., Anwar, J., Snee, L.W., and Hussain, S., 1986, Mingora emerald deposits (Pakistan): suture associated gem mineralization: Economic Geology, v. 81, p. 2022-2028.

Kozlowski, A., Metz, P., and Jaramillo, H.A.E., 1988, Emeralds from Somondoco, Columbia: chemical composition, fluid inclusions and origin: Neues Jahrbuch für Mineralogie Abhandlungen, v. 159, p. 23-49.

Metzger, F.W., Kelly, W.C., Nesbitt, B.E., and Essene, E.J., 1977, Scanning electron microscopy of daughter minerals in fluid inclusions: Economic Geology, v. 72, p. 141-152.

Ottaway, T.L., Wicks, F.J., Bryndzia, L.T., and Spooner, E.T.C., 1986, Characteristics and origin of the Muzo emerald deposit, Columbia: International Mineralogical Association, Abstracts with Programs, Stanford, California, p. 193.

Potter, R.W., II, 1977, Pressure corrections for fluid-inclusion homogenization temperatures based on the volumetric properties of the system $NaCl-H_2O$: U.S. Geological Survey, Journal of Research, v. 5, p. 603-607.

Renders, P.J., and Anderson, G.M., 1987, Solubility of kaolinite and beryl to 573K: Applied Geochemistry, v. 2, p. 193-203.

Roedder, E., 1963, Studies of fluid inclusions II. Freezing data and their interpretation: Economic Geology, v. 58, p. 167-211.

Roedder, E., 1984, Fluid Inclusions: Mineralogical Society of America, Reviews in Mineralogy, v. 12.

Geology of World Emerald Deposits:
A Brief Review

Ali H. Kazmi and Lawrence W. Snee

INTRODUCTION

Emerald has been known to man since prehistoric times. The earliest known records of the geologic occurrence of emerald come from Egypt, where the so-called Cleopatra's Mines were being worked as early as 2000 B.C. Owing to subsequent political vicissitudes, these mines went into oblivion for many centuries until Mohammad Ali Pasha, king of the Ottoman Empire, organized an expedition to find the lost mines. Frederic Cailliand, a French explorer and member of this expedition, rediscovered the Cleopatra's Emerald Mines in 1818. Attempts were made to reopen these mines but they could not be run commercially. Meanwhile better quality emeralds were discovered and introduced in the world market from other parts of the world, mainly Colombia, and the Egyptian emeralds could not compete with these.

Emeralds from the Habach Valley in Austria were known to the Romans nearly 2000 years B.P. However, no effort was made to mine them until the Middle Ages when the Archbishop of Salzburg established a mine at Habachtal.

In Colombia and Brazil, emerald was known to the American Indians prior to the fifteenth century, much before the first Spaniards arrived in the New World. Emerald was worshipped and used in jewelrey and was traded as far south as Peru and Chile and as far north as Mexico. The first operating emerald mines in Colombia were seen by the Spaniards in 1537 at Turqmeque, Boyaca. Soon thereafter the Spaniards began to work the mines in the Chivor area (Keller, 1981). Since then the Colombian emeralds have found their way all over the world, particularly in the Mediterranean countries, the Middle East, and India, where emeralds have been held in the highest esteem.

Other emerald deposits of the world were discovered much later, most of them within the past 150 years or so. At present, emeralds are known to occur in several countries of the world mainly Afghanistan, Australia, Austria, Brazil, Bulgaria, Colombia, India, Madagascar, Mozambique, Nigeria, Norway, Pakistan, South Africa, Tanzania, United Arab Republic, USSR, USA, Zambia, and Zimbabwe (fig. 8.1). Recently emerald has been reported from Ghana also. With the exception of the Colombian, Panjsher (Afghanistan), Austrian (Habachtal), Brumado (Brazil), and Pakistani emeralds (except Khaltaro) almost all other emerald occurrences are in or along pegmatite veins which have cut mafic schists. A brief continent-wise account of the geology of these emerald deposits is given in the following pages.

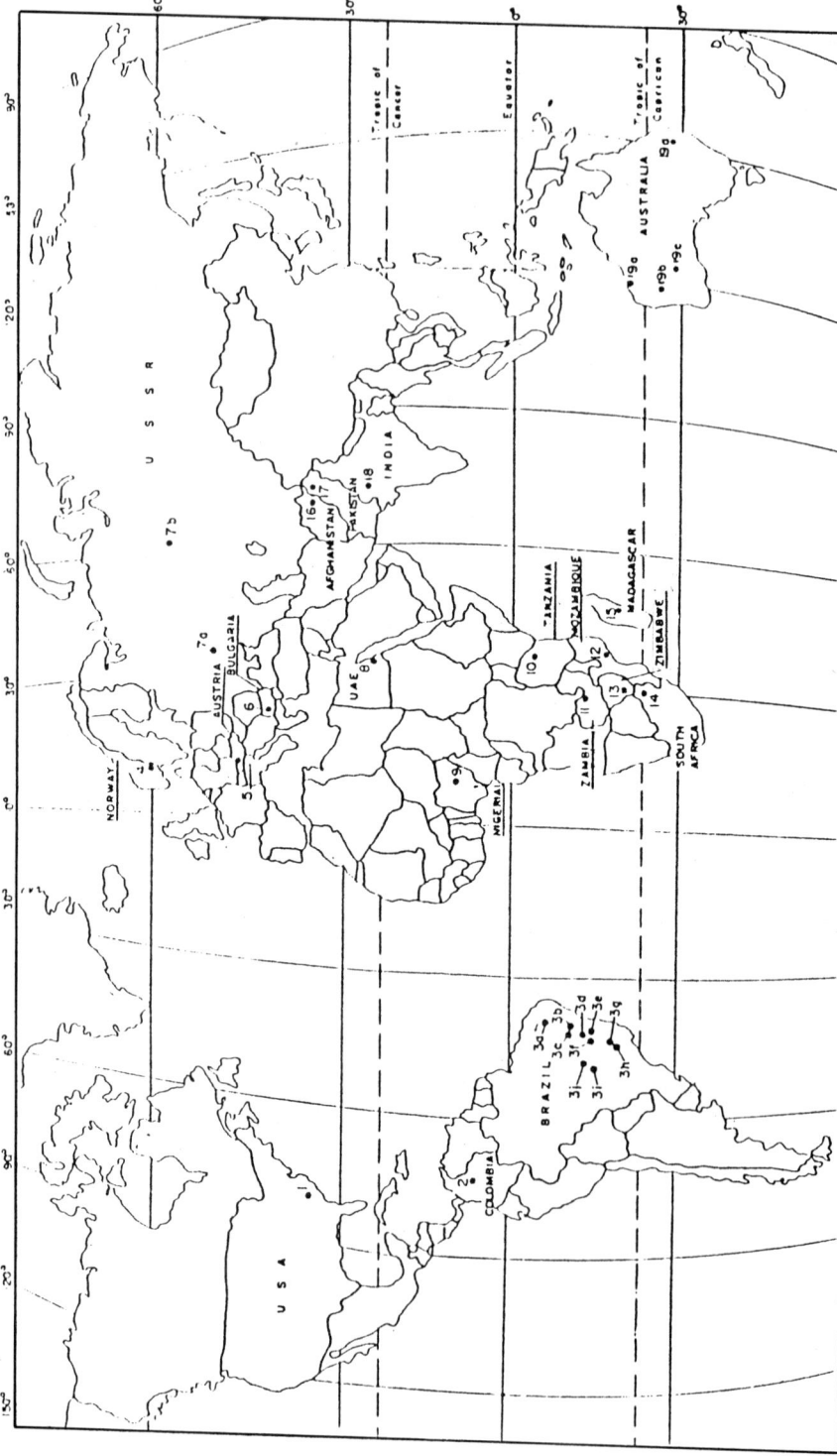

Figure 8.1. 1. USA, N. Carolina emerald; 2. Colombia, Muzo and Chivor emeralds; 3. Brazil, (a) NE. Bahia-Salininha and Carnaiba green beryls, (b) SE. Bahia-Brumado emeralds, (c) Minas Gerais, Emerald dos Ferros and Itabira emeralds; 4. Norway, Eidsvoll emerald; 5. Austria, Habachtal emerald;. 6. Bulgaria, Rila emerald; 7. USSR, (a) Ukrain and (b) Notlshoi Reft River emerald; 8. Egypt, Gebel Sikait and Gebel Zabara emeralds; 9. Nigeria, Jos emerald; 10. Tanzania, Lake Manyara emerald; 11. Zambia, Miku emerald; 12. Mozambique, Morrua emerald; 13. Zimbabwe, Ft. Victoria, Sandawana and Filabusi Emeralds; 14. South Africa, Gravelotte emerald; 15. Madagascar, Ankadilalana emerald; 16. Afghanistan, panjsher and Badel emeralds; 17. Pakistan, Gandao, Nawe Kili, Pranghar, Bucha, Khanori, Aman Kot, Maimola, Mingora, Charbagh, makhad, Gujurkili and Khaltaro emeralds; 18. India, Ajmer-Udaipur meralds; 19. Australia, (a) Wodgina emerald, (b) Warda Wara, Melville and Poona emeralds, (c) Menzies emerald, (d) Emmaville and Torrington emeralds.

Photo. 8.1. A fine example of Arabic calligraphy engraved on an emerald. The inscription consists of a verse from Quran. *Courtesy National Museum of Natural History, photo G. 3920.*

Photo. 8.2. Emerald and diamond necklace of the Spanish Inquisition period. *Courtesy National Museum of Natural History, photo No. 77-10583.*

AFRICA

Emeralds are mainly found on the eastern side of Africa; deposits occur in Egypt, Tanzania, Zambia, Madagascar, Mozambique, Zimbabwe, and South Africa. The sole presently known exception is the Jos deposit of Nigeria which is located in the western part of the African continent. The eastern part of Africa largely comprises a number of Proterozoic to Archean cratonic blocks (2600—2500 m.y) such as the Nubian (northeast Africa), the Tanzanian, the Rhodesian, and the Kaapvaal (South Africa) cratons. These cratons are separated by intervening mobile belts such as the Mozambique, Kibaride, Ubendian, Irumide, Zambezi, Katangan, Damarides, Limpopo, and Namaqualand—Natal belts (fig. 8.2). These ancient metamorphic fold belts range in age from Early Proterozoic (2100—1950 m.y) to Pan-African (730-600 m.y.). As would be seen from the following account, the African emeralds are found both in the cratonic areas as well as the mobile belts.

Emerald Deposits of United Arab Republic.

In the United Arab Republic (UAR), emerald deposits are found in the vicinity of the ancient emerald workings known for ages as "Cleopatra's Mines". These are located about 161 km northeast of the Aswan Dam in the Zabara Mountains, which run along the west coast of the Red Sea. The main deposits occur near Gebel Sikait (24°40'N; 34°48') which is 24 km from the sea coast and Gebel Zabara (24°45'N; 34°42'E), 16 km north of Gebel Sikait (fig. 8.3).

Regional geological setting

The Zabara Mountains, which contain the emerald deposits, form a part of the Arabian-Nubian shield along the west side of the Red Sea. West of the River Nile, amphibolite, gneisses, and schists are exposed. These rocks probably form the eastern margin of the pre-Pan African craton and represent the deformed foreland of the African continental plate. Eastward these crystalline rocks have been faulted and thrust over by a greenschist assemblage consisting of layered volcanic, sedimentary, and ophiolitic rocks. This whole sequence was later intruded by batholithic and plutonic syn- to post-orogenic and anorogenic masses of diorite, granodiorite, granites, gabbros, and syenites (Vail, 1983).

In the Eastern Desert of southern UAR, ultramafic bodies are scattered and isolated (fig. 8.3). These have been interpreted as dismembered fragments in a regional allochthonous ophiolitic melange (Shackleton and others, 1980). Farther eastward, in Saudi Arabia, the ophiolitic complexes occur in linear belts (e.g., Hulayfah-Hamdah belt) which are interpreted to indicate subduction zones that have been disrupted by later igneous intrusions and offset by faults and represent different stages of development or levels of preservation (Vail, 1983).

Geology of the emerald mines

According to Soliman (1986), the Gebel Sikait and Gebel Zabara area, where the emerald mines are located, largely comprise Late Proterozoic metasediments (1200 to 850 m.y.), ophiolitic melange (800 to 600 m.y.), and associated gabbros, diorites, tonalites, and granodiorites (fig. 8.3). There is no apparent relationship between these older intrusions and young tin- and beryllium-bearing granites. A variety of metamorphic rocks including mica, talc, hornblende, actinolite, chlorite, tourmaline, quartz, muscovite, and graphite schists are

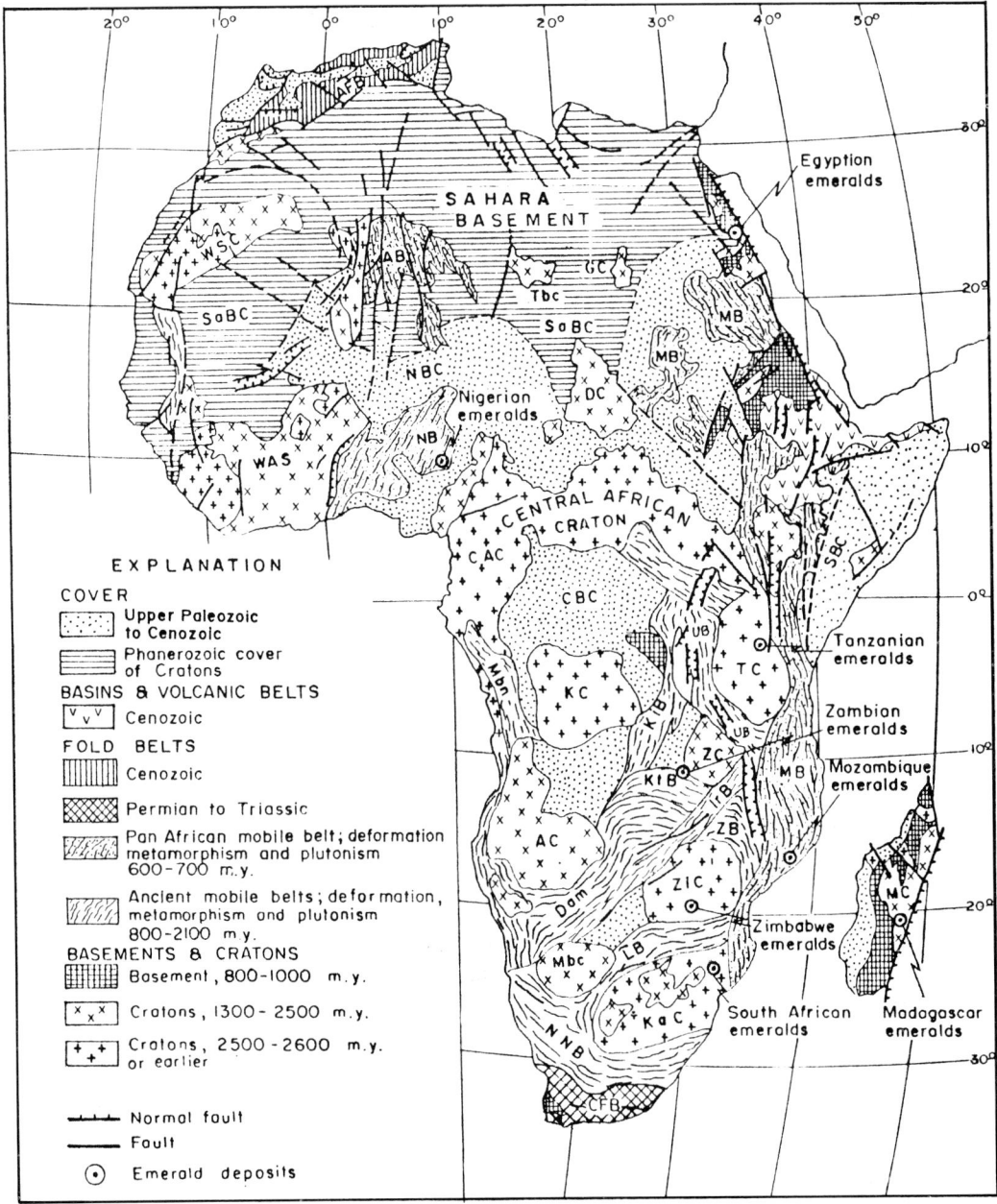

Figure 8.2. Geological sketch map of Africa showing location of emerald deposits (Geology after Leonov and Khain, 1982). AB-Ahaggar belt; AC-Angolan Craton; AFB-Atlas fold belt; CAC-Central African Craton; CBC-Congo basement; CFB-Cape fold belt; Dam-Damaride belt; DC-Darfur Craton; GC-Gilf Kebir Craton; IrB-Irumide belt; KaC-Kaapvaal Craton; KC-Kasai Craton; KiB-Kibaran belt; Kt-B-Katangan copper belt; LB-Limpopo belt; MB-Mozambique belt; Mbn-Maiyumbian belt; MC-Madagascar Craton; MhC-Maltahohe Craton; NB-Niger belt; NBC-Niger basement cover; NNB-Namaqualand-Natal belt; SaBC-Sahara Basement cover; SBC-Somali basement cover; TbC-Tibesti Craton; TC-Tanzanian Craton; UB-Ubendian belt; WAS-West African Shield; WSC-West Sahara Craton; ZB-Zambezi belt; ZC-Zambia Craton; ZiC-Zimbabwi Craton.

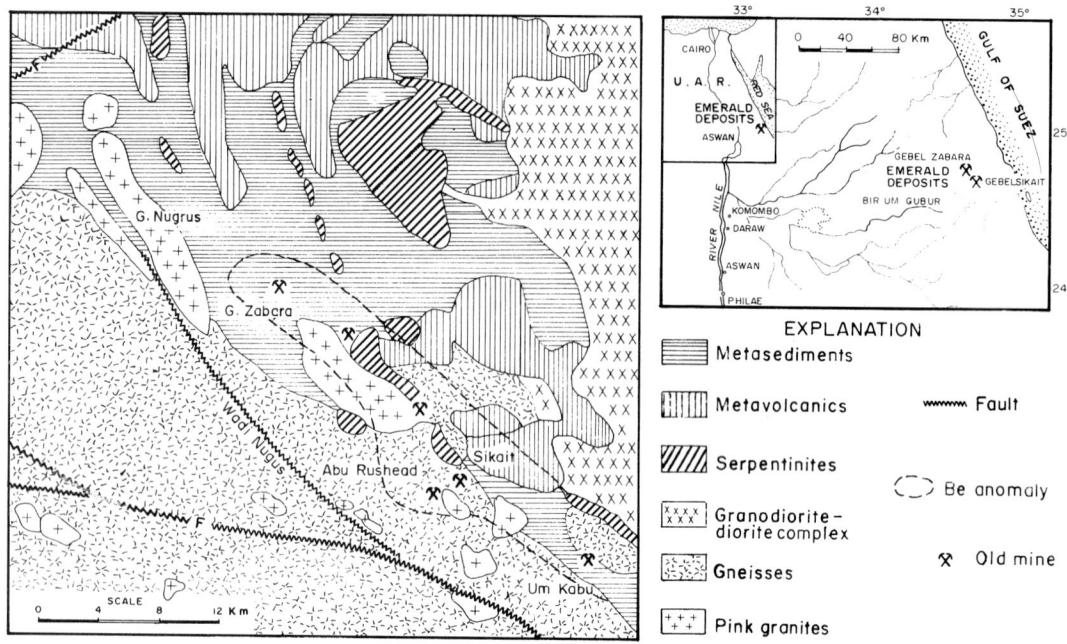

Figure 8.3. Map showing location and geology of Gebel Sikait and Gebel Zabara emerald deposits (adapted from Soliman, 1986).

exposed at the mines. They have been intruded by 640 to 480 m.y. biotite and muscovite-biotite, granite, greisen, and quartz veins (Soliman, 1986). These younger granites and the associated pegmatites and quartz veins contain beryl.

According to Sinkankas (1981) emeralds occur in three types of schists: (a) biotite-quartz schist which is exposed at the mine on the east side of Wadi Sikait. It has been cut by quartz veins in which emerald occurs in association with biotite, plagioclase, calcite, and iron oxides. Biotite contains inclusions of zircon and allanite. (b) Actinolite-biotite schist that is exposed at the mines at the base of Gebel Sikait. In this schist, emerald is associated with actinolite, biotite, beryl, quartz, calcite, magnetite, and hematite. (c) Tourmaline-biotite schist that crops out at Um Harba mines east of Wadi Sikait also contains emerald.

Aeromagnetic survey of the emerald-bearing region reveals three "deep-seated linear tectonic zones" or faults extending deep into the crust (fig. 8.3). According to Soliman (1986) they have been possibly rejuvenated periodically and have provided pathways for pneumatolitic and hydrothermal mineralizing fluids associated with tin-bearing granite. The fluids caused the desilicification of the granite and silicification of the metaultramafic rock and produced extensive metasomatism, greisenization, albitization, silicification, tourmalinization, and mineralization (Sn, Nb, Be, W, Mo, Bi, B and F). Geochemical survey of this region has highlighted the presence of a northwest-trending beryllium-rich zone, parallel to one of these deep-seated faults (the Wadi Nugrus fault). This zone is situated at the contact of the older granite gneisses and the metasediments. The ultramafic rocks are also strung bead-like fashion along this beryllium enriched zone and it is characterized by the outcrops of the younger tin-bearing granites and occurrence of emeralds.

Emerald Deposits of Tanzania

In Tanzania an emerald deposit, known as Manyara, is located near the village of Maji Mota, 110 km west-southwest of Arusha, along the southwestern margin of Lake Manyara, about 2.4 km west of the shore of this lake (fig. 8.4).

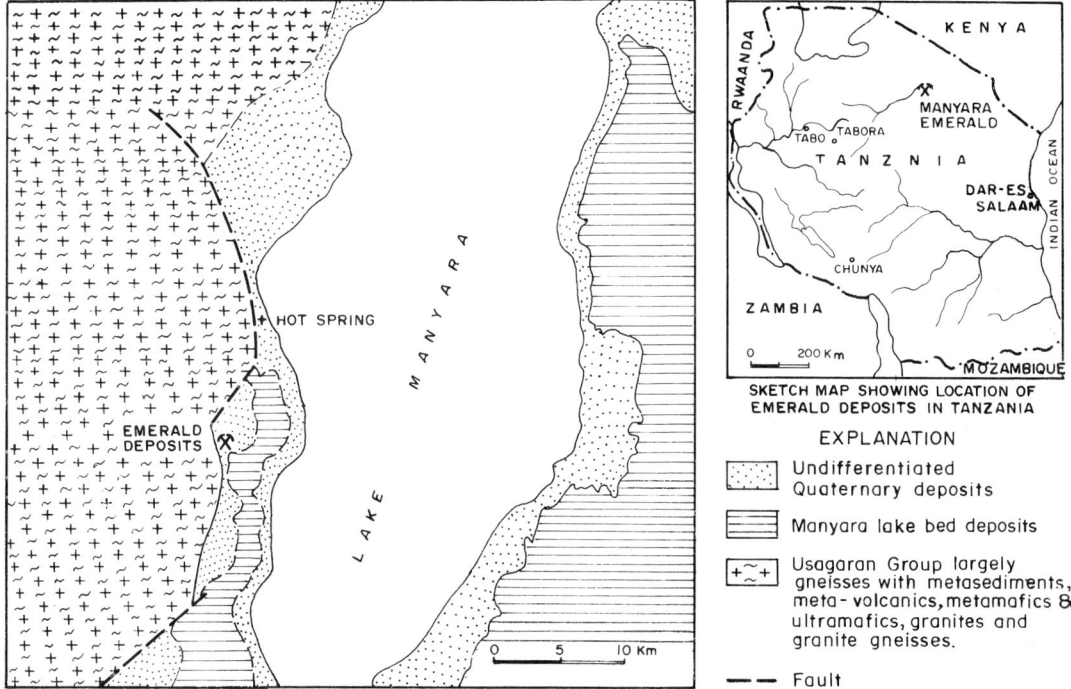

Figure 8.4. Map showing location and regional geological setting of the Manyara emerald deposits, Tanzania.

Regional geological setting

This region is part of the Proterozoic Mozambique orogenic belt (Holms, 1951). The salient features of this belt include Pan-African K/Ar ages, easterly dipping thrusts of the western orogenic front, upper amphibolite facies metamorphism, presence of granulite, charnockite and anorthosite (representing lower crustal environment ?), and polyphase folding. Migmatization is common. There are post-tectonic granites. The late Proterozoic metasediments are pelites and monotonous eugeoclinal metagreywakes (Prochaska and Pohl, 1983).

Within the Mozambique Belt there are numerous small ultramafic bodies associated with mafic rocks. They are noted in the vicinity of Lake Manyara where they commonly occur in the form of a string of pearl which may be interpreted as a thrust or suture zone (Pohl and others, 1980). High-grade metamorphism and intrusion of pegmatites resulted in gem mineralization, including emeralds.

According to Prochaska and Pohl (1983) the orogenic model favored for the major imprint within the Mozambique Belt is a continental collision setting, partly contemporaneous with the late Proterozoic island arcs in the evolving Arabian-Nubian shield.

Petrochemistry of some of the mafic and ultramafic rocks in the vicinity of Lake Manyara reveals high chromium and large-ion-lithophile element contents, which indicates that these rocks were mantle derived but were contaminated by crustal material during their upward migration. This feature suggests that a tensional regime affected the continental crust resulting in a rift-type tectonic setting and a passive continental margin (Prochaska and Pohl, 1983).

Geology of the Manyara emerald deposit

The rocks exposed in the vicinity of the Lake Manyara emerald deposit comprise a sequence of biotite schist, amphibolite with mica bands, chlorite schist, amphibolite, kyanite-almandine-amphibolite gneisses, and highly granitoid gneisses (Gübelin, 1974).

Emeralds are commonly found along the contacts of biotite schist and the intruding pegmatite veins. They occur partly in the pegmatite but mainly in the biotite schist (Thurm, 1972). The emerald mineralization is structurally controlled by two distinct east-trending strike-slip faults.

Emerald Deposits of Mozambique

Until recently emeralds were unknown in Mozambique, though as early as 1957 Behier reported fine, green emerald crystals with biotite in a block of gneiss. A specimen, No. 65-522, is preserved in the Museu Freire d'Anderade with a doubtful locality of Porto Amelia (Behier, 1957). It was however only in early 1970's that gem quality emeralds were found in Morrua district.

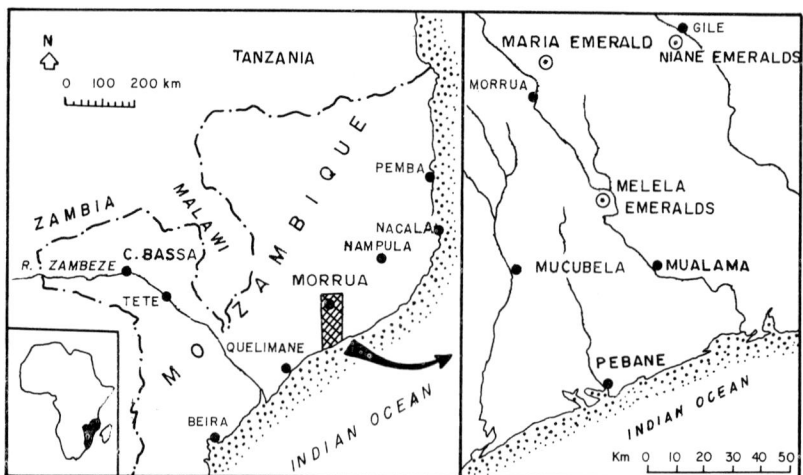

Figure 8.5. Map showing location of Morrua emerald deposits, Mozambique (after Neves, 1978).

Regional geological setting

In Mozambique, emeralds are found within the Mozambique Belt, which is characterized by high-grade metamorphism, presence of granulites, charnockites, anorthosites, polyphase folding, magmatism, and post-tectonic Pan-African granites.

In southeastern Mozambique between Mocuba and Nacala, the Mozambique Belt largely comprises Precambrian gneisses and metasediments which have been extensively intruded by beryl-rich granitic pegmatites. Though beryl occurs commonly in these pegmatite fields, emerald has been reported only from Morrua area about 110 km northeast of Mocuba. Emeralds occur in a 6-m-thick zone of schist along the contact zone of a pegmatite vein. This pegmatite has been mined for beryl, tantalite, and mica (Mumme, 1982).

Geology of the Morrua emerald deposits

According to Neves (1978), emeralds are found in the Morrua area of Mozambique at three localities, namely Rio Maria III, Melela, and Niane (fig. 8.5). He mentions that emerald occurs in hydrothermal veins cutting Precambrian ultrabasic rocks which include amphibolite and talc-actinolite schist. The emeralds are associated with biotite, plagioclase, and quartz. The emerald-bearing pegmatites also contain scheelite, apatite, stilbite, molybdenite, pyrite, fluorite, and calcite.

Emerald Deposits of Madagascar

Emeralds have been relatively recently discovered in Madagascar and reported by Hanni and Klein (1982). The emerald mines are located at Ankadilalana (47°55'E, 21°21'S), northeast of Fianarantsoa. The emeralds occur in biotite schist and according to Hanni and Klein this deposit is of a similar nature as the Manyara (Tanzania) and Habachtal (Austria) emerald deposits.

Regional geological setting

Geologically Madagascar may be conveniently divided into two segments, an eastern Proterozoic to Archean basement complex which largely comprises granites and gneisses with some schists and western platform cover and shelf deposits ranging in age from upper Paleozoic to Cenozoic (fig. 8.2). The basement rocks have been extensively intruded by pegmatite dikes and there are several large pegmatite fields, scattered throughout the entire length of the island. These pegmatites are rich in beryl and contain blue, green, golden, pink as well as bicolored green rose beryls. Besides beryl they contain topaz, garnet, amethyst, chrysoberyl, tourmaline, columbite, euxenite, xenotime, fergusonite, monazite, ilmenite, martite, ampangabeite, and samarskite (Lacroix, 1922).

Geology of the Ankadilalana emerald deposit

The Ankadilalana emeralds are found in biotite schist where it has been intruded by pegmatite veins. Emeralds occur in the contact zone between the pegmatites and the biotite schist. The biotite schist is largely comprised of biotite with smaller amounts of quartz, plagioclase, hornblende, apatite, and zircon. The pegmatites are probably related to the strongly foliated leucogranitic gneiss which crops out in the area.

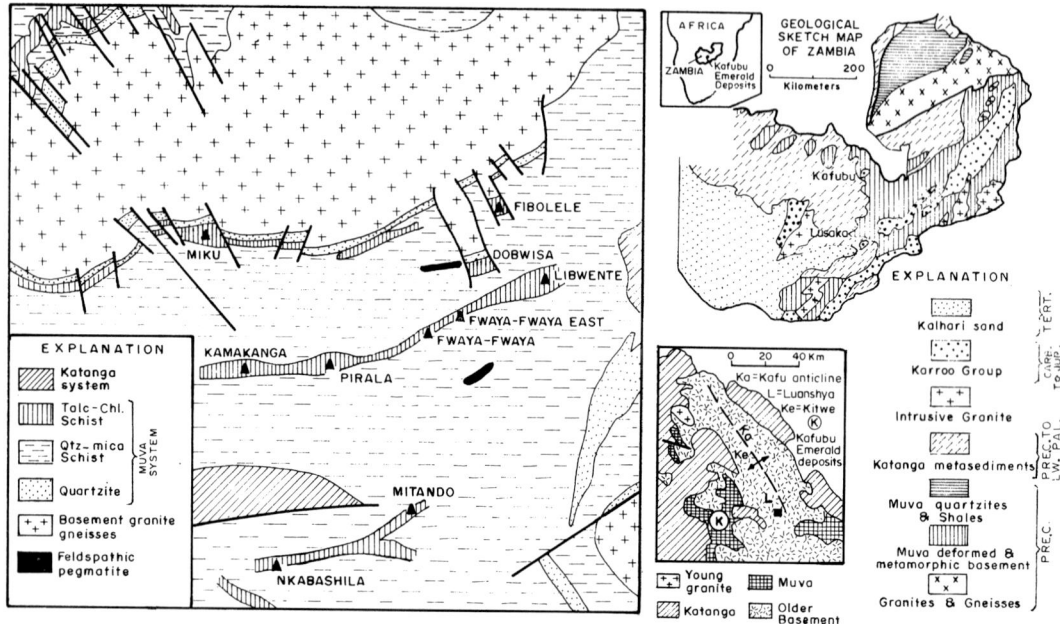

Figure 8.6. Maps showing location and geological setting of Zambian emerald deposit (after Tether and others, undated).

Emerald deposits of Zambia

Emeralds were discovered in Zambia in 1931, but until the early sixties they were of little commercial interest (Bank, 1974). However since 1977, Zambia has become one of the important producers of emerald in the world. The Zambian emerald deposits are located in northern Zambia about 30 km southwest of the town of Kitwe and about the same distance west of the town of Luanshya (fig. 8.6) in the Miku River valley, near its confluence with the Kafubu River. This emerald locality is commonly known as the Kafubu area. In this region the best known and better studied emerald deposit is located at Miku (28°03′30″ E, 13°04′01″S). Other emerald prospects occur south, southeast of Miku, within a distance of 12 km (fig. 8.6) at Nkabashila, Mitando, Kamakanga, Pirala, Fwaya-Fwaya, Libwente, Dabwisa, and Fibolele.

Regional geological setting.

The Kafubu emerald deposits are located at the edge of the Kantangan copper belt near its contact with older basement complex (fig. 8.6). The Kafubu area is a part of the overturned southern limb of a pre-Lufilian antiform. Lufilian tectonism produced northwesterly shearing. The area is underlain by talc-chlorite amphibolite, magnetite schist, quartz-mica schist, and quartzite of the Pre-Katangan Muva system.

Geology of the emerald deposits.

The emeralds are found in westerly striking talc-chlorite amphibolite and magnetite schist, which according to Hickman (1972) was derived from an ultrabasic igneous rock. Within this schist there are lenses and bands of a dark-colored schist which is entirely composed of biotite and phlogopite with scattered crystals of tourmaline. Emeralds mainly occur in this biotite-phlogopite schist. According to Hickman, the emerald-bearing biotite-phlogopite schist band commonly contains an inner core of tourmalinite and an outer transition zone between the biotite-phlogopite host rock and the country rock, which is the talc-chlorite amphibolite and magnetite schist (Fig. 8.7).

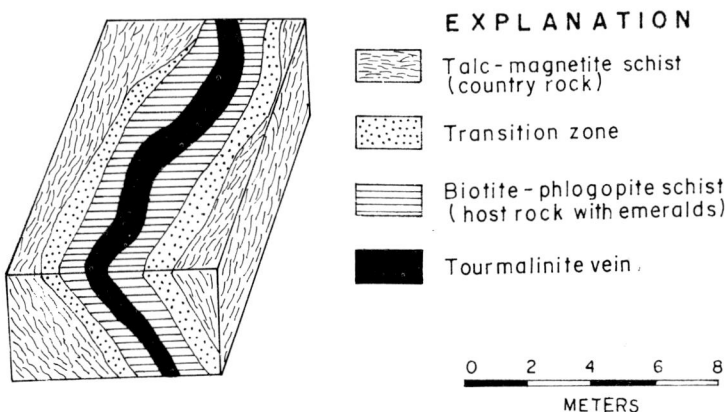

E X P L A N A T I O N

Talc-magnetite schist (country rock)

Transition zone

Biotite-phlogopite schist (host rock with emeralds)

Tourmalinite vein

0   2   4   6   8
METERS

Figure 8.7. Block diagram showing geology of the Miku emerald deposit (adapted from Hickman, 1972).

The country rock has been intruded by pegmatites and tourmaline-rich veins which are probably associated with the foliated granite in the core of "Chitita-Makobo Culmination" to the north (Tether and others, undated). According to Tether and others the tourmaline veins and biotite-phlogopite schists are the likely source of beryllium and chromium respectively. Intrusion of pegmatitic and tourmaline veins resulted in the alteration of the country rock into the biotite-phlogopite schist and the transition zone. The emeralds formed due to pneumatolitic interaction between the intruding veins and the biotite-phlogopite schist.

Emerald Deposits of Zimbabwe.

Emeralds occur in the southern part of Zimbabwe associated with three distinct greenstone belts namely the Fort Victoria greenstone belt, the Mweza greenstone belt, and the Filabusi greenstone belt (fig. 8.8). Each of these belts contains a number of mines or mining claims. Emeralds were first discovered in Zimbabwe in 1956 in the Mweza belt. Emeralds from this belt are commonly referred to as the Sandawana emeralds. Emeralds were discovered in the Filabusi belt in 1958 and in the Fort Victoria greenstone belt in 1960. Apart from these three gemstone belts, emerald occurs in other parts of the country also but the stones are very small and of poor quality (Anderson, 1976 a, b; 1978).

Regional geological setting

The emeralds occur on the Zimbabwean craton, which largely comprises an Archean granite-greenstone terrain. This region contains one of the oldest known (2,700 to 3,500 m.y.) fragments of the Earth's surface. These rocks were tectonized and metamorphosed and presently crop out in the form of relatively narrow, linear greenstone belts. An almost 500-km-long linear feature, by far the most conspicuous geological feature in Zimbabwe, the Great dike (2,500 m.y.), which is of mafic to ultramafic composition, cuts through the craton in a northerly direction. The Zimbabwean craton is bounded by high-grade, polymetamorphic mobile belts, namely the Mozambique to the east, the Zambezi to the north, the Damara to the west, and the Limpopo to the south (fig. 8.2). These belts represent reworked Archean cratonic material and infolded younger supracrustal rocks. Some geologists believe that these belts represent ancient sutures (Burke and others, 1977).

The volcanic and sedimentary rocks of the greenstone belts have been subdivided into three lithostratigraphic divisions namely the Sebakwian, Bulawayan, and Shamvaian Groups. The basal Sebakwian Group largely consists of mafic and ultramafic rocks interlayered with sedimentary rocks and it is intruded by ultramafic rocks. The Sebakwian Group is overlain by the Bulawayan group which mainly consists of mafic to felsic lavas and pyroclastic rocks interlayered with ferruginous phyllites. The Shamvaian Group tops the greenstone belt sequence and predominantly contains sandstones and shales with subordinate volcanic rocks. These rock units of the greenstone belt have been intruded by a variety of granitic rock types (Anhaeusser, 1976).

The emerald deposits are associated with pegmatites at the margins of greenstone belts, mainly in the metaultramafic rocks of the Sebakwian Group. Some of these pegmatites such as the Bikifu pegmatite, close to the Chikwanda emerald deposit, are approximately 2,650 m.y. old (Cooper, 1964).

Geology of the Fort Victoria Emeralds

The main deposits in the vicinity of Fort Victoria (fig. 8.8) are Novello (including Twin Star mines, 17 km northwest of Fort Victoria), Mayfield Farm (12 km northeast of Ft. Victoria) and Chikwanda (80 km east of Fort Victoria), Popoteke, and Renders.

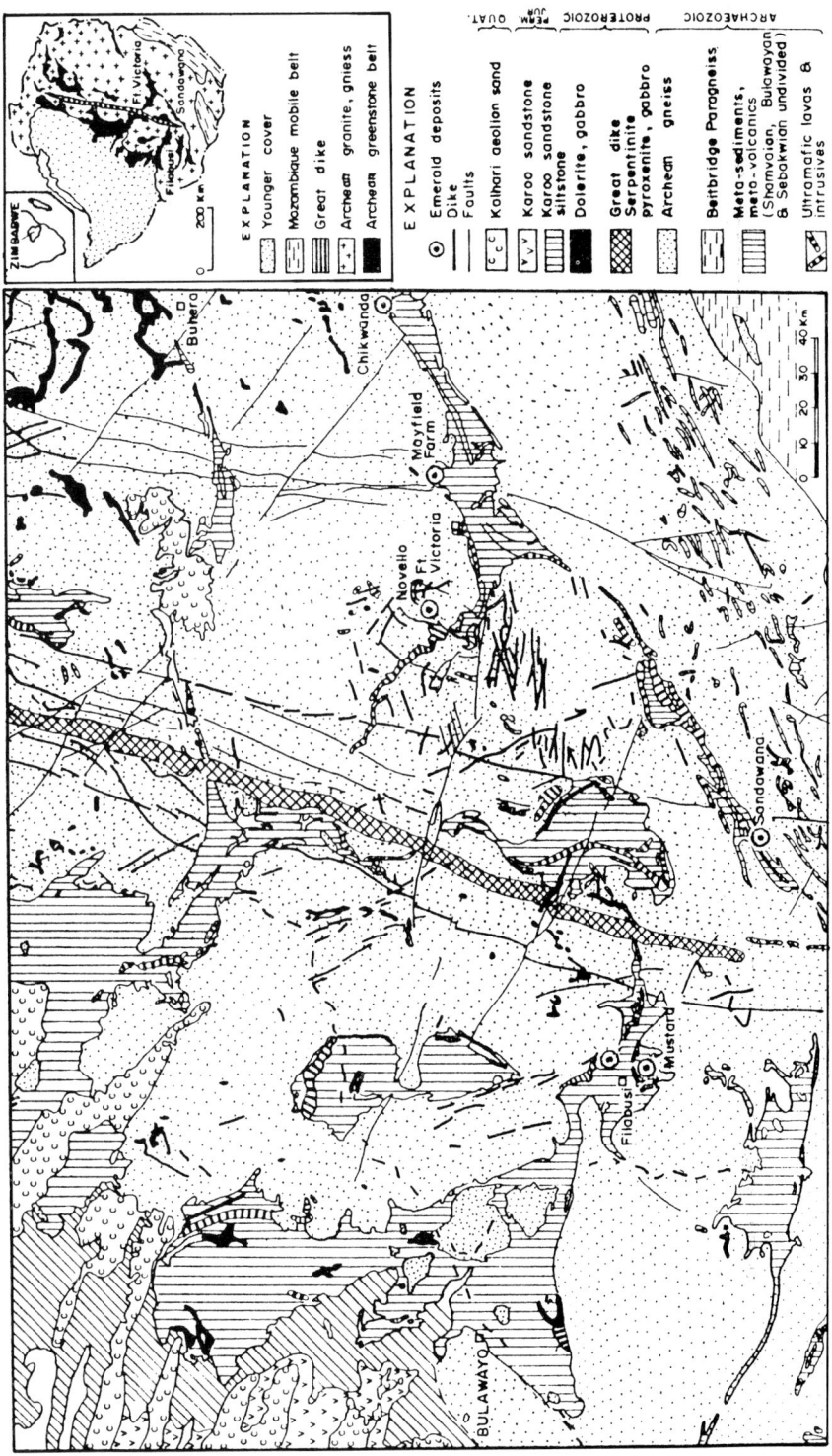

Figure 8.8 Map showing location and regional geological setting of the Zimbabwe emerald deposits.

Gneissic granite crops out in the Novello area and contains a large body of serpentinite intruded by pegmatite and quartz veins (Sinkankas, 1981). These intrusions follow major shear zones and tend to form large lenticular bodies which are scattered randomly over a distance of 5 km. Emeralds tend to be associated with tightly folded and contorted mica schists, particularly along fold axes (Mumme, 1982). According to Martin (1962), emeralds occur in close proximity to the pegmatites and are also found in the pegmatite walls.

The Mayfield Farm deposit also occurs in an environment where serpentine rock and ultramafic schist is associated with the granitic basement complex. Here emerald is found in contorted biotite-phlogopite schist and biotite-talc-chlorite schist along the contact of quartz veins (Mumme, 1982).

The Chikwanda mines have a similar geological setting. Here also, the emeralds are associated with serpentinite in an old basement complex which includes banded iron ores and greenstones (Anderson, 1976a).

Geology of the Mweza (Sandawana) emeralds

The Mweza greenstone belt contains a number of emerald occurrences, the principal ones being the Sandawana, Kanya Halza, Athens, Maharani, and Vidan. The Sandawana deposit is however the most important in Zimbabwe and is located on the southern side of the Mweza Hill Range, 5.2 km southwest of the confluence of the Nuanetsi and Sungai (Mutsume) rivers. The Mweza Range contains a core of tightly folded metasediments (phyllites, sericite-quartz schists, quartzites, and banded ironstone). There are narrow sills of altered ultramafic rocks. The metasediments are flanked by greenstones and epidiorites. The metasediments and associated rocks have been intruded by granites and associated pegmatites. Emeralds occur in tremolite schists adjacent to the granite pegmatites (Anderson, 1978; Gübelin, 1960).

Geology of the Filabusi emeralds

The Filabusi greenstone belt also contains a number of emerald occurrences, the more significant ones being Flame Lily, Mustard, and Coen's Luck. These are located about 90 km southeast of Bulawayo and southeast of Filabusi Post Office. The geological setting of these deposits is similar to the Sandawana deposit.

Emerald Deposits of the Republic of South Africa

In South Africa, emeralds occur in the Murchison Range in northeast Transvaal. They are located in a 36-km-long emerald belt near the town of Leysdorp in the Gravelotte district (fig. 8.9) and often are referred to as Gravelotte emerald (Le Grange, 1929; Van Eeden and others, 1939).

Regional geological setting

The Republic of South Africa mainly consists of two major types of geotectonic terrains, namely cratonic areas and younger mobile belts. The northern half of the country is covered by the large Kaapvaal Craton (> 2500 m.y.) followed by the relatively younger (< 2,500 m.y.) Maltehohe Craton to the west. The Kaapvaal Craton is surrounded by the Mozambique belt to the east, the Limpopo belt to the north, and the Namaqualand—Natal belt to the south and west. The Namaqualand belt is followed to the south by the Cape fold belt.

The emeralds occur on the Kaapvaal Craton. This craton comprises four main structural units (fig. 8.9). The main craton is largely granitic and contains gneissic rocks which were folded or metamorphosed 2500-2600 m.y. ago or earlier. In the northeast corner there are four main zones of east-trending greenstone belts (3000 to 3500 m.y.) which are similar to the greenstone belts of Zimbabwe. In the central part of the Kaapvaal Craton there is a relatively small sedimentary basin filled with sedimentary rocks of the Witwatersrand Supergroup (2500 to 2750 m.y.). A short distance to the north the Kaapvaal Craton is covered by the Bushveld complex (2,000 m.y.) which mainly consists of volcanics, diabase sheets, felsites, granophyres and interbedded sedimentary rocks, mafic and ultramafic rocks, and late Bushveld granites.

As in the case of Zimbabwe, the emeralds in South Africa also occur in the schists of the greenstone belt. In this country the Archean greenstone belts comprise a basal ultramafic—mafic group of rocks (Onverwacht Group), which pass upward into mafic to felsic volcanic rocks interbedded with sedimentary rocks (Fig Tree Group) followed by an arenaceous sedimentary sequence of rocks (Moodies Group).

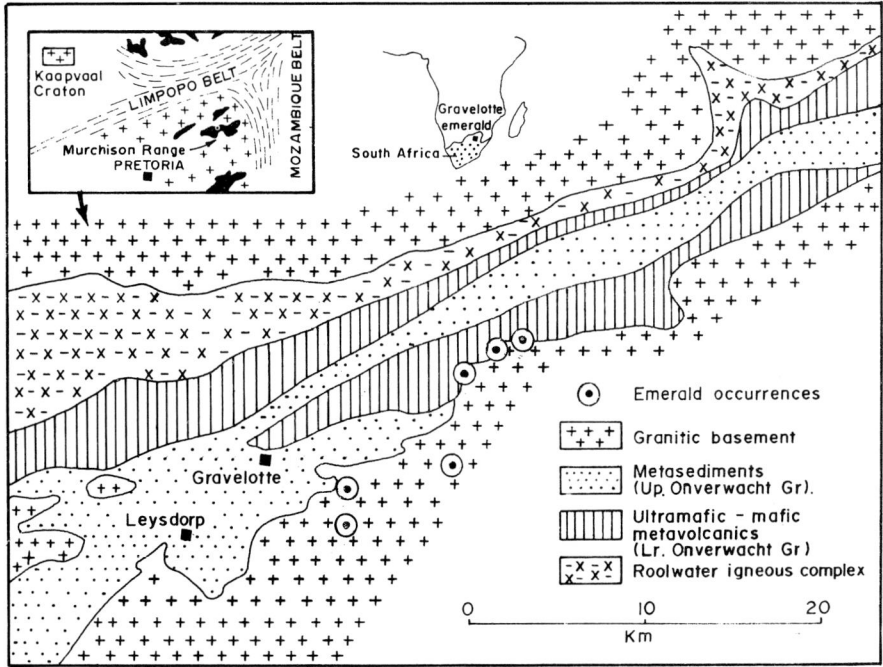

Figure 8.9. Map showing location and regional geological setting of the south African emeralds (geology after Van Eeden and others, 1939).

Geology of the Gravelotte emeralds

The South African emeralds occur in the Murchison Range greenstone belt, in the form of an emerald belt east of Gravelotte (fig. 8.9).

The emerald belt is located in a 6-8-km-wide east-northeast-trending exposure of Murchison schists, which form a folded mass of schists, amphibolites, quartzites, slates, and calcsilicate rocks, resting on granites. Emerald deposits are aligned along the contact of the granite and the Murchison schists.

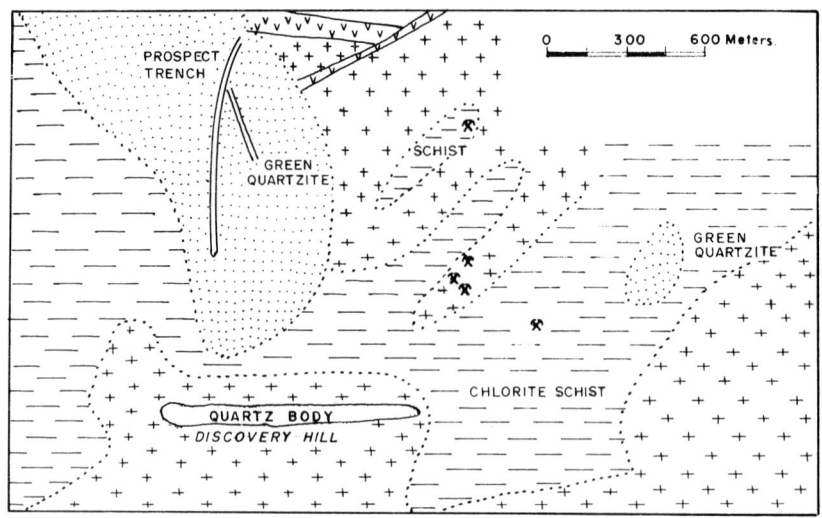

Figure 8.10. Geological sketch map of Germania Koppies (Hills), near Leysdorp, Murchison Range, Transvaal (after O.R. Van Eeden and others, 1939).

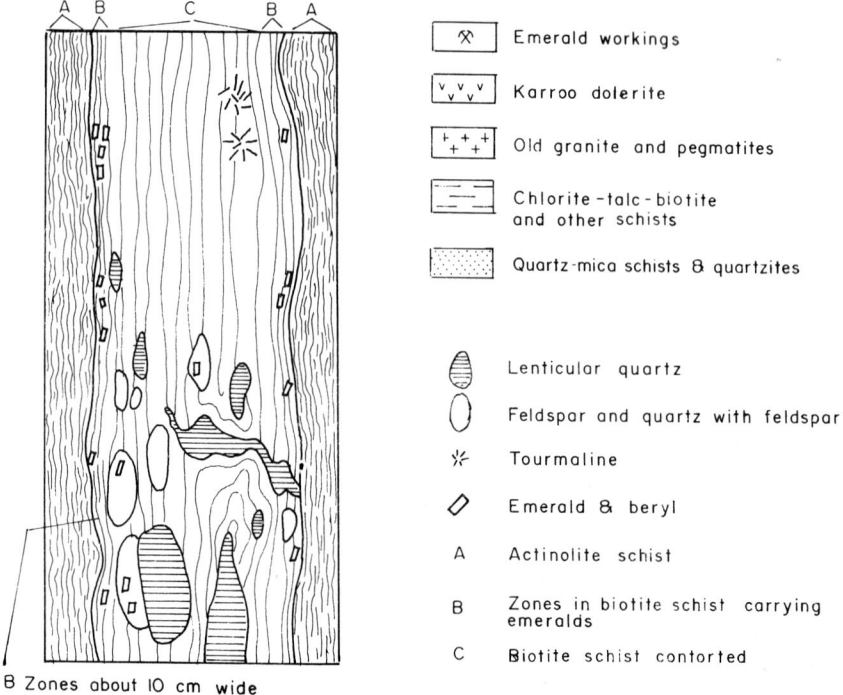

Figure 8.11. Geological cross-section of emerald bearing biotite vein, Barbara Farm, Leysdorp District, Murchison Range, Transvaal (after Le Grang, 1929).

The emerald deposits are found in lenticular masses of amphibole-muscovite schist and talc-biotite schist enclosed in old granites (fig. 8.10). Pegmatite veins intrude the schists, which are believed to be the alteration product of ultramafic rocks. Emerald occurs in biotite schist (fig. 8.11) and in sugary vein quartz and is associated with fluorapatite, aquamarine, beryl, biotite, chalcopyrite, clinochlore, almandine-molybdenite, muscovite-sericite, nontronite, plagioclase, pyrite, quartz, scapolite, and tourmaline (Sinkankas, 1981).

Emeralds and Green Vanadium Beryls in Kenya.

Although there are no available reports on the occurrence of an emerald deposit in Kenya, Schmetzer and Bank (1981) have given a partial microprobe chemical analysis of an emerald reported from Kwale District in Kenya. The information given by them is as follows:

Locality	Color	V	Cr	Fe
Kwale Dist. Kenya	green	0.01	0.14	0.32

The chromium-vanadium ratio indicates that the stone must have been an emerald.

Ghera and Lucchesi (1987) have on the other hand reported the occurrence of a green vanadium beryl from the Taita District of southern Kenya. In this region beryl occurs commonly as a constituent of granitic pegmatites intruded in the Precambrian gneisses, schists, marbles, and quartzites, which form the basement rocks. The basement rocks contain significant amounts of graphite (5-20 volume % in calcsilicates and 0-2 vol. % in marbles).

According to Ghera and Lucchesi (1987) the Kenyan vanadium beryl formed due to metasomatism by magmatic fluids caused by intrusions of pegmatites in vanadium-rich calcsilicate rocks and calcareous metasediments, presumably at 380° to 475°C and 2-3 kbar. The vanadium-rich green beryls of Salininha, Bahia, Brazil are believed to be of a similar origin (Graziani and Lucchesi, 1979).

A S I A

In Asia, emerald deposits are found on the eastern side of the Ural Mountains in the USSR, in eastern Afghanistan, in northern Pakistan, and in western India (fig. 8.12). As in the case of African emeralds, these deposits are also located near the margins of cratonic areas close to mobile belts or suture zones. Geologically the Asian mainland largely consists of several stable cratonic or shield areas (rocks ranging in age from 2600 to 1600 m.y.) separated by intervening fold belts or suture zones. The Ural Mountains mark one such zone. During the upper Paleozoic, Europe and Asia collided resulting in the formation of the Urals. The Ural emeralds formed in this suture zone.

The Afghan emeralds are located in the Panjsher suture zone which is the site of the closure of the Paleotethys Sea. The Pakistan emeralds are located in the Indus suture zone which is the site of closure of the Neotethys. The Indus suture zone marks the collisional margin between the Indo-Pakistan subcontinent and Asia. The Indian emeralds formed in a linear belt in the Precambrian Arravali group of rocks. This belt also represents a probable ancient Archean suture.

The geology and gemology of the Pakistani emeralds have been discussed in detail in previous chapters. The other Asian emerald deposits are briefly described below in a geographic sequence beginning with the deposits in the Urals and followed by the Indian and Afghan occurrences.

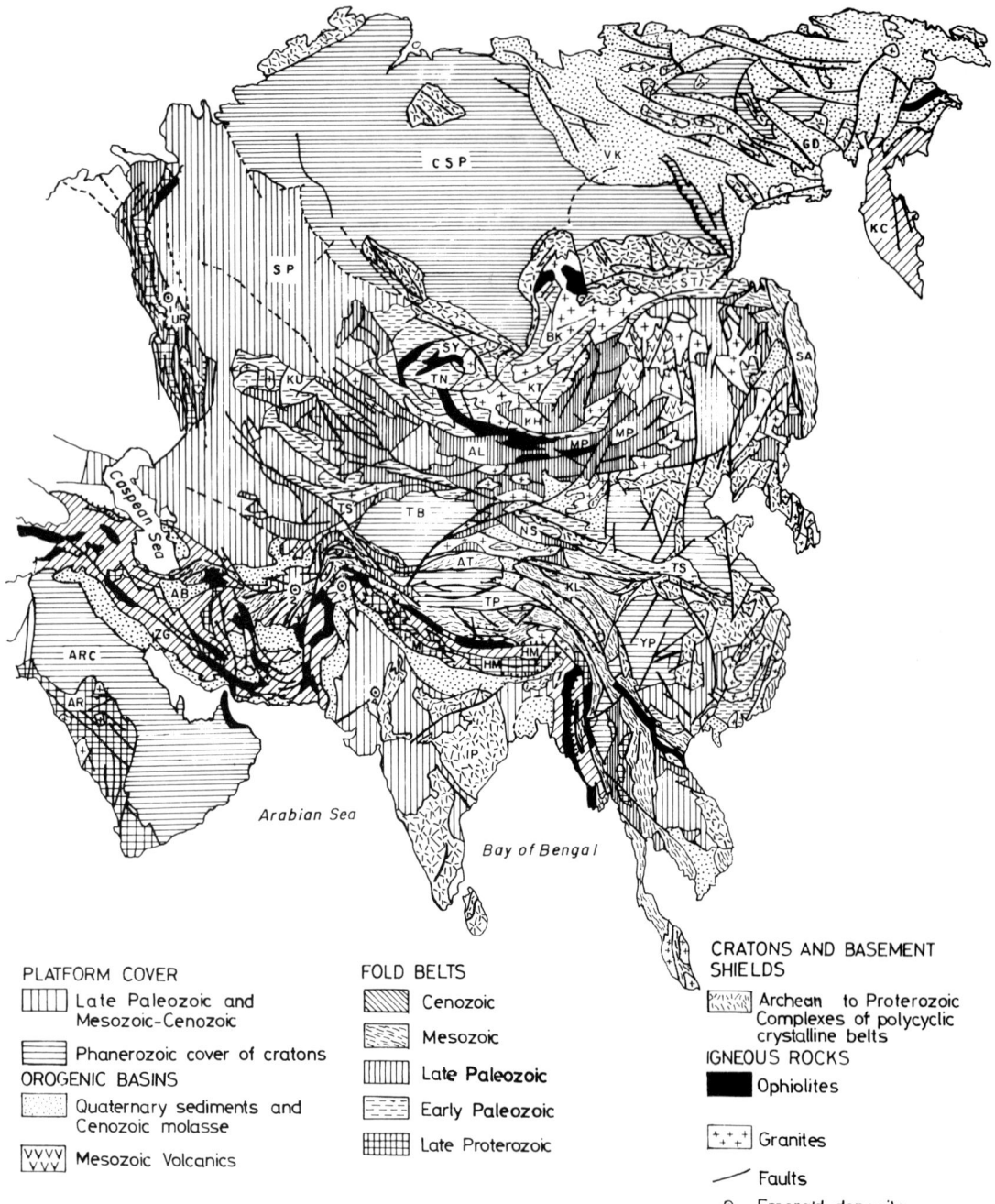

PLATFORM COVER

|||| Late Paleozoic and Mesozoic-Cenozoic

▤ Phanerozoic cover of cratons

OROGENIC BASINS

Quaternary sediments and Cenozoic molasse

VVVV Mesozoic Volcanics

FOLD BELTS

▧ Cenozoic

Mesozoic

|||| Late Paleozoic

Early Paleozoic

▦ Late Proterozoic

CRATONS AND BASEMENT SHIELDS

Archean to Proterozoic Complexes of polycyclic crystalline belts

IGNEOUS ROCKS

■ Ophiolites

+++ Granites

╱ Faults

⊙ Emerald deposits

Figure 8.12. Geological sketch map of Asia showing location of emerald deposits (geology after Leonov and Khain, 1982). 1. Ural emeralds; 2. Panjsher emeralds; 3. Pakistan emeralds and 4. Rajasthan emeralds. A-Arakan; AB-Alburz; AL-Altai; AR-Arabian shield; ARC-Arabian-Shield cover; AT-Alynt Tagh; BK-Baikalian Mts.; CK-Cherskiy; CSP-Central Siberian Plateau; GD-Gydan; HM-Himalayas; IP-Indian Peninsular Shield; KC-Kamchat Ka; KH-Khangai Mts.; KL-Kwen Lun; KT-Kentai; KU-Kazakh uplands; MP-Mongolian Plateau; NS-Nan Shan; SA-Sikhote Alin; SP-Siberian Plain Cover; ST-Stanovoy; SY-Soyan; TM-Takla Makan; TN-Tannu; TP-Tibetan Plateau; TS-Tsingling Shan; TIS-Tien Shan; UR-Urals; YP-Yunkwci Plateau; ZG-Zagros; VK-Ver Khayansk.

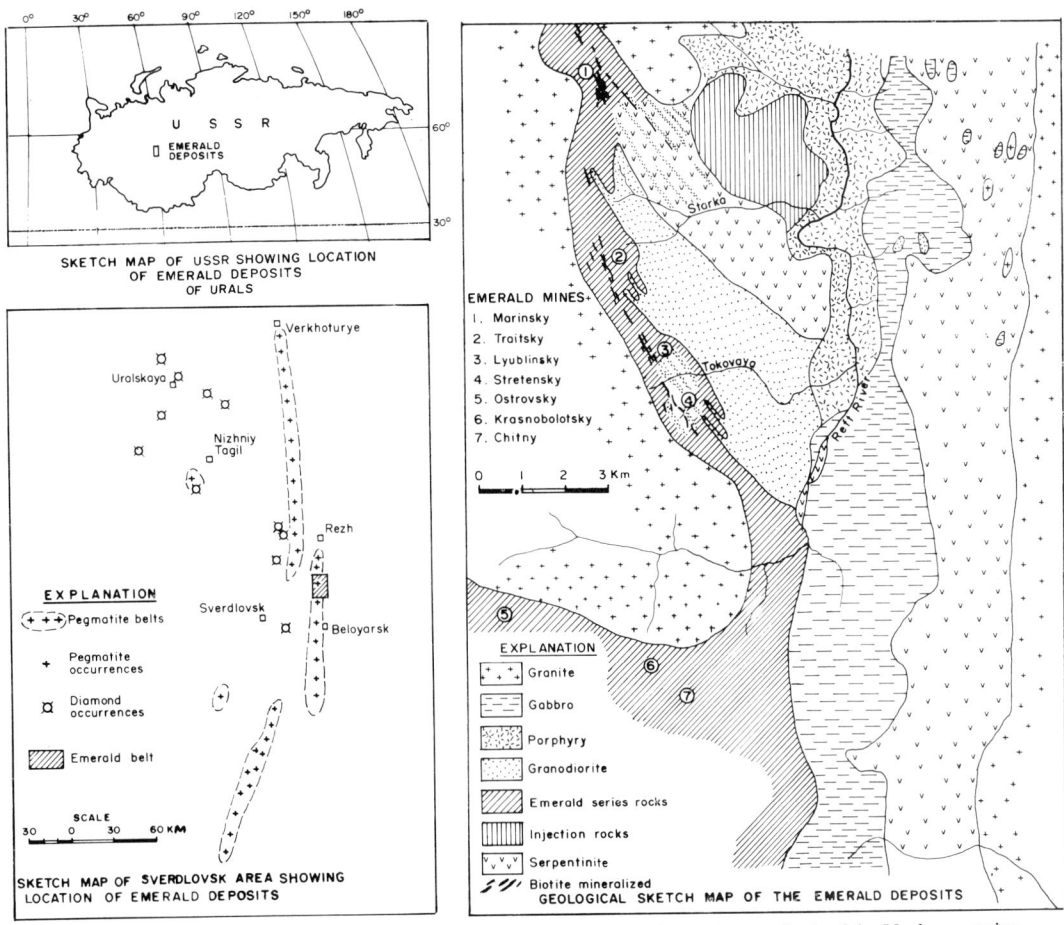

Figure 8.13. Geological sketch map of the Reft River emerald deposits on the east flank of the Ural mountains (after Fersman, 1929).

Emerald Deposits of USSR.

In the USSR, emeralds have been reported from the eastern side of the Urals in the Bolshoi Reft River Valley, near the town of Bagdanovich, which is located about 58 km east of Sverdlovsk; emeralds have also been found in the Ukraine (Lavrinenko and others, 1971). (The Ukrainian emeralds are discussed with European occurrences). The Bolshoi Reft River deposits are the most important ones and have been mined intermittently ever since they were discovered in 1830 or 1831 (Sinkankas, 1981).

Regional geological setting

The Ural Mountains have a north-south trend and divide Asia from Europe. The Urals include two main tectonostratigraphic belts. The western belt is largely composed of Late Proterozoic to early Cambrian metamorphic and sedimentary rocks which are covered by late Carboniferous to Permian rocks. The eastern belt, which comprises late Carboniferous to

Permian metamorphic rocks is in thrust contact over the western belt. The metamorphic rocks of the eastern belt were intruded by Late Paleozoic granites. Emerald deposits are found in the eastern belt, in the Bolshoi Reft River valley. In this region the main emerald mines (from north to south) are Marinsky, Troitsky, Lyublinsky, Sretenskey, Ostrovsky, Krasnobolotsky, and Chitny (Fig. 8.13). Emeralds occur in a north-trending band of biotite and amphibole schists associated with granite pegmatite (Mumme, 1982). This band includes three distinct north-trending geologic zones as described below (Sinkankas, 1981; Fersman, 1923, 1929).

The Bolshoi Reft River Emerald Belt

Western Zone: This zone largely comprises biotite granite, muscovite granite, and muscovite-biotite granite with pegmatite and inclusions of amphibolite.

Central Zone: This zone constitutes a relatively narrow belt (0.5—0.7 m wide) of metamorphic rocks which are in contact with the granitic rocks to the west for a distance of about 25 km. The metamorphic rocks largely consist of serpentine, talc, chlorite, talc-phlogopite and tremolite rocks, amphibolites, amphibole gneisses, and quartzites. The quartzites contain small bodies of serpentinite. The metamorphic rocks are intruded by many quartz diorite and diorite porphyry dikes. The emerald deposits are located in this Central zone.

Eastern zone: This zone includes the Asbostovskiy intrusive complex which contains basic and ultrabasic rocks such as peridotites, dunites, pyroxenites, and gabbros.

Emerald mineralization

According to Fersman (1929) the emeralds are found exclusively in the biotite schist of the Central zone in association with quartz or granitic pegmatite veins and lenses. Emeralds also occur in actinolite-talc schist or talc schist. Vlasov and Kutakova (1960), on the other hand, have reported that the emeralds are found only in desiliconized pegmatites and tend to occur mainly in the phlogopite zone and at its contact with plagioclase bodies and cores. Emeralds are commonly present in plagioclase itself. It is considerably rarer that the emerald crystals are observed in the phlogopite-talc zones in which the talc has formed with the phlogopite. In isolated cases the fine emerald crystals occur in quartz separations in the plagioclase cores, as well as in actinolite lenses and phlogopite-tremolite zones. Minerals closely associated with emerald are phlogopite, plagioclase, fluorite, topaz, apatite, and beryl.

Emerald Deposits of India.

Regional geological setting

India contains two main geological provinces, namely the peninsular region to the south, which is almost entirely composed of Archean to late Precambrian igneous and metamorphic rocks and late Cretaceous volcanics, and the extra-peninsular region to the north, which comprises the Himalayan ranges and the Gangetic trough. The Himalayan ranges are composed of a vast Tethyan fold and thrust belt which contains sedimentary and metamorphic rocks ranging from Precambrian to the Neogene.

The northwestern margin of the peninsular region is characterized by the northeast-trending Aravalli hill range which largely comprises granites, granite gneisses, metasediments and metavolcanics ranging from Archean to late Precambrian. The Indian emeralds occur in the Aravallis.

Geology of the Emerald deposits

In India, emeralds are found along a 200-km-long and relatively narrow northeast-trending emerald belt between the cities of Ajmer and Udaipur, in the Rajasthan Province (fig. 8.14). This belt is traversed by several northeast-trending parallel faults and is largely made up of a 20- to 60-km-wide faulted block of Precambrian metasediments of the Delhi System; the Delhi System largely comprises isoclinally folded quartzite and mica-schist. To the west the rocks of the Delhi System are unconformably overlain by the rocks of the Vindhyan Group. At places this contact is faulted and it is characterized by extensive northeast-trending intrusions of mafic rocks.

To the southeast of the emerald belt the rocks of the Delhi System have been faulted against Archean banded gneisses and metasediments of the Aravalli System. The Aravalli System is mainly mica-schist and gneissic granite with quartzite, amphibolite, calc-schist, and crystalline limestone. These rocks have been tightly folded and faulted.

Within the emerald belt, the metasediments of the Delhi System contain several northeast-trending linear intrusions of mafic rocks which are aligned along major faults. These sediments

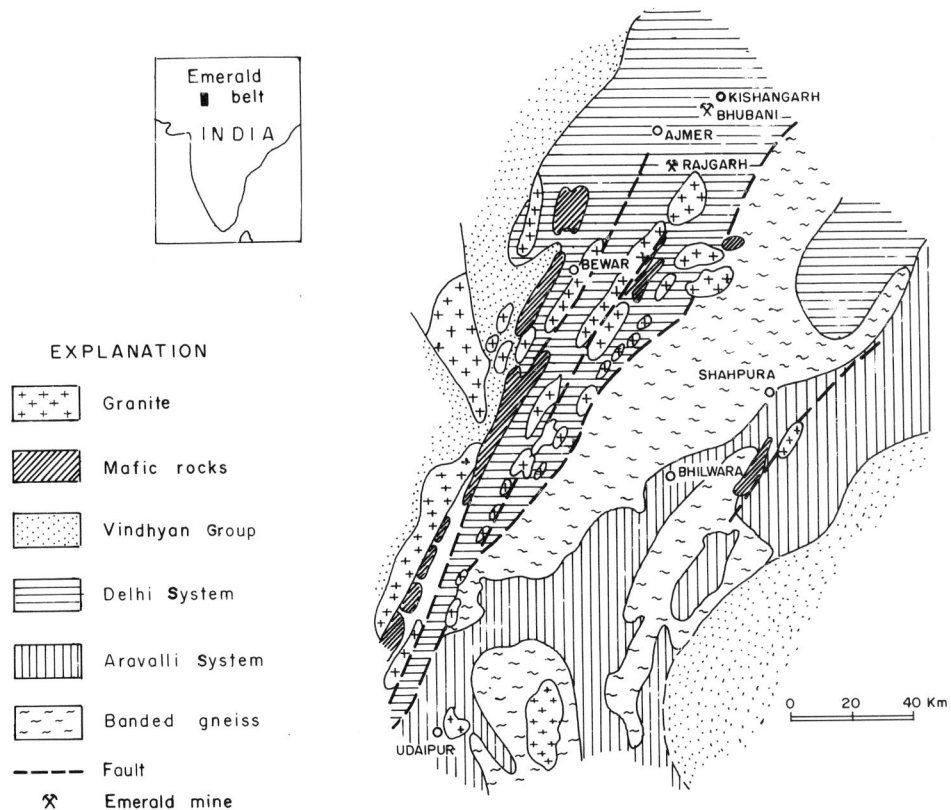

Figure 8.14. Geoloical sketch map of the Rajasthan emerald belt, India (after Mumme, 1982).

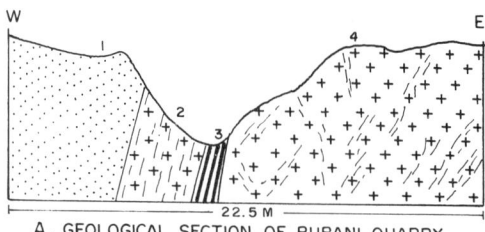

A. GEOLOGICAL SECTION OF BUBANI QUARRY

A. Geological section across Bubani emerald quarry. 1. Phyllites and schists with pegmatites; 2. Biotite, biotite-muscovite schist with pegmatite and tourmaline; 3. Talc schist with emeralds; 4. Pegmatite and tourmalinite with biotite schist zenoliths.

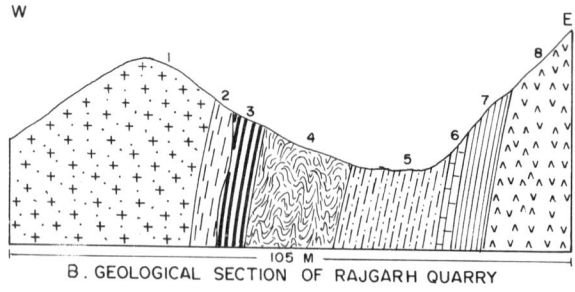

B. GEOLOGICAL SECTION OF RAJGARH QUARRY

B. Geological section across Rajgarh emerald quarry. 1. Pegmatite with tourmalinite; 2. Biotite-muscovite schist; 3. Pegmatite with emerald-bearing schistose xenoliths; 4. Biotite schist, highly contorted; 5. Phyllites and schists; 6. Calc-silicate rock; 7. Hornblende-chlorite-actinolite schist with pegmatite; 8. Peridotite (altered).

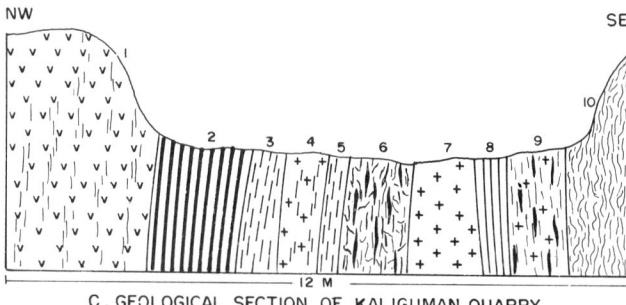

C. GEOLOGICAL SECTION OF KALIGUMAN QUARRY

C. Geological section across Kaliguman emerald quarry (adapted from Roy, 1955). 1. Altered peridotite with actinolite and tremolite (15 ft.); 2. Actinolite-tremolite-biotite schist with emeralds (4½ ft.); 3. Biotite schist (1 ft.); 4. Pegmatite with patches of biotite schist (1¼ ft.); 5. Biotite schist (9 inches); 6. Hornblende schist with quartz vein (3 ft.); 7. Pegmatite (4 ft.); 8. Hornblende schist with vein quartz (1 ft.); 9. Feldspathic biotite schist with pegmatite (3½ ft.) and quartz vein; 10. Hornblende-chlorite schist.

Figure 8.15. Geological sections across Bubani, Rajgarh and Kaliguman emerald mines Rajasthan, India (adapted from Roy, 1955).

also contain talc, biotite, and actinolite schists in which emeralds are found. The emerald belt rocks of the Delhi System have been intruded by tourmaline granite and associated pegmatite veins, which also have a northeast orientation and which caused the emerald mineralization. The emeralds are associated with beryl, tourmaline, apatite, quartz, feldspar, muscovite, and biotite. The host rocks contain kaolin, sericite, chlorite, albite, tourmaline, vermiculite, calcite, talc, serpentinite, tremolite, and anthophyllite. The main emerald deposits in the Rajasthan emerald belt are as follows (Roy, 1955):

Bubani Mines: These are located between Bubani and Muhami villages (26°31′N; 74°48′E). Here emerald is sporadically distributed in biotite and talc-schists (fig. 8.15).

Rajgarh Mines: These are located about 1.6 km southeast of Rajgarh village (26°17.5′N; 74°38′E). At this place highly folded and contorted biotite schists have been intruded by a beryl-bearing pegmatite. Emeralds occur in a 6-m-wide biotite schist zone adjacent to pegmatite (fig. 8.15).

Chat Mines: These are located near Chat village (26°18′; 74°28′45″E) about 1.2 km south of Rajgarh mines. At these deposits muscovite-biotite, biotite, and hornblende schists have been intruded by pegmatite. Here emerald occurs in the biotite schists.

Tekhi Mines: These are located near the city of Deogarh, 1.6 km southeast of Tekhi village (25°31′N; 73°57″E). Here biotite schist with talc and actinolite schist were invaded by siliceous veins (pegmatites?). Emeralds are found in talc schist and actinolite schist.

Kaliguman Mines: These are located near Kaliguman village (25°20′N; 73°50′E) about 89 km north-northeast of Udaipur city. At these deposits hornblende schist was intruded by numerous granite pegmatite veins. Emeralds occur in veinlike bodies of soft talcose biotite schist emplaced between hornblende schist and an altered peridotite rock (fig. 8.15).

Gum Gurha Mines: These are the southernmost deposits in the emerald belt and are located about 45 km southwest of Kaliguman. Here emerald occurs in talcose biotite schist. This schist is a part of a sequence of actinolite, talc, and biotite schists and altered peridotites intruded by pegmatites.

Emerald Deposits of Afghanistan

Regional geological setting

In a very generalized manner Afghanistan may be divided into four geotectonic regions (fig. 8.16). Southeastern Afghanistan is formed of frontal elements of the Indo-Pakistan crustal plate; here, the Indo-Pakistan plate largely consists of a marginal fold and thrust belt. The southwestern part of Afghanistan is the Seistan microcontinental crustal block which is largely covered with unconsolidated sediments. The northern regions of the country comprise parts of the Asian crustal plate. The southern margin of the Asian plate is covered by Mesozoic folded shelf carbonates, volcanic rocks, and carboniferous ophiolites and deep-sea sediments. The latter comprise an early Mesozoic suture zone. The suture zone rocks have been thrust southward along an extensive fault complex, the Hindukush fault (Stocklin, 1977). South of this suture zone, between the Asian, Indian, and the Seistan crustal blocks lies the relatively narrow Farah-Hilmand-Nuristan-Pamir fold belt of Paleotethyan affinity. The Afghan emeralds are found in the early Mesozoic suture zone north of Nuristan and in the Paleotethyan zone in eastern Nuristan.

Location of emerald deposits

In Afghanistan, emerald deposits occur in the Panjsher valley of the Kapisa District in Parwan Province and at Badel (34°50′20″; 70°50′30″E) in Konar District of Nangarhar Province (fig. 8.16). In the Panjsher valley, according to J.F. Shroder Jr. (personnel communication, 1984), emeralds are found at the following six localities:—

— Darun, Parwan Province, Kapisa District,
    35°29′15″N: 69°54′15″E.
— Dahane Revat, Parwan Province, Kapisa District,
    35°29′N: 69°50′E.

— Buzmal, Parwan Province, Kapisa District,
  35°28'35"N; 69°30'00"E.
— Riwat, Parwan Province, Kapisa District,
  35°28"N: 69°52'30"E.
— Mikeni, Parwan Province, Kapisa District,
  35°25'20N; 69°46'45"E.
— Dar Khenj, Parwan Province, Kapisa District,
  35°24'50"N; 69°45'30"E).

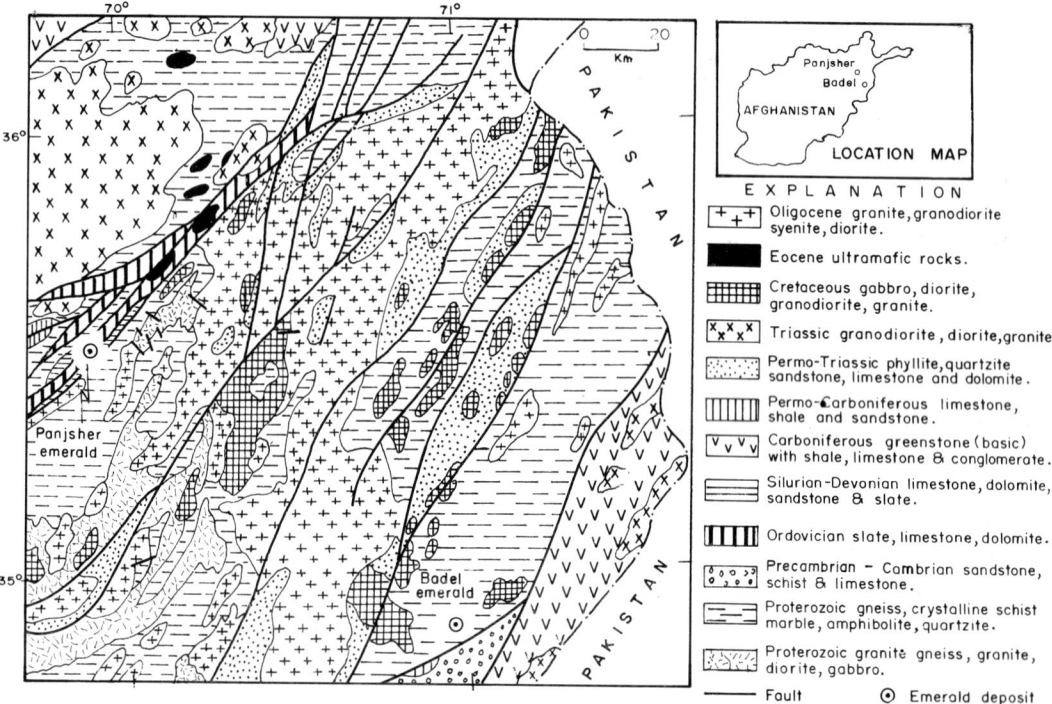

Figure 8.16. Map showing geological setting of Panjsher and Badel emerald deposits in Afghanistan, (geology from Kafarskiy and others, 1976).

Geology of the Panjsher emerald deposits

All the emeralds emanating from Afghanistan are commonly known as Panjsher emeralds, which is of course not correct because besides Panjsher, emeralds occur at Badel in Konar District. The Herat-Panjsher suture zone has a northeast trend along the Panjsher valley. In this valley emerald mineralization is confined to a 16-km-wide area on the southern side of the valley. This area consists of predominantly Ordovician carbonate rocks intruded by diorite-gabbro and quartz-porphyry dikes and/or sills (fig. 8.16). These occurrences are along shear zones occupied by siliceous rocks containing biotite, albite, tourmaline, and pyrite. Beryllium- and emerald-bearing rock are confined to small quartz-ankerite and dolomite veinlets in metasomatically altered diorite-gabbro, dolomitic marble, quartz-biotite schist, and quartz-porphyry. Commonly emerald crystals occur intergrown with quartz, ankerite, or calcite.

The better quality emeralds are found in the Darkhenj and Buzmal areas in highly altered diorite-gabbro dikes or sills. The alteration has resulted in the biotitization (phlogopitization), albitization, carbonation, and silicification of these rocks and production of a network of quartz-ankerite veins which enclose the emeralds (J.F. Shroder Jr., personnel communication 1984).

From the above account it appears that the Panjsher emeralds, like the Swat emeralds are suture-related.

Geology of the Badel emerald deposits

Badel emeralds are located at Badel (34°50′20″N; 70°56′30″E) in the Konar District. At this locality Proterozoic amphibolite, marble, and gneiss are exposed. Associated with the latter and presumably cutting the amphibolite there is a 20-m-long, 20-50-cm-thick pegmatite dike. At the outer contact of this dike, in thin biotite schists, grass green emeralds have formed (Rossovskiy, 1980).

The Konar emeralds, like the Khaltaro emeralds (Pakistan) and emeralds from several other localities of the world are of pegmatitic origin.

AUSTRALIA

Geological Framework and the Emerald Occurrences.

The greater part of the Australian continent comprises Precambrian to Mesozoic platform cover which is characterized by a mildly deformed sequence of shallow water continental sediments, small plutons, and basaltic sheets (fig. 8.17). This vast platform cover contains several scattered masses (inliers) of ancient orogenic and metamorphic belts and cratons. The eastern part of the continent contains the coastal (Tasmanides) Paleozoic fold belt. The central part is traversed by Proterozoic (1730 m.y. to 1040 m.y.) metamorphic fold belts, fault blocks, and deformed basins such as the Gawler block, Musgrave block, Almadeus basin, Arunta block, Creek Province fold belt, and McArthur basin. The Western part largely comprises the Archean cratons, the Yilgarn and the Pilbara, with the intervening Hamersley basin fold belt and the Bangemall basin.

Emerald deposits are located on the Pilbara and Yilgarn cratons (fig. 8.17). The Pilbara craton hosts emeralds in the Pilbara gold field area at Wodgina, Calverts White Quartz Hill, and McPhees Patch southeast of Port Hedland (fig. 8.18). The Yilgarn craton bears emeralds at two principal localities. One of these is in the vicinity of the town of Cue, at Poona and Warda Wara in the Murchison gold field and at Melville in the Yalgoo gold field (fig. 8.19). The other deposits occur near Menzies (fig. 8.20) associated with another gold field. Emeralds have been reported to occur in the eastern Paleozoic fold belt also at Torrington and Emmaville in New South Wales (figs 8.21 and 8.22). The Emmaville stones however are deficient in chromium and are essentially green beryls rather than emeralds (Brown, 1984). The emerald-producing mines of Australia are largely located in the Poona and Menzies region.

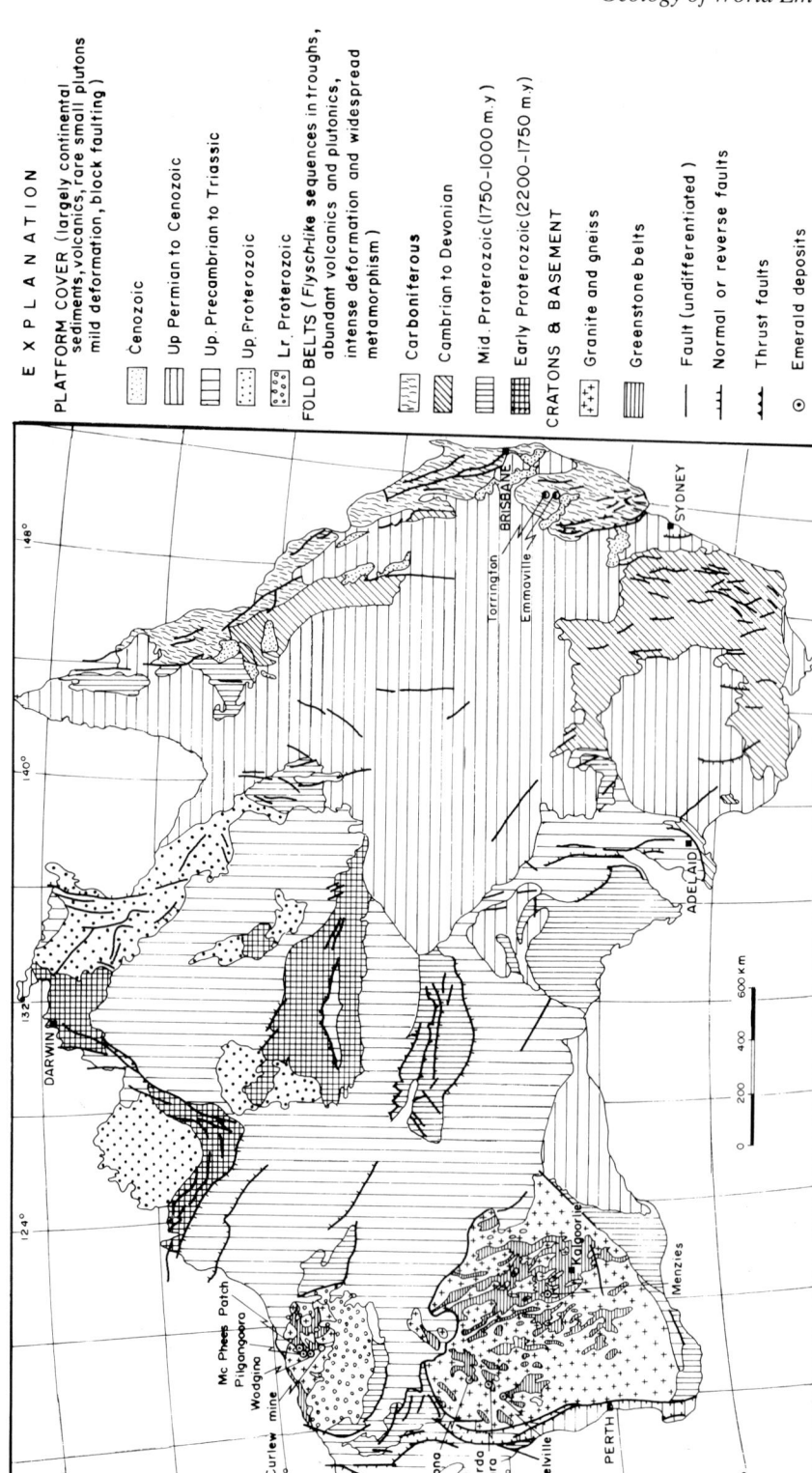

Figure 8.17. Geological sketch map of Australia showing location of emerald deposits (adapted from Geol. Soc. Australia, 1971, Tectonic map of Australia).

Analogous to some of the South African Archean cratons, the Pilbara and Yilgarn blocks are largely granites and gneisses which contain linear to arcuate greenstone belts. These belts largely consist of low- to medium-grade metamorphosed volcanic and sedimentary rocks. The greenstones either have intrusive contacts with surrounding plutonic rocks or are bounded by steep shear zones. They are characterized by tight upright folds, steep foliations, and greenschist to amphibolite facies metamorphism. At places, the greenstone belts were formed during two or more volcano-sedimentary cycles separated by an unconformity. Each of these cycles mainly consists of a basal ultramafic to mafic extrusive and hypabyssal rock sequence capped by acid volcanics and sedimentary rocks (Williams, 1973; Gemuts and Theron, 1973).

The lower greenstones are ultramafic lavas, pyroclastics, and massive peridotites and pyroxenites. The lavas consist of komatiites and komatiitic basalts. These rocks have been metamorphosed to tremolite-chlorite-antigorite schists and amphibotites. The lower greenstones have been intruded by Na-rich 2700-3000 m.y.-old granites, whereas the upper greenstones unconformably overlie the granites. The upper greenstones have been dated at 2700 to 2670 m.y. while the upper age limit of the lower greenstones is believed to be older than 3000 m.y. (Glikson and Lambert, 1976). In the Pilbara region a portion of the lower greenstones (Talga-Talga Subgroup) yielded a date of 3500 m.y. (Hickman, 1980; Nisbet and Chinner, 1981). The Archean volcano-sedimentary sequence in Pilbara and Yilgarn cratons was later intruded by 2600 m.y. potassium-rich granites (table 8.1).

Table 8.1. Stratigraphic sequence of rocks in the Pilbara and Yilgarn cratons (after Glikson and Lambert, 1976).

Rock units	Pilbara Craton	Yilgarn Craton
Potassic granites	Moolyolla Granite, 2670 ± 95 m.y.	Mungari Granite 2615 ± 15 m.y.
Molasse		Kurrawang Beds Unconformity
Turbidites	Mosquito Creek Succession.	Mungari Beds, Association 4.
Upper Greenstones	Upper part of Warrowoona Succession	Mulgabbie Formation, Association 3.
		Unconformity.
Na-rich granites.	Plagioclase rich granites,	Na-rich granites, 2800-2700 m.y. and possibly older.
Lower greenstones	Warrawoona Succession	"Western greenstones".

Besides emeralds, the Australian greenstone belts host important nickel, copper, cobalt, and gold deposits. Platinum, palladium, and irridium group minerals are also associated with these mineral deposits.

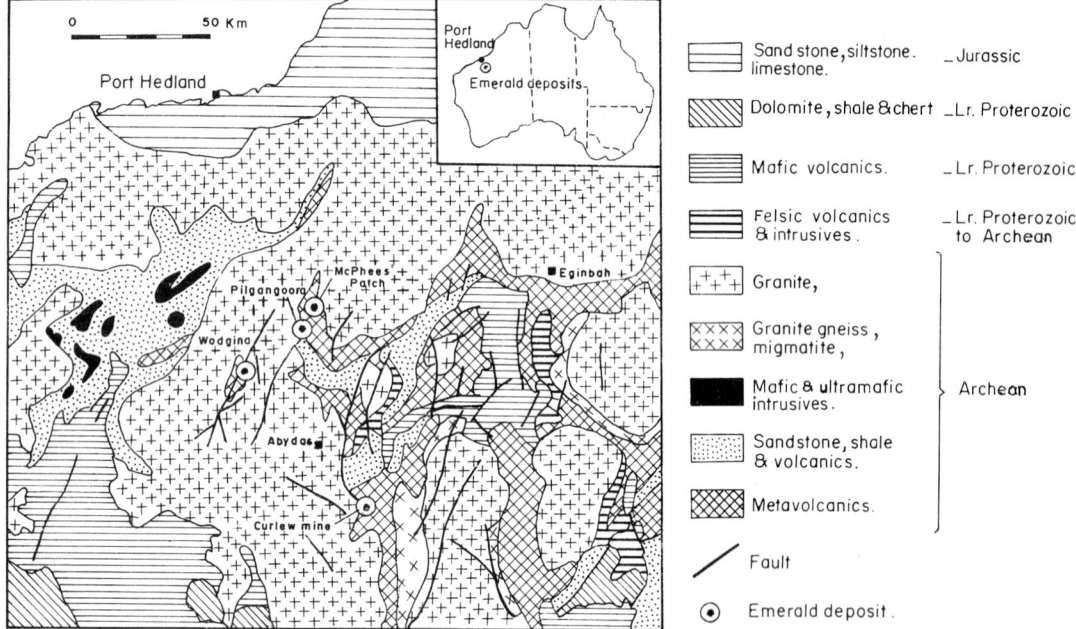

Figure 8.18. Geological map of the emerald deposits of the Pilbara Craton, Australia, (adapted from Geol. Soc. Australia, 1971, Tectonic map of Australia).

Emerald Deposits of Pilbara Craton

On the Pilbara Craton, emeralds occur at four localities as summarized below:

Pilgangoora

This deposit is located about 4 km south-southeast of Fort Hedland in Western Australia. It is located near the contact of the greenstone belt and the granites (fig. 8.18) and is associated with beryl pegmatites. Here emerald occurs with quartz, mica, and feldspar (Mumme, 1982).

Wodgina

Emeralds have been reported to occur near Wodgina, south of Port Hedland, 11 km east of Roebourne. This region consists of folded greenstones and metasediments intruded by a granitic complex and unconformably overlain by Late Proterozoic rocks which have been intruded by diabase dikes and sills. Emeralds were formed in pegmatites which cut through the schistose rocks (Mumme, 1982).

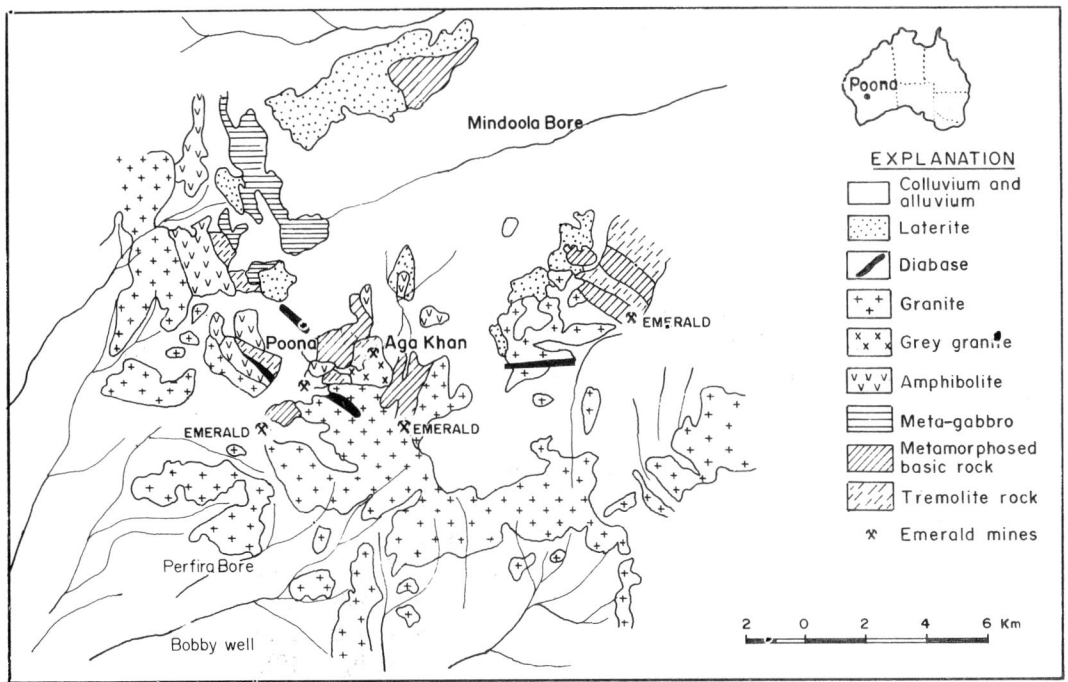

Figure 8.19. Geological sketch map of the Poona emerald deposits, Western Australia, (modified from Mumme, 1982; adapted from Geol. Soc. Australia, 1971, Tectonic map of Australia).

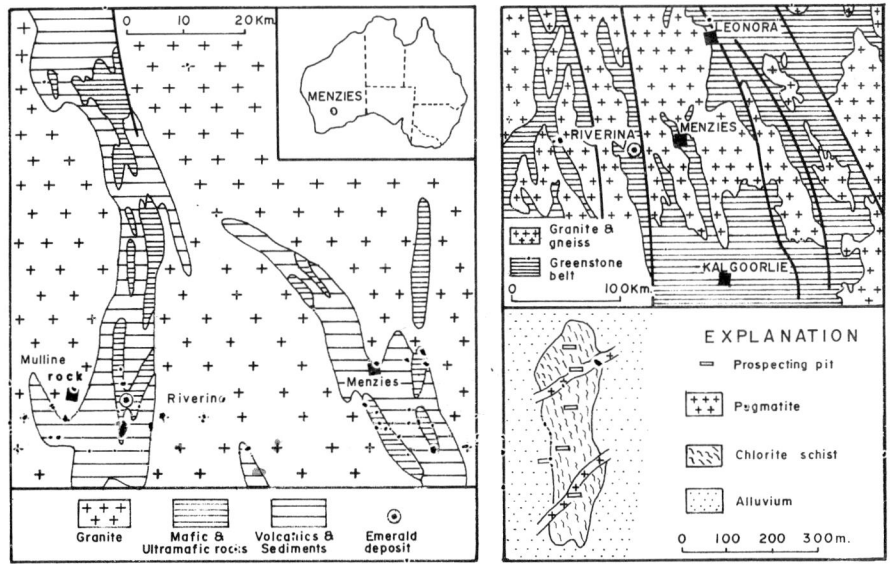

Figure 8.20. Sketch map showing location and geology of the Menzies emerald deposit, (adapted from Whitfield, 1975; Platt, 1980; and Garstone, 1981).

Calvert White Quartz Hill

Also known as Curlew mine (Mumme, 1982), this deposit is located about 180 km southeast of Port Hedland and occurs in the same greenstone belt as the Wodgina and McPhee's Patch emeralds (fig. 8.18).

Emerald Deposits of the Yilgarn Craton

Emerald deposits are concentrated at two principal localities on the Yilgarn Craton. One of these is in the vicinity of the town of Poona, southeast of Meekatharra, where emerald occurs at Poona, Warda Wara, and Melville. The other main locality is centered near Menzies, northwest of Kalgoorlie. Some of these deposits have been recently studied in detail and a brief account of their geology is given in the following pages.

Poona emerald

The Poona emerald deposit is located near Poona (27°10'S: 117°25'E) in Murchison gold field, 64 km northwest of the town of Cue. It occurs in granitic pegmatites which have intruded Archean greenstones. Greenstone xenoliths occur in the pegmatites (Graindorge, 1974). The greenstones were metamorphosed to amphibolites and hornblende-chlorite schist. The latter are intruded by epidiorites, diabase, and serpentinite dikes (fig. 8.19). Along a 60-cm-wide contact zone with the pegmatites the hornblende schist is commonly altered to biotite schist. The best emeralds are found in the mica schist adjacent to beryl-bearing pegmatite (Simpson, 1948). In many cases, porphyroblastic textures are seen in the mica schist. The pegmatite dikes can be traced to their parent granite which has been dated as 2590± 23 m.y. old (Muhling and de Laeter, 1971). Besides emeralds the pegmatites contain biotite, muscovite, lepidolite, zinnwaldite, topaz, tourmaline, fluorite, cassiterite, manganocolumbite, and monazite.

According to Graindorge (1974), the Poona emeralds are the product of contact metamorphism and pegmatitic hydrothermal processes associated with the granite intrusion into an Archean amphibolite.

Warda Wara emerald

Emerald has been found at Warda Wara, 70 km south of Poona in the Murchison gold field. Kalgoorlie Series rocks crop out in the area and emerald occurs in quartz (Mumme, 1982).

Melville emerald

Emerald occurs in pegmatites 4 km north of Melville in Yalgoo gold field, Murchison. In this region several emerald- and beryl-bearing veins of microcline-quartz granite intrude amphibolite. The emeralds are largely concentrated in biotite schist along the contacts of pegmatite veins and amphibolites.

Menzies emerald

Emeralds were discovered in 1974 in a greenstone belt in an old gold mining lease on Riverina station about 46 km west of Menzies, in Western Australia (fig. 8.20). This deposit lies in the Norseman-Wiluna nickel sulfide belt in the eastern half of the Yilgarn Craton. In the Menzies area the following geologic sequence is exposed (Kriewaldt, 1970).

1. Basalt, quartz-mica schist, slate, greywacke, and banded iron formation; all intruded by ultramafics.
2. Talc-chlorite schist, pyritic chert, conglomerate with chert pebbles, basalt, slate, siltstones, grits, and volcanics.
3. Quartz porphyry, ultramafics, and abundant mafic sills which intrude (4) and (5).
4. Intrusive granite.
5. Pegmatite, felsite.

The granite which intruded the metamorphosed sedimentary and volcanic rocks is about 2600 m.y. old (Kriewaldt, 1970).

The Menzies deposit was briefly described by Whitfield (1975) who also published a geological sketch map of the deposit (fig. 8.20). According to him the Menzies emeralds occur in two pegmatite veins (about 7 m wide) that cut a lenticular body of chlorite schist (500 m x 120 m). The protolith of this schist is a peridotite. The pegmatites contain abundant scattered pale green opaque beryl crystals. The emeralds are found along the contact zone between the pegmatites and the chlorite schist. According to Whitfield this mineralized zone is 6 m wide and occurs on either margin of the pegmatites.

Following Whitfield a more detailed account of the geological setting and origin of Menzies emeralds was published by Garstone (1981). According to him the Menzies emeralds are found in northwest-trending pegmatite veins which have intruded a body of metamorphosed ultramafic rock at Riverina Station (fig. 8.20). The pegmatite intrusion caused potassium metasomatism of the ultramafic rock which occurs in the form of a chlorite schist near the surface. Near its contact with pegmatites this schist is interfoliated with mica schist. The mica schist is largely characterized by crenulated tremolite/actinolite-phlogopite/biotite assemblage. Bladed spinifex texture is preserved by grains of phlogopite cross-cutting a lineated tremolite/ actinolite groundmass. Whole rock analyses of the ultramafic host rock by Garstone have indicated a composition intermediate between basaltic komatiite and peridotitic komatiite (Cr_2O_3 = 0.18-0.31%).

The pegmatites have concordantly intruded the ultramafic schist along the foliation and are either wholly composed of feldspar, blocky quartz, or intergrowths of quartz and feldspar. These pegmatites are similar to the complex "granitic pegmatites of the cross-line" as described by Beus (1965) and their intrusion has resulted in the formation of three contact reaction zones in the ultramafic host rock. These are (1) hornblende zone, (2) plate phlogopite zone, and (3) actinolite zone. Talc and chlorite schist are also found in the reaction zone surrounding the pegmatites.

Green Gem Beryl Deposits in the Eastern Paleozoic Fold Belt

Geological setting

The eastern Paleozoic fold belt of Australia extends along the east coast from Tasmania northwards to the Torres Strait. The central part of this belt, south of Brisbane, in the northeastern corner of New South Wales, is often referred to as the New England fold belt (Broughton, 1979). Green gem beryls have been mined from this belt near the town of

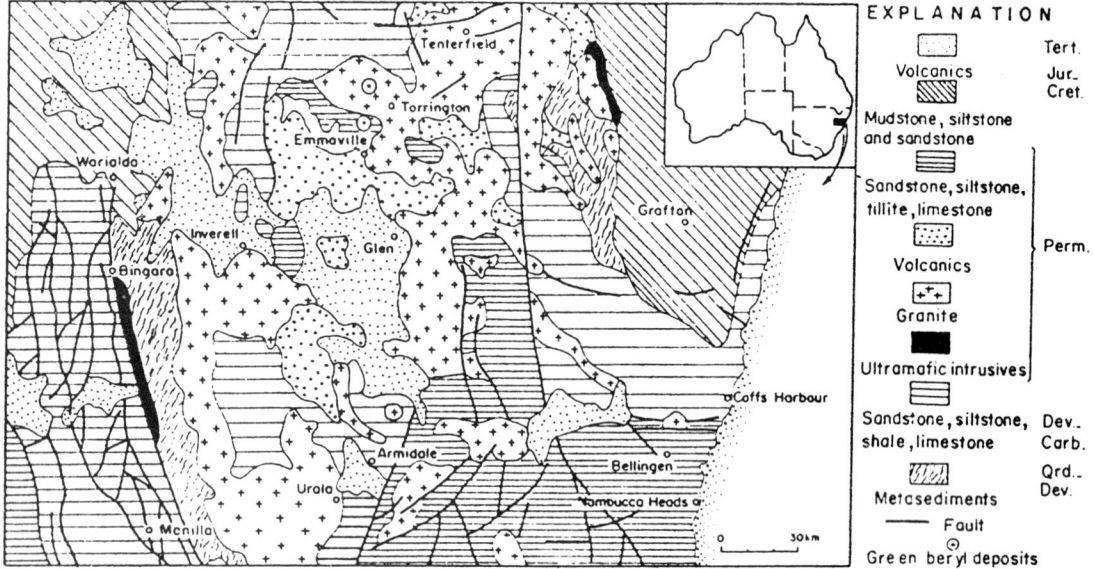

Figure 8.21. Map of a part of the New England fold belt showing the regional geological setting of the Emmaville green beryl deposit, Australia.

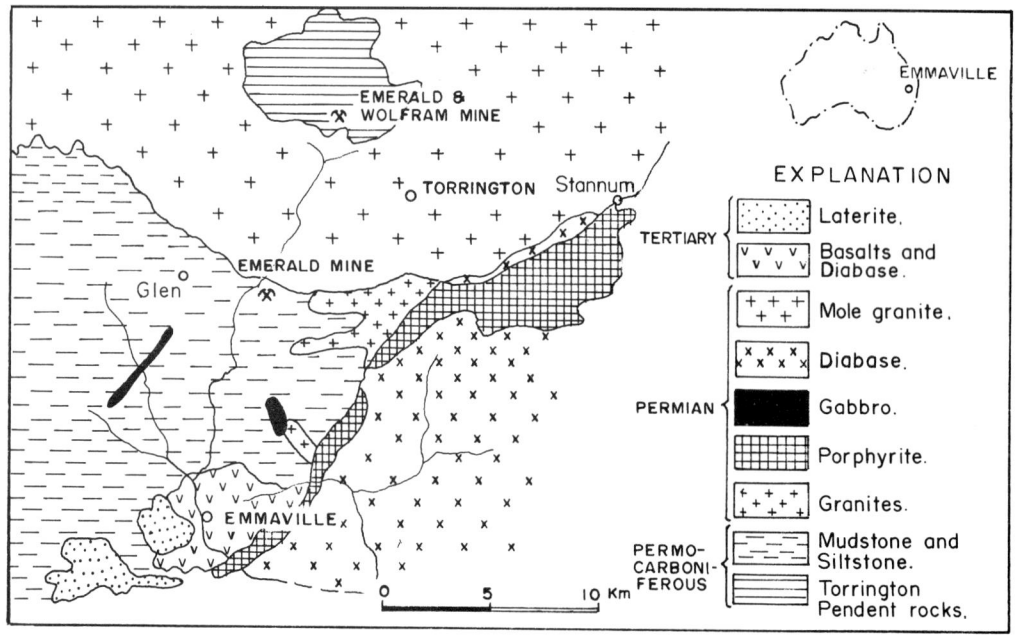

Figure 8.22. Geological sketch map showing location of emerald deposits in the Emmaville-Torrington area, New South Wales, Australia (adapted from Mumme, 1982).

Emmaville since 1890. For decades these stones have been mined and sold as emeralds and until the publication of Brown's paper in 1984, they have been known as Australia's first emeralds.

The New England fold belt includes three structural elements which consist of a central granitic batholith straddled by two fold and thrust belts on the its eastern and western sides. The contact zones between the granite and the fold belts are characterized by two tectonized serpentine belts (fig. 8.21). One would have expected the emeralds to occur along the thrust belt of the ultramafic rocks on either side of the New England batholith. Strangely enough there are no reports of emeralds from these serpentine belts. The so-called Emmavile "emeralds", which are truly green beryls occur in the New England batholith in a terrain that is largely Permian sedimentary and volcanic rocks intruded by Permian granite. This region also contains deposits of tin, tungsten, bismuth, molybdenum, silver, gold, copper, lead, and zinc as well as alluvial diamond, sapphire, and zircon.

Geology of the Emmaville green beryl deposit

The Emmaville "emerald" deposit which is now known to be actually green beryl, was discovered in 1890 and its exploitation started in 1891. The deposit is located at the Glen, formerly known as DeMilhau's reef or Cleary's lode, 9 km (5.5 miles) north of Emmaville (Sinkankas, 1981). The green beryl is found in quartz vein with cassiterite, topaz, fluorite, ferberite, arsenopyrite, and quartz (Barrie and Kalix, 1959) or embedded in kaolinized feldspathic rock, in places, surrounded by massive arsenopyrite, rarely encrusted with crystals of cassiterite or contained in plates of mica (Pittman, 1901). Green beryl generally occurs in pegmatitic rocks related to the nearby late Permian Mole granite. The pegmatites have intruded Permian metasediments (fig. 8.22).

The pegmatite is formed of quartz, topaz, feldspar, and mica. According to Brown (1984), the green beryl occurs irregularly as bunches embedded in cavities and often surrounded by dickite. The green beryls range in color from light green to bright emerald green with a yellowish rather than bluish secondary line. This green color is largely due to 1000 ppm vanadium and perhaps due to 450 ppm copper content in the beryls rather than due to their 350 ppm chromium content.

Torrington emerald

Emeralds have been reported from Heffernan's Wolfram mine 5 km northwest of Torrington, about 40 km north of Emmaville (fig. 8.22). Emeralds occur in quartz and pegmatite veins and are associated with quartz, feldspar, biotite, and wolframite (Mumme, 1982). The authors have not been able to obtain the chemical analyses of Torrington emeralds from the available literature. Considering the fact that this deposit is located in the same geological terrain as the Emmaville deposit it would not be surprising if the Torrington stones also turn out to be green beryls.

EUROPE

Geological Framework

In a very generalized manner it may be said that Europe includes the following six geological subdivisions (fig. 8.23).

1. *Eastern Phanerozoic cratonic cover.* Vast areas of eastern Europe between the Urals and the Carpathians are dominated by a thick mass of Phanerozoic cratonic cover. North of the Black Sea an Archean fold belt crops out through this cover.

2. *The Baltic shield.* This shield is an Archean metamorphic and plutonic complex and covers almost the entire area of Sweden and Finland and small adjacent areas of USSR.

3. *The Caledonide fold and thrust belt.* This is an early Paleozoic fold belt which covers the northern parts of Ireland, Scotland and most of Norway.

4. *The Hercynian fold and thrust belt.* This is a late Paleozoic fold belt exposed in central and western Europe and southern Great Britain.

5. *Central Late Paleozoic to Early Cenozoic cover of young platforms and segments of cratons.* Late Paleozoic to Cenozoic sedimentary rocks cover large areas in northern Europe extending from western Spain to Germany and Czechoslovakia. At several places inliers of older rocks (tectonized blocks) are seen protruding out of this cover (fig. 8.23).

6. *Southern Late Cretaceous to Early Cenozoic Alpine fold and thrust belt.* This belt is of Tethyan affinity and was thrust northwards over the Paleozoic rocks of Central Europe. It marks the collision zone between Eurasia and North Africa.

In Europe, emeralds occur in the Habach Valley in Austria, in the Rila Mountains in Bulgaria, and in the Ukraine in the USSR. Emeralds have been reported from Eidsvoll, north of Oslo in Norway, but truly these stones are vanadium green beryls rather than emeralds. These deposits are briefly described below.

"Emerald" Deposit of Norway

Regional geological setting

In Norway the Eidsvoll emerald deposit is located near the railway station of Eidsvoll, at the southern end of Lake Mjosen on its west bank, about 60 km north-northeast of Oslo (fig. 8.24). It has been also referred to as "Byrud Minnesund" emerald deposit.

The Eidsvoll emerald deposit is located on the Precambrian Baltic shield. A series of Proterozoic crystalline rocks crop out north, east, and south of Eidsvoll roughly east of the line which may be drawn linking Oslo and Hamar (fig. 8.24). These Proterozoic rocks fall in two distinct geotectonic zones separated by a northwest-trending thrust fault that runs parallel to a line between Minnesund and Magnoi. Burke and others (1977) and Zeck and Malling (1976) believe that this line marks a 1100 m.y. old Sveco-Norwegian suture characterized by tectonized ultramafic and mafic volcanic rocks. According to them, reactivation is to the west and it appears to have been a site of continental collision.

The Proterozoic rocks to the south of this line of thrust faulting are largely various kinds of northwest-trending gneissic rocks ranging from granite gneiss, augen gneiss, granodiorite gneiss, quartz-dioritic gneiss, tonalitic gneiss to migmatites. North of the Minnesund—Magnoi thrust line a thick sequence of Proterozoic metavolcanics, amphibolites, metagabbros, metadiorites, and metamorphosed porphyritic granites occurs. However, the southern part of

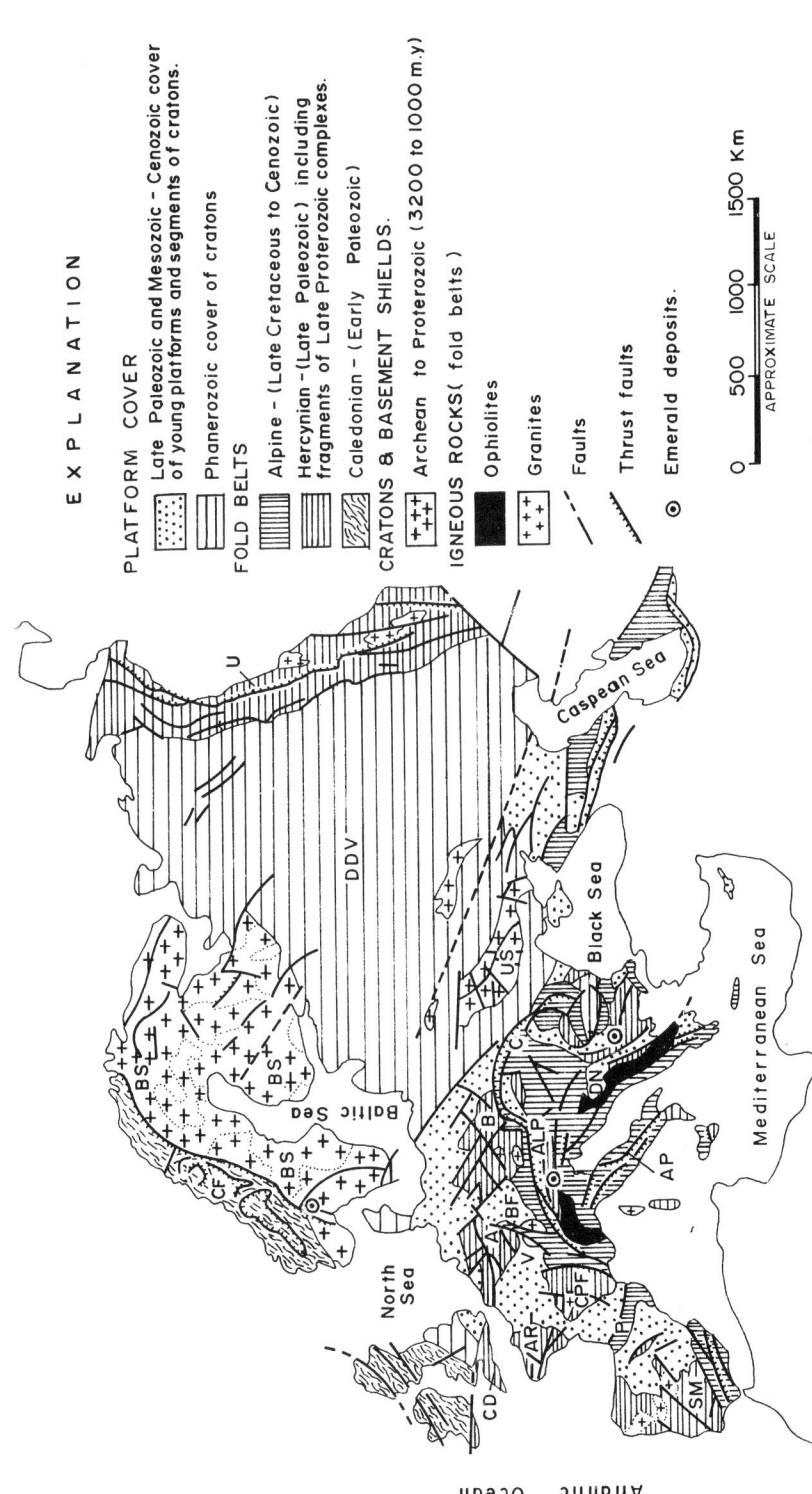

Figure 8.23. Geological sketch map of Europe showing location of emerald deposits (geology from Leonov and Khain, 1982). A-Ardennes; ALP-Alps; AP-Apennines; AR-Armorica; B-Bohemia; BF-Black Forest; BS-Baltic shield; C-Carpathians; CD-Cornwall and Devon; CF-Caledonian fold belt; CPF-Central Plateau of France; DDV-Dnepr, Don,& Volga basins; DN-Ninaric Alps; P-Pyrenees; SM-Spanish Meseta; U-Urals; US-Ukrainian shield; V-Vosges.

to its higher concentration, is the main coloring agent rather than chromium. Recent studies (Wood and Nassau, 1968; Schmetzer and others, 1974; Schmetzer, 1982) indicate that vanadium and chromium play a common role in imparting green color to the emerald as in the case of Salininha (Brazil) green beryls (Draper, 1963) and Emmaville (Australia) green beryls (Brown, 1984), both of which were previously considered to be emeralds. The Eidsvoll gem beryls would thus be classified as vanadium green beryls rather than emeralds.

Since the pegmatite has intruded bituminous schists, it appears that as in the case of Colombian emeralds, vanadium and chromium have been abstracted by the pegmatitic fluids from the bituminous schists.

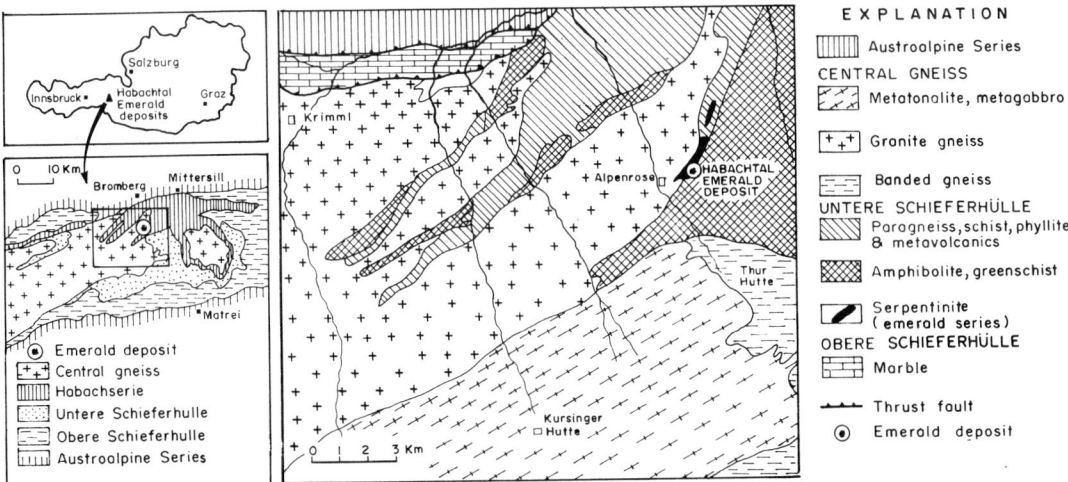

Figure 8.25. Location and geological sketch maps of the Habachtal emerald deposit, Austria, (adapted from Okrusch and others,1981 and Grundmann, 1985).

Emerald Deposit of Austria

Location and regional geological setting

In Austria, emeralds have been known since Roman times in the Habach Valley, which carries a tributary of the Salzach River. The emerald deposits are at Leckbachscharte in the Hohe Tauern hill range at an altitude of 2290 m above sea level, about 89 km southwest of Salzburg (fig. 8.25). The deposit is situated in the Lechbach-Rinne ravine, on the eastern slope of the Habach Valley. Emeralds were sporadically mined from 1860 until 1914.

The Habachtal emeralds are light to dark green in color and largely occur in the ½-cm-size range. They are associated with biotite, oellacherite, tourmaline, quartz, talc, dolomite, pyrite, galena, bertrandite, chrysoberyl, and phenacite.

A wide range of mineral inclusions characterize the Habachtal emeralds. These include biotite, plagioclase, epidote, allanite, clinozoisite, titanite, rutile, quartz, apatite, muscovite, chlorite, talc, tremolite, actinolite, pyrite, pyrrhotite, molybdenite, scheelite, calcite, dolomite, pentlandite, chalcopyrite, magnetite, phenacite, tourmaline, zircon, and ilmenite.

The Habachtal emeralds, often also referred to as Leckbachscharte emeralds, are located in the thrust belt which forms the northern segment of the Alpine thrust and fold belt. The Hohe Tauern hill range, where the emeralds occur, comprises two nappe zones. The Austroalpine nappes form the upper sheet which is underlain by the Pennine nappes. Erosion of the Austroalpine nappes has exposed the rocks of the Pennine nappes and this structure has been commonly referred to as the "Tauern window". The tectonostratigraphic rock sequence from youngest to oldest which is exposed in the Tauern window is as follows (Grundmann, 1985):

3. Obere Schieferhülle   — Permian—Mesozoic
2. Zentrale Gneise Untere   — Late Hercynian
1. Untere Schieferhülle   — Cambrian—Ordovician.

The Zentrale Gneise is exposed in the core of the Tauern window and is formed from late Hercynian granitoids ranging in composition from granites and granodiorites to tonalites. These rocks also have been referred to as "Augen und Flasergneise" and comprise coarse-grained biotite-muscovite-augen gneisses which, according to Morteani (1974), are orthogneisses and are believed to be 240 m.y. old (Jäger and others, 1969).

The Zentrale Gneise is in tectonic contact with the surrounding Untere Schieferhülle. The latter is comprised of two units, the "Penninic Altristallin" which consists of gneisses and migmatites and the "Habachserie" which is largely metasediments interlayered with metavolcanics (greenschists and amphibolites) and serpentinites (Grundmann, 1985; Okrusch and others, 1981).

The Obere Schieferhüle includes Permian to Jurassic metasediments and ophiolites which unconformably cover both the Zentrale Gneise and the Untere Schieferhülle (fig. 8.25).

The emeralds formed at the contact between the "Augen und Flasergneise" and the Habachserie. This is a highly tectonized contact zone characterized by layers and lenses of serpentinite which range in length from a few meters to over 100 meters. They have been metasomatically altered to talc-carbonate, talc-actinolite-tremolite, chlorite, or biotite schists. Serpentinite lenses are enclosed in banded biotite-epidote-plagioclase gneisses interlayered with amphibolite.

Geology of the Habachtal emeralds

The Habachtal emeralds are largely found in biotite schist, chlorite schist, or in tremolite-actinolite-talc schists, which formed at the margins of metasomatically altered serpentinites (Morteani and Grundmann, 1977). This schist zone often has been referred to as a "Blackwall sequence" which was initially highlighted by Fersman (1929) based on his investigations of the Tokowoja emeralds in the Urals. In the Urals, "blackwall sequence" has formed by reaction between beryllium-rich pegmatitic fluids and serpentinite.

From the genesis standpoint the Habachtal emeralds are of considerable interest because pegmatites do not occur at the Habachtal emerald deposits. Okrusch and others (1981) have therefore concluded that in the Tauern window the blackwall assemblages were not formed by the influx of pegmatitic fluids into the ultramafics, but due to "metasomatic exchange between ultramafics and adjacent silicic country rocks during metamorphism". Their work confirmed the earlier views of Morteani and Grundmann (1977) that the beryllium for the Habachtal emeralds was derived from the metavolcanics in the Habachserie through metamorphic processes.

Over the years different views have been expressed regarding the origin of Habachtal emeralds. Some authors believed that they were the product of contact metamorphism (Weinschenk, 1896) others were of the view that they were formed by hydrothermal fluids from aplitic dikes injected into metabasic rocks (Leitmeier, 1937) or that they were formed late during Alpine tectonometamorphism (Höll, 1975). Konigsberger (1913) and more recently Morteani and Grundmann (1977), have suggested that the emeralds are porphyroblasts that were deformed during Alpine metamorphism and that during this process beryllium was derived from the volcanic rocks. They have highlighted the fact that the emeralds are pretectonic or Alpidian but there are no exposed Alpidian granites in the Habach area, which could provide beryllium-bearing hydrothermal fluids. Okrusch and others (1981) have shown that some of the metavolcanic Habachserie rocks in the vicinity of Leckbachscharte emerald deposits contain anomalous amounts of beryllium as indicated below:

Serpentinites	1.8 — 4.2 ppm
Talc schists	1.5 — 5.8 ppm
Talc-actinolite-biotite schist	2.8 — 3.6 ppm
Chlorite-biotite schist and biotite schist	4.7 — 66 ppm

The Alpine metamorphism that affected the Habachtal emeralds is believed to be a two-stage event, a low-grade syntectonic stage and a high-grade post-tectonic stage (Morteani and Rasse, 1974). It is believed that this metamorphism produced a maximum temperature of 500°C (Hoernes and Friedrichsen, 1974). Isotopic age data on biotite indicates the cooling age of rocks to be 25 m.y. (Kreuzer and others, 1973). Based on kyanite-grade metamorphism Morteani and Grundmann (1977) have assumed a 4-6 kbar pressure. Because of the high-grade condition of the metamorphism that affected the pre-tectonic emeralds, it is important to note that critical geologic evidence for the origin of the Habachtal emerald may have been obscured during metamorphism and deformation.

Emerald Deposit of Bulgaria

An emerald deposit, the Mt. Rila emerald deposit, occurs on the flanks of Damga Peak, in the Rila Mountains, near Urdini Lakes, about 50 km south of Sofia (fig. 8.26).

The Rila emeralds are associated with a 20-m-long and 2.5-m-wide pegmatite vein that occurs in the contact zone of biotite gneiss with serpentinized ultrabasic igneous rock.

The pegmatite dike exhibits distinct zoning. From the center of the dike to its periphery the mineral zoning is in the following order—plagioclase, phlogopite, tremolite, chlorite, and talc. The emeralds mainly occur in the plagioclase-phlogopite zones (Mumme, 1982).

According to Mumme, the emerald was formed during desilicification of pegmatite solutions which occurred when solutions rich in Be, Si, Al, Ca and K penetrated into the ultrabasic rocks. The addition of chromium that was derived from the basic igneous rocks resulted in the production of emerald during crystallization of beryl.

Emerald is associated with beryl, feldspar, amphibole, quartz, muscovite, garnet, apatite, zircon, allanite, spinel (?), fuchsite, pyrite, chalcopyrite, molybdenite, bismuthinite, bismutite, calcite, guembelite, and iron hyroxides.

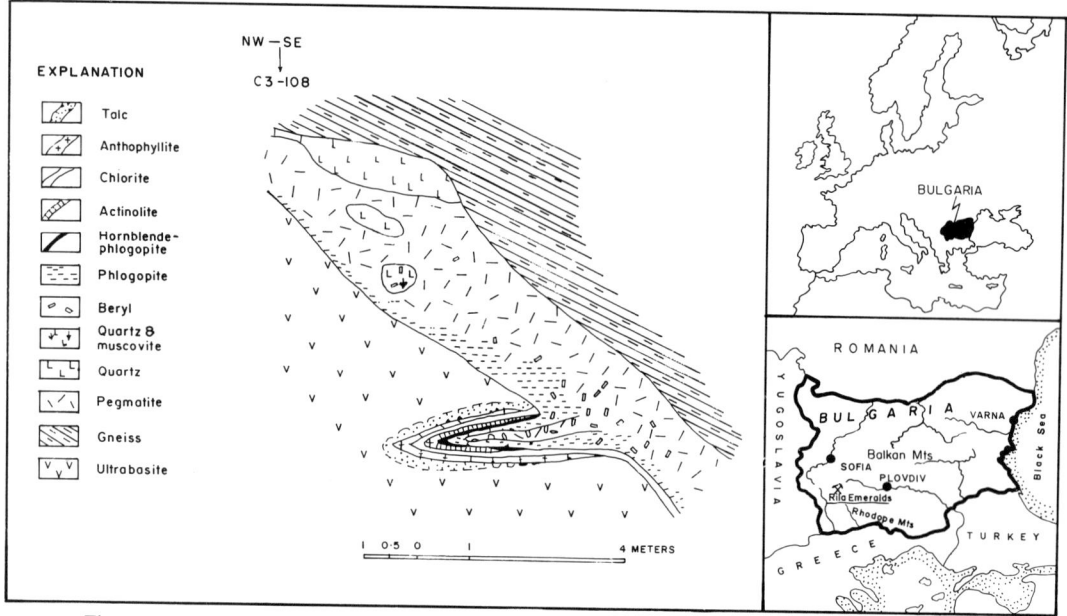

Figure 8.26. Location and geological sketch map of the Rila emerald deposits, Bulgaria, (adapted from Petrusenko and others, 1980).

Emerald deposits of Ukraine, USSR

According to Lavrinenko and others (1971) emeralds were discovered in the Ukranian crystalline shield in the course of exploration for rare metals in the albite granite. This granite contains spodumene, beryl, tantalite, tapiolite, and other minerals.

The emeralds occur in the "exomorphosed" zone of a pegmatite that has intruded mafic and ultramafic rocks.

NORTH AMERICA

Geological Framework

The North American continent comprises a central Archean crystalline massif whose northern part is known as the Canadian Shield whereas its southern part is buried under a thick pile of Phanerozoic sedimentary platform cover which forms the great plains and the interior lowlands (fig. 8.27). This central crystalline zone is bounded to the west by the complex Cordilleran fold and thrust belt which extends from Alaska to Panama and farther southwards links up with the Andean fold belt. The Cordilleran fold belt includes rocks ranging in age from Precambrian to Neogene.

The central crystalline shield is bounded to the east by another north-trending fold and thrust belt. The inner part of the belt is known as the Grenville belt with a reactivation history going back to 1400-1500 m.y. (emplacement of anorthosite), followed by rifting, Grenville orogeny (1100 m.y) and continental suture (Burke and others, 1977). Eastwards the Grenville belt is followed by the Appalachian fold and thrust belt which largely includes rocks ranging in age from Early Paleozoic (Cambrian) to Neogene. This complex belt has gone through

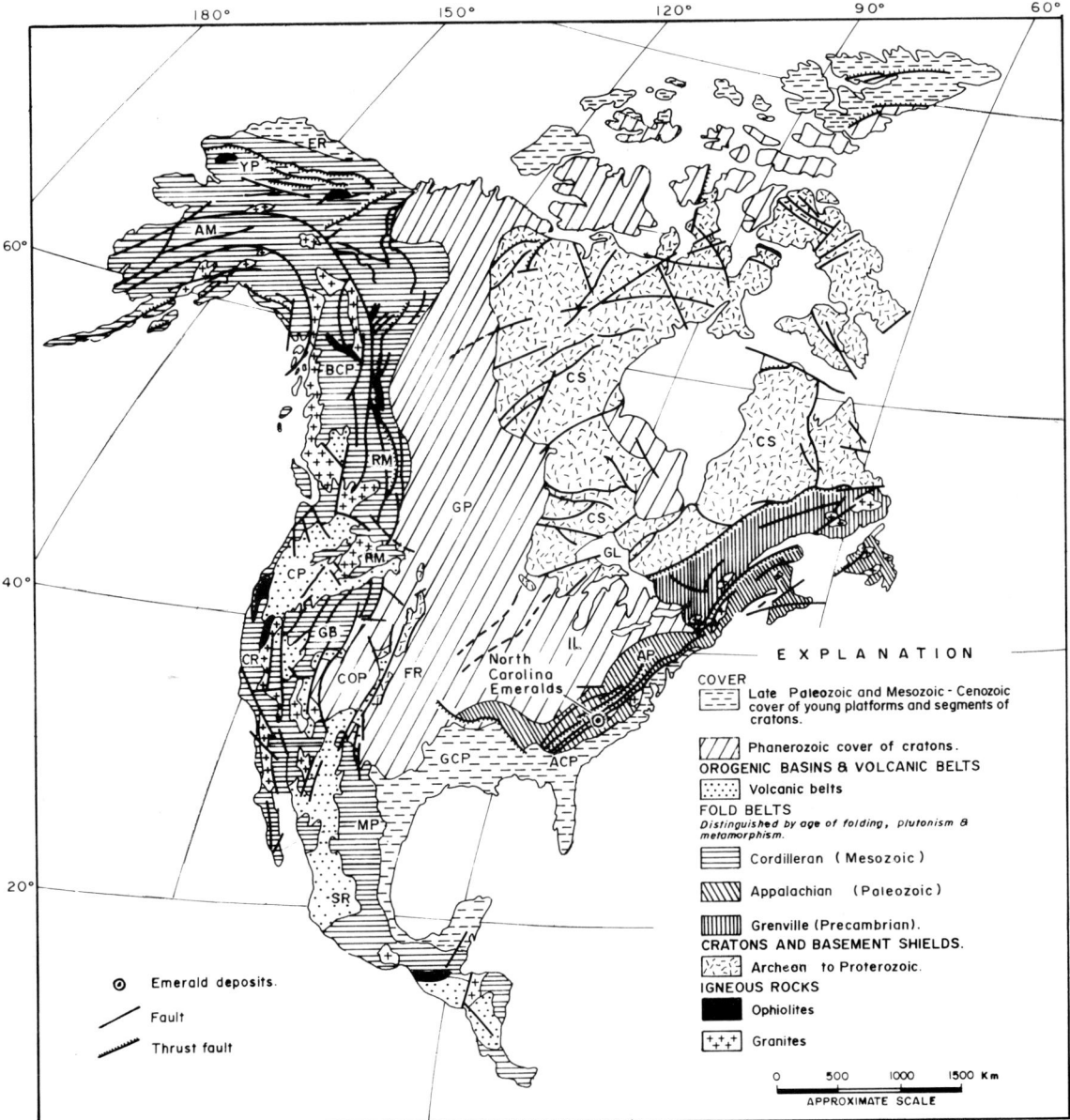

Figure 8.27. Geological sketch map of North America showing location of emerald deposits (geology from Leonov and Khain, 1984). ACP-Atlantic Coastal Plain; AM-Alaska Mts.; AP-Appalachian Range; BCP-British Colombia Plateau; COP-Colorado Plateau; CP-Colombia Plateau; CR-Coastal Ranges; CS-Canadian Shield; ER-Endicott Range; FR-Front Ranges; GB-Great Basins; GCP-Gulf Coastal Plain: GL-Great Lakes; GP-Great Plain (Platform); IL-Interior Lowlands (Platform); MP-Mexican Plateau; RM-Rocky Mts.; SR-Sierra Madre; YP-Yukon Plateau.

numerous periods of deformation and an intricate history of rifting, opening of oceans, continental or arc collision and suturing (Burke and others, 1977).

In the North American continent, emeralds occur in the southern Appalachian and thrust fold belt and are found only in North Carolina, USA. These are briefly described below.

North Carolina Emeralds

Location and geological setting

In North Carolina, emeralds have been mined from the following three localities (Broughton, 1974):

1. Near Spruce Pine on Big Crabtree Mountain, 20 km south of Bakersville in Mitchell County.

2. Near Shelby (8 km to the west) in Cleveland County, near the highly mineralized King's Mountain, adjacent to the South Carolina state line.

3. At Hiddenite, in Alexander County, 8 km southeast of Taylorsville, 20 km northwest of Statesville, on Route 90, at the foot hills of the Blue Ridge Mounains (fig. 8.28).

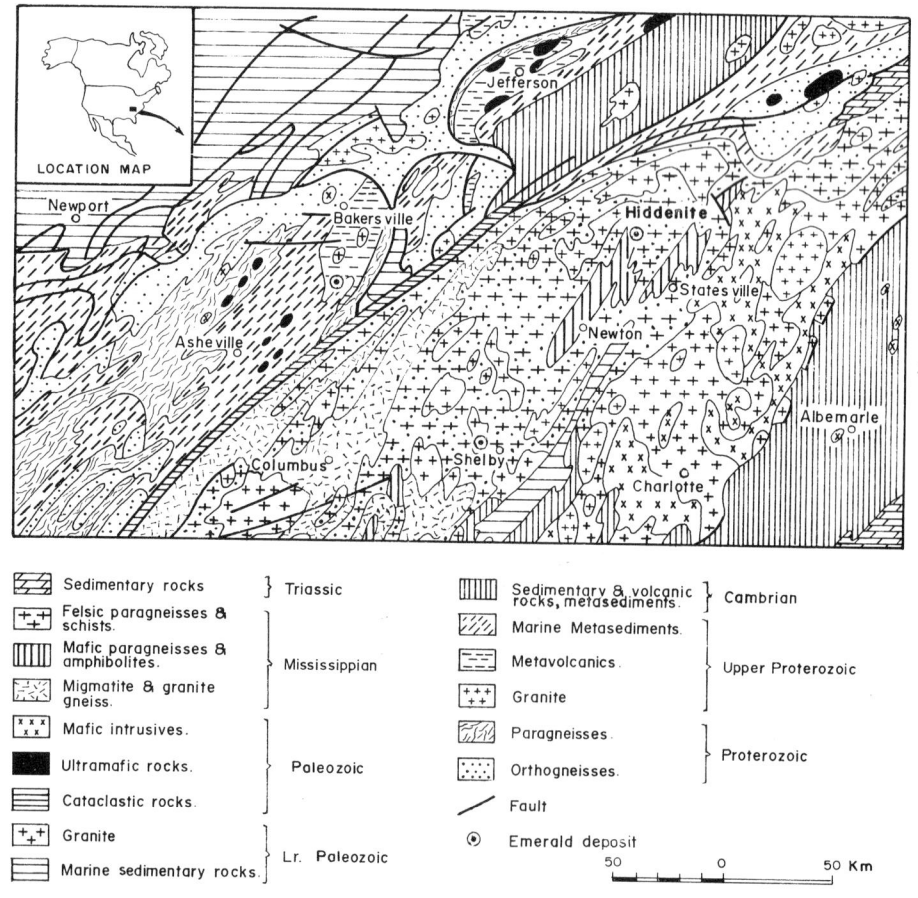

Figure 8.28. Map showing location and geology of North Carolina emeralds, USA (geology from King and Beikman, 1974).

The Spruce Pine deposit occurs in Precambrian metavolcanics which have been intruded by Middle Paleozoic (350-400 m.y.) granitic rocks. The Shelby and Hiddenite deposits formed in felsic to mafic paragneisses and schists of uncertain age but, in this region of the Piedmont province, believed to be of Precambrian age (King and Beikman, 1974; Palache and others, 1930). These schists and gneisses have been intruded by Lower to Middle Paleozoic granitic rocks.

Geology of the North Carolina emerald deposits

Spruce Pine emerald deposit:—This deposit was discovered by J.L. Rorison and D.A. Bownan in 1894 near Bakersville (Kunz, 1894). It has been sporadically mined since then. The emeralds are found in a 5-ft-thick pegmatite vein which is made of coarse-grained quartz, feldspar, garnet, and schorl. The pegmatite vein intruded dark grey to black biotite-muscovite schist and gneiss. The emeralds are best developed at the contact of the pegmatite and the country rock. They become paler or colorless in the pegmatite vein (Broughton, 1974).

Shelby emerald deposit:—This deposit was discovered in 1908 by W.B. Turner who found loose fragments of emeralds in the top soil (Broughton, 1974). At this site emeralds occur in a pegmatite which cuts "hornblende hypersthenite and olivine gabbro". According to Sterrett (1912) both these rocks have been intruded and surrounded by biotite granite.

Hiddenite emerald deposit:—This is the most important of the North Carolina emerald deposits and is believed to have been discovered by W.E. Hidden in 1880 (Hidden, 1883; Broughton, 1974). At this deposit emeralds occur in weathered cavities in pegmatite veins. These strike east-west and form a swarm with veins of "considerable number and variable size". They weather into a red clayey subsoil. Emeralds are found in the decomposed weathered soil as well as in the cavities of the unweathered rock (Broughton, 1974).

The pegmatites have intruded Precambrian greyish colored, intensely crumpled and folded, banded paragneisses which contain quartz, plagioclase, biotite, garnet, zircon, and apatite. According to Palache and others (1930), the protolith of this gneiss was an argillaceous sandstone. They envision the following three stages of pegmatite formation:

1. *Lit-par-lit pegmatite:* This stage followed the metamorphism of the argillaceous sandstone; the pegmatite contains quartz, andesine, orthoclase, microcline, bronzite, tourmaline, apatite, and pyrite.

2. *Hiddenite pegmatite:* The lit-par-lit pegmatite stage was followed by tectonism and renewed metamorphism, shearing of country rock, and injection of the second pegmatitic stage, the Hiddenite pegmatite stage (syntectonic). These pegmatites parallel the gneissose structure and have sharp contacts with the county rock. This stage is characterized by quartz, andesine, microcline, hiddenite, tourmaline, garnet, dumortierite, sillimanite, zircon, biotite, sericite, rutile, apatite, pyrite, and calcite.

3. *Hiddenite "cavities" stage:* The Hiddenite syntectonic stage was followed by shearing and formation of vugs and cavities in which a variety of minerals were formed at relatively lower temperatures through percolating mineralized solution. According to Palache and others (1930), the following suite of minerals has been found in these cavities:

Quartz, amethyst, albite, adularia, hiddenite, holmquistite, beryl, tourmaline, garnet, muscovite, nontronite, rutile, apatite, monazite, pyrite, arsenopyrite, calcite, ankerite, siderite, and aragonite.

Photo. 8.3. Emerald crystals from Hiddenite, North Carolina, U.S.A. *(photo © 198, Harold & Erica Van Pelt).*

Photo. 8.4. Emerald crystal, 11.5x6 cm, 2800 cts, from Tokovaya, Ural Mountains, American Museum of Natural History collection, *(photo © 198, Harold & Erica Van Pelt).*

The Hiddenite deposits are unique for their mineralized "cavities" and in several respects resemble the "Alpine clefts" of the Austrian Tauern Window in their form, structure, habit, and mineral contents.

From the above account of the geology of the deposits it appears that the North Carolina emerald deposits are of the classical schist-pegmatite type wherein older Precambrian mafic/ultramafic chromium-bearing schists have been intruded by younger (Middle Paleozoic) granites; associated with these granites are the later emerald-bearing pegmatites. However, Sinkankas (1982) has pointed out that the beryl-bearing veins at Hiddenite (Hiddenite "cavities" stage of Palache and others) are not pegmatites but hydrothermal veins deposited along fractures in the gneissic country rock. He has also compared them with the Alpine clefts and has suggested that they "possibly represent a unique mode of emerald occurrence".

SOUTH AMERICA

Geological Framework

The eastern portion of the South American continent is primarily Precambrian crystalline basement which is flanked on its western margin by the wide Andean Cordilleran and thrust fold belt. The Precambrian basement includes two main shield complexes, the Guiana shield in the northern part and the Brazilian shield in central and eastern Brazil. These complexes are comprised of Archean to Proterozoic metamorphosed igneous, volcanic, and sedimentary rocks. They are surrounded by a thick platform cover ranging in age from Precambrian to Cenozoic (fig. 8.29).

The western fold and thrust belt that comprises the Andean Cordillera comprises rocks ranging in age from the Paleozoic to Cenozoic, including tectonized blocks of Precambrian crystalline rocks. This belt has a complex history of multiple deformation in the lower Paleozoic, upper Paleozoic, Mesozoic, and Cenozoic.

In South America, emeralds are found in Colombia in the northern part of the fold belt of the Andean Cordillera and in Brazil in the Brazilian Precambrian shield. These are briefly described in the following pages:

Emerald Deposits of Colombia

Location and geological setting

The best and the largest known deposits of emerald in the world undoubtedly occur in Colombia. Along with the Pakistani emeralds, these are also geologically one of the better studied deposits.

The Colombian emerald deposits occur in the fold and thrust belt of the Andean Cordillera in the northern part of Colombia, where the Andes bifurcate into three subparallel north-trending hill ranges, namely (from west to east) the Cordillera Occidental, Cordillera Central, and Cordillera Oriental. The core of these hill ranges are igneous and metamorphic rocks, overlain on their flanks by folded and faulted Cretaceous sedimentary rocks. The western range is largely Mesozoic mafic volcanic rocks, with Cenozoic intrusive rocks of intermediate composition. The central range is mainly Paleozoic to Mesozoic felsic to intermediate (largely granitic) intrusive rocks. The eastern range mostly consists of sedimentary rocks, mainly

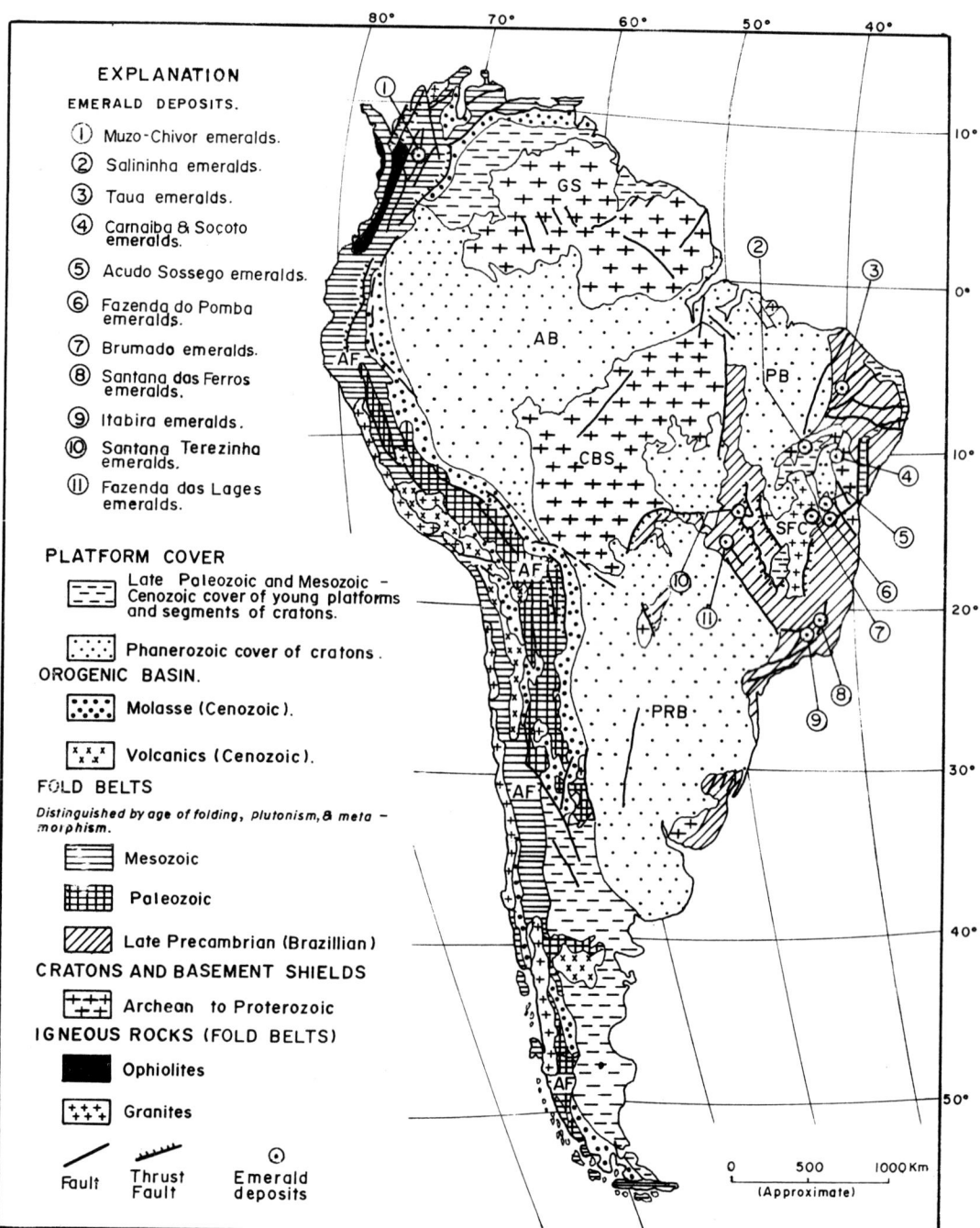

Figure 8.29. Geological sketch map of South America showing location of emerald deposits (geology from Leonov and Khain, 1982). AB-Amazon Basin; AF-Andean fold belt; CBS-Central Brazillian Shield; GS-Guiana Shield; PB-Parnaiba Basin; PRB-Parana Basin; SFC-Sao Francisco Craton.

Photo. 8.5. Chivor emerald mines, Colombia *(photo © 198, Harold & Erica Van Pelt)*.

Photo. 8.6. Emerald crystal from Coscuez, Colombia, 32x22 mm, 120 ct, *Courtesy AMGAD Inc. (photo © 198, Harold & Erica Van Pelt)*.

limestone and shale. There are however, outcrops·of Mesozoic metamorphic and felsic to intermediate intrusive rocks exposed in the northern part of this range (Beus, 1979).

The emerald deposits are confined to the eastern range, the Cordillera Oriental, where they occur in a 50-70-km-wide belt which extends in a northwest-southeast direction for about 200 km. It is situated about 50 km northeast of Bogota (fig. 8.30). From north to south, there are at least twenty mining sites, namely Jesus, Maria, Otanche, Muzo-Coscuez, Briceno, Maripi-Coper, Chiquinquira, Yacopi-Pacho Paime, Susa-Fuquene, Raquira, Suesca-Lenguazque, Supata, Ramiriqui-Sotaquira-Siachoque, Macanal-Campoher Moso, Somondoco-Guateque, Chivor, Almeida, Nemocon-Zipaquira, Buenavista, Ubala-Junin, Vega de San Juan, and Gachala. In recent years, mining has been extended to the southeast and southwest of Bogota where the Guayaketal-Quetame deposit has been found near the headwaters of Rio Meta, 50 km southeast of Bogota and the Andalucia (Valle) deposit has been located in the Rio Cauca valley, about 240 km west of Bogota (Hintze, 1979). However, these deposits may be grouped into two principal emerald-producing districts, namely Muzo and Chivor. The Muzo group of mines are located near the town of Muzo, 105 km north of Bogota, whereas the Chivor group of mines are located about 85 km northeast of Bogota.

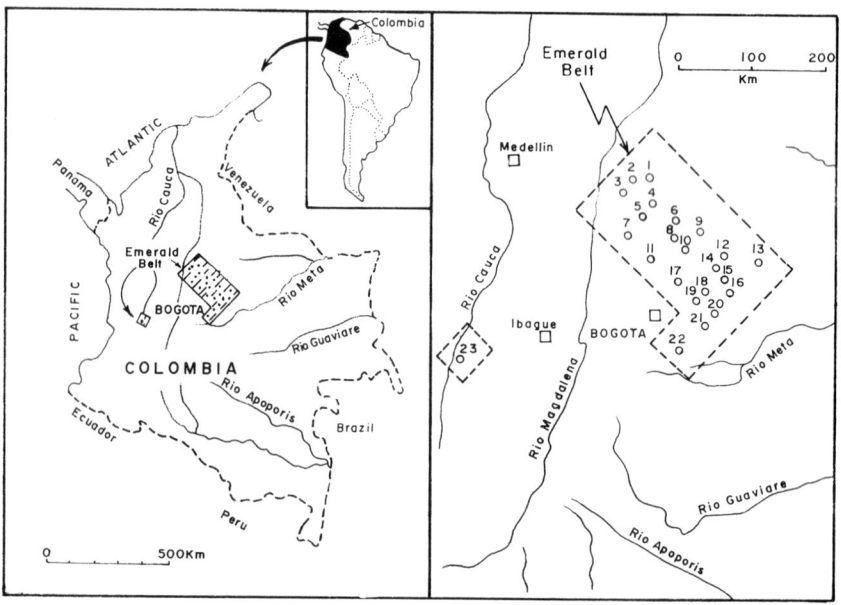

Figure 8.30. Map showing location of Columbian emerald deposits, (adapted from Hintze, 1979). 1. Jesus Maria (Santander); 2. Otanche; 3. Muzo-Coscuez; 4. Briceno; 5. Maripi-Coper; 6. Chiquinquira; 7. Yacopi-Pacho-Paime; 8-Susa-Fuquene; 9. Raquira; 10. Suesca-Lenguazaque; 11. Supata; 12. Ramiriqui-Sotaquira-Siachoque, 13. Macanal-Compoher Moso; 14. Somondoco-Guateque; 15. Chivor; 16. Almeida 17. Nemocon-Zipaquira; 18. Buenavista; 19. Ubala-Junin; 20. Vega de San Juan; 21. Gachala; 22. Guayaketal-Quetame; 23. Andalucia (Valle).

Most of the emerald deposits are in steep, rugged terrain covered with impenetrable tropical rain forest. Mines are located at altitudes ranging from 600 m to 1200 m above sea level. The major streams and their tributaries have a northeasterly or northwesterly orientation. From this fact, as well as the northwest trend of the emerald belt, Beus (1979) has concluded that there

Photo 8.7. Emerald crystal in matrix, from Muzo, Colombia. *(photo © 198, Harold & Erica Van Pelt).*

is a deep-seated northwest regional tectonic zone which intersects the principal structural lineaments of the northeast-trending Cordillera Oriental. This conclusion is supported by the structural analysis of Kutina (1970).

In the Colombian deposits, the emeralds occur in fracture-filling calcite veins which have a north-northeast orientation, parallel to the folding and the main fault system. The emerald deposits are located close to the intersections of these faults with the northwest-striking faults. It has been concluded therefore that the emerald mineralization is structurally related to intersections of northeast- and northwest- striking regional fault zones.

Geology of the Muzo emerald deposits

Sedimentary rocks of Cretaceous age are exposed in the Muzo District. These largely consist of limestone and shale of the Villeta Formation of mid-Cretaceous age. The Villeta Formation has been divided into a lower Cambiado member and an upper Capas esmeraldiferas member (fig. 8.31). These two members are separated by two thin agglomeratic layers of calcite crystals known as Cama and Cenicero (Oppenheim, 1948).

The shales in the Capas esmeraldiferas or the emerald beds are at places highly carbonaceous. They are fractured and have been cut by numerous white fracture-filling calcite veins. The emeralds occur in these calcite veins and are associated with calcite, dolomite, quartz, pyrite, chalcopyrite, albite, fluorite, apatite, and parisite (Keller, 1981).

Geology of the Chivor emerald deposits

The Cretaceous rock sequence at Chivor almost entirely consists of shales with minor limestone and sandstone. These sediments are at least 1000 m thick and are intensely faulted and folded. The emeralds occur in quartz and pyrite veins in blue-gray argillite in the middle part of the section along axes of tight folds. Unlike Muzo there has been no report of emeralds occurring in calcite or dolomite at Chivor (Keller, 1981). The veins run parallel to the bedding of the sediments and emeralds are commonly found where two veins intersect.

There are three parallel pyritic and limonitic "iron bands" in the stratigraphic section, about 50 m apart. According to Keller (1981), these bands appear to control the distribution of emeralds to some extent because emeralds are most abundant underneath the lowest iron band and between the lowest and middle band. Very few emeralds occur above the middle or upper iron bands. It is likely that these bands may have acted as barriers to the upward movement of the mineralizing fluids.

Geochemistry and origin of Colombian emeralds

Geochemical studies of Colombian emeralds by Beus (1979) reveal that the usual relationship between sodium-metasomatism and the concentration of beryllium in hydrothermal processes is maintained even in the unique Colombian environment. In the Colombian deposits the emerald mineralization generally occurs in the most altered facies of the black shales in which the mineralized zones have been metasomatically altered showing strong carbonization and sodium metasomatism. There has been widespread development of calcite and dolomite, followed by sodium plagioclase. Within the tectonic zone the black shales are enriched in CO_2, Ca, Na, Mg, Mn, S, F, and P and depleted in K, Si, and Al. There has been intense hydrothermal leaching of almost all trace elements mainly Li, Mo, Ba, Zn, V, and Cr.

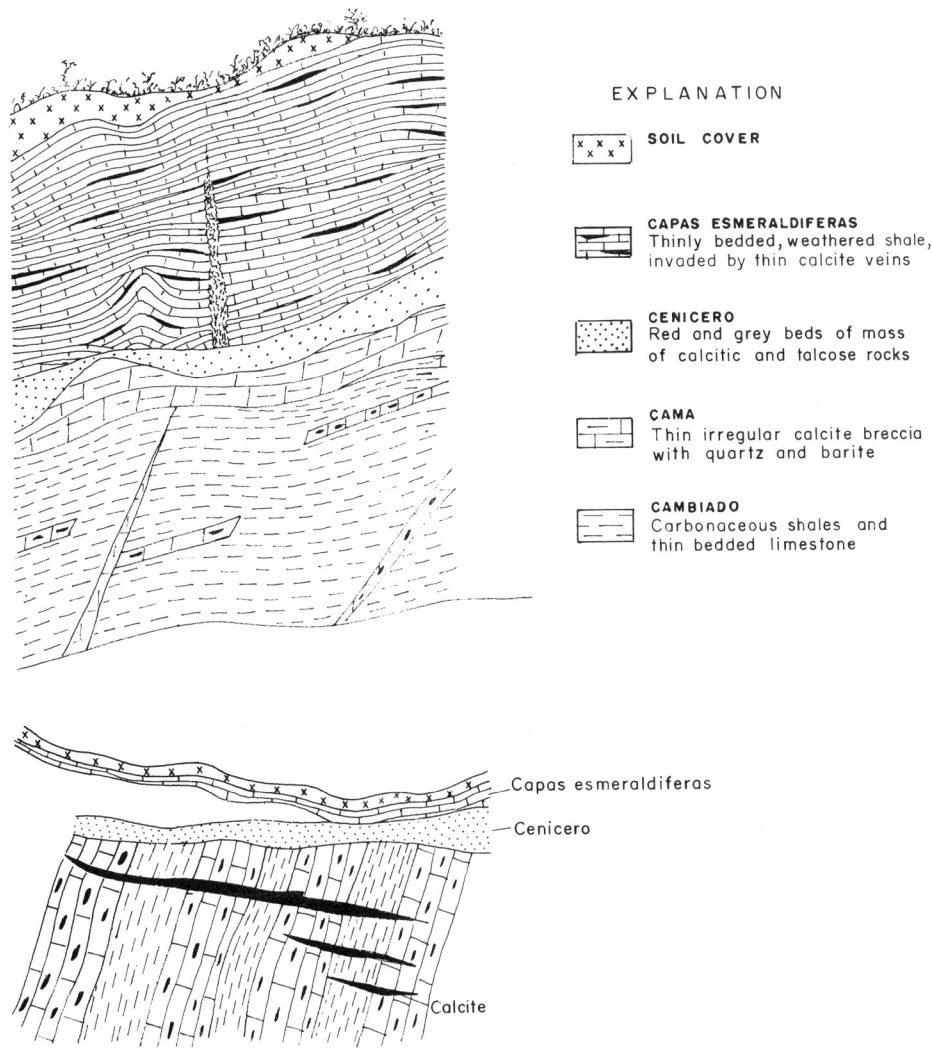

EXPLANATION

☒ SOIL COVER

CAPAS ESMERALDIFERAS
Thinly bedded, weathered shale,
invaded by thin calcite veins

CENICERO
Red and grey beds of mass
of calcitic and talcose rocks

ÇAMA
Thin irregular calcite breccia
with quartz and barite

CAMBIADO
Carbonaceous shales and
thin bedded limestone

Capas esmeraldiferas

Cenicero

Calcite

Cambiado limestone and shale

Figure 8.31. Geological cross sections of Muzo emerald deposit (adapted from Pratt and Texpet, 1979).

As established by Beus (1979), the distribution of sodium and potassium in the altered and unaltered black shales in the Colombian emerald belt and the distribution pattern of the trace elements can serve as a reliable guide for distinguishing mineralized tectonic zones which are otherwise difficult to recognize visually.

According to Beus (1979), low-temperature, sodium-rich carbonate hydrothermal solutions with some chloride ion have deposited the emeralds and associated minerals in the Colombian deposits. These solutions contained beryllium probably in the form of complex carbonate-chloride compounds. Chromium has been extracted from the black shales (average Cr content about 100 ppm) as a result of the interaction of mineralizing solutions with rocks in the fractured tectonic blocks. In this hypothesis the main unresolved problem is to nail down a plausible source for beryllium. So far no magmatic rocks have been found in the emerald belt region. The black shales outside the mineralized zone contain about 4 to 4.5 ppm beryllium. In the mineralized zone there has been leaching of the beryllium and the average content here is about 2 to 3 ppm. Beus has therefore suggested the possibility that beryllium may have been leached by carbon dioxide-rich connate waters from the surrounding rocks followed by its reprecipitation in the form of emerald.

Pratt and Texpet (1979) and Beus (1979) have drawn attention to the presence of a number of salt domes in the emerald-bearing region of Cordillera Oriental and have hinted that these may have been the source of the sodium-rich carbonate hydrothermal solutions responsible for emerald mineralization. More recently, Ottaway and others (1986) suggested that these carbonate-rich solutions were derived from the evaporites and initially contained beryllium along with other elements. They interpret the Cenicero to be the remains of an evaporite which conducted hydrothermal fluids into the Muzo deposit and may have been the source of hypersaline fluids found in the fluid inclusions in emerald, quartz, and parisite. They have suggested that evaporite dissolution either regionally and/or locally mobilized other elements such as Be, Mg, Ca, Ba, F, and LREE.

Different interpretations for the origin of Colombian emeralds have been presented by Renders and Anderson (1987) and Kozlowski and others (1988). From fluid inclusion data, Kozlowski and others (1988) showed that the beryls were formed at high temperatures (probably greater than 470°C). Based on chemical arguments, these authors concluded that the origin of the emeralds involved only chemical components mobilized from the country rocks and no deep-seated magmatic source was necessary. In contrast to Beus (1979), Renders and Anderson (1987) concluded that Colombian emeralds formed at high temperature but beryllium was transported as hydroxy-complexes in hydrothermal fluids to the sites of emeralds deposit.

Emerald Deposits of Brazil

Location and regional geological setting

Though search for emeralds began in Brazil in the 16th century it was not until 1963 that emeralds were commercially mined in Brazil. Emerald mineralization and mining has been reported from five principal regions of Brazil. From north to south (and east to west), these regions are as follows (fig. 8.29):

1. Taua, State of Ceara.
2. Northern part of the State of Bahia, which contains the emerald deposits of Carnaiba and Socota and the Salininha green beryls.
3. Southern part of the State of Bahia, where emeralds occur near Fazenda do Pombo, Acude Sossego, and Brumado.
4. State of Minas Gerais where the Itabira and Santana Dos Ferros deposits are located.
5. State of Goias which contains the more recent emerald finds at Fazenda Das Lages and Santa Terezinha De Goias.

All these emerald deposits are located on the Precambrian Brazilian shield complex. They are however located in Upper Proterozoic to Cambrian (Brazilian-Pan African, 600-500 m.y. old) metamorphic fold and thrust belts which contain metasediments, metavolcanics, and metaultramafic rocks intruded by relatively younger granitic rocks. The Ceara, Bahia, and

Photo 8.8. Emerald crystals in matrix, Brazil, collection Los Angeles County Museum of Natural History. *(photo © 198, Harold & Erica Van Pelt).*

Minas Gerais deposits are located in the eastern (coastal) tectonic belt, whereas the Goias deposits are located in the Central tectonic belt (fig. 8.29).

The eastern part of the Brazilian Shield, south of the Amazon, which contains these two tectonic belts has been involved in several orogenic events (Brazilian-Pan African, 900-550 m.y. ago; Ferreira, 1972). This region contains both reactivated older basement as well as rocks formed during rifting (ocean-opening) and suturing (ocean-closing events) during the Late Proterozoic. The Central tectonic belt which runs from Brasilia to the Amazon Delta, apparently follows a Proterozoic suture line as indicated by a series of dismembered ophiolites that are aligned in a north-south direction for approximately 600 km along the 49th longitude (Burke and others, 1977). Similarly a series of ophiolitic rocks are strung out in a north-south direction in a necklace-bead-fashion along a fault line on the northeastern margin of the large Sao Francisco craton, which lies to the west of Salvador and which is an isolated fragment of the unreactivated Archean basement (fig. 8.29). These features may also represent an ancient suture zone (Jordan, 1972).

Geology of the Taua emerald deposit

The Taua deposit, which is also known as the Boa Esperanca emerald deposit is located north of Campos Sales, in Taua County of the State of Ceara, in northeastern Brazil (fig. 8.29). It is believed to have been discovered in 1954 and was intermittently mined until 1973. The Taua region is largely comprised of Proterozoic metamorphic basement rocks interspersed with Precambrian rocks deformed by the Brazilian-Pan African event. These rock formations have a fan-shaped orientation, with the outcrops and the structures radiating northeastward and southward from the Campos Sales region (fig. 8.29).

The Taua emerald deposit is associated with intrusion of aplite veins and lenticular pegmatitic bodies which intruded Precambrian mylonitized gneisses, amphibolites, migmatites, mica schists, and greenschists. The pegmatites are upto one meter thick and largely consist of quartz, feldspar, apatite, and schorl. The emeralds were formed in pegmatites and in the contact zone between pegmatites and greenschists. The greenschists mainly contain phlogopite-tremolite schist (Sauer, 1982).

Geology of the Carnaiba emerald deposits

The Carnaiba emerald was discovered in 1964 (Jordan, 1972). The deposits occur near the Carnaiba village, 30 km south of the city of Campo Formoso in the northern part of the State of Bahia. The deposits are spread over an area of over 40 sq. km and have been mined extensively since 1965. Hundreds of shafts and adits have been made and abandoned over the 25 years of mining. Currently the main mines are centered at Munde, Carnaiba de Cima, Carnaiba de Baixo, Bode, Lagarto, Gaviao, Jatoba, Queimada Grande, Sambaiba, Marota, and Braulia (fig. 8.32).

The Carnaiba emeralds occur in the northern part of the Archean Tequie' Complex (2600-3000 m.y.) in the northern part of the Sao Francisco craton (fig. 8.32) and are associated with the relatively small Carnaiba granitic pluton. This pluton is located in an antiform within the narrow north-trending Sierra de Jacobina mountain range. The granite intruded an Early Proterozoic volcano-sedimentary complex of the Sierra de Jacobina, which is largely fuchsitic quartzites, pelitic schists, graywackes, amphibolites, and chromite-bearing ultramafic rocks. The structure is highly imbricated and the whole sequence has been thrust over the Archean

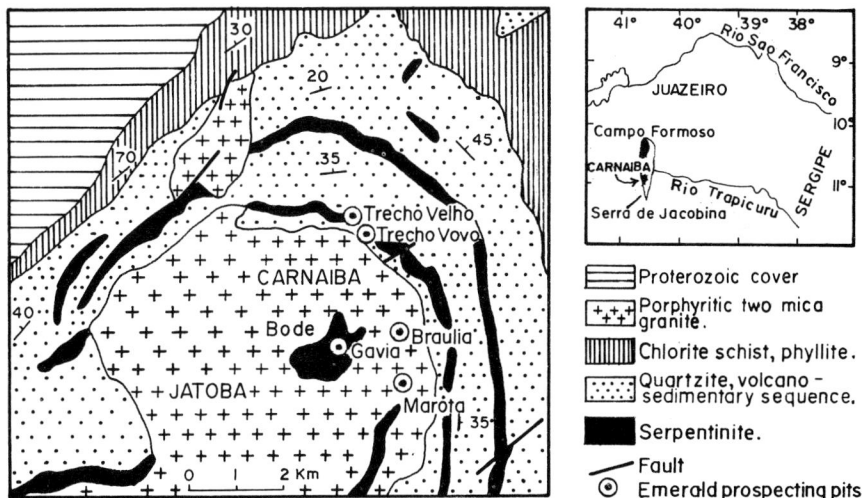

Figure 8.32. Map showing location and geology of the Carnaiba emerald deposits, Brazil (after Rudowski and others 1987).

basement along a décollement localized in the ultrabasic horizon (Rudowski and others, 1987; Couto and Almeida, 1982).

The Carnaiba granite is a porphyritic muscovite-biotite granite. Extensive pegmatite bodies developed from this granite. They are of two kinds, one with potassium feldspar, quartz, muscovite, and garnet and the other with plagioclase, quartz, mica, and tourmaline. The pegmatites also contain emerald, molybdenite, scheelite, alexandrite, phenacite, chalcopyrite, pyrrhotite, pyrite, and fluorite. Light green beryl-emerald occurs in pegmatite, while deeper green emerald occurs in the phlogopite zone of the "biotite schist" at the contact of the pegmatite and the ultrabasic rocks. The emerald-bearing schist largely comprises phlogopite, very litte quartz, apatite, biotite, molybdenite, and trace scheelite. Metasomatic zoning has formed at the contact of the pegmatites and serpentinites and the following sequence is commonly observed—a central plagioclase zone, followed by phlogopite zone, phlogopite-chromite zone, phlogopite zone with talc, chromite and magnetite, and talc-chromite-magnetite zone with some phlogopite (Rudowski and others, 1987; Sauer, 1982).

Geology of the Socoto emerald deposit

The Socoto emerald deposit is one of the new emerald finds in Brazil, and it was discovered in 1983 (Read, 1984). It is located near the Socoto village about 50 km north of the Carnaiba emeralds, approximately 20 km northeast of the city of Campo Formoso in northern Bahia and occurs in the Sierra de Jacobina mountain range (fig. 8.33).

The Socoto emeralds were formed in the same geological terrain as the Carnaiba emeralds and have a similar geological setting. The Socoto deposit lies on the Transamazonian Early Proterozoic Campo Formoso granite. This pluton includes a central biotite-muscovite mesocratic granite surrounded by an outer muscovite-biotite leucogranite. The granites exhibit extensive alteration; alteration phases include muscovite, epidote, potassium feldspar, and tourmaline (Rudowski and others, 1987). The granites intruded Early Proterozoic metasediments and metavolcanics which include amphibolites and ultrabasic rocks and the whole sequence was thrust westwards over an Archean gneissic complex (fig. 8.33).

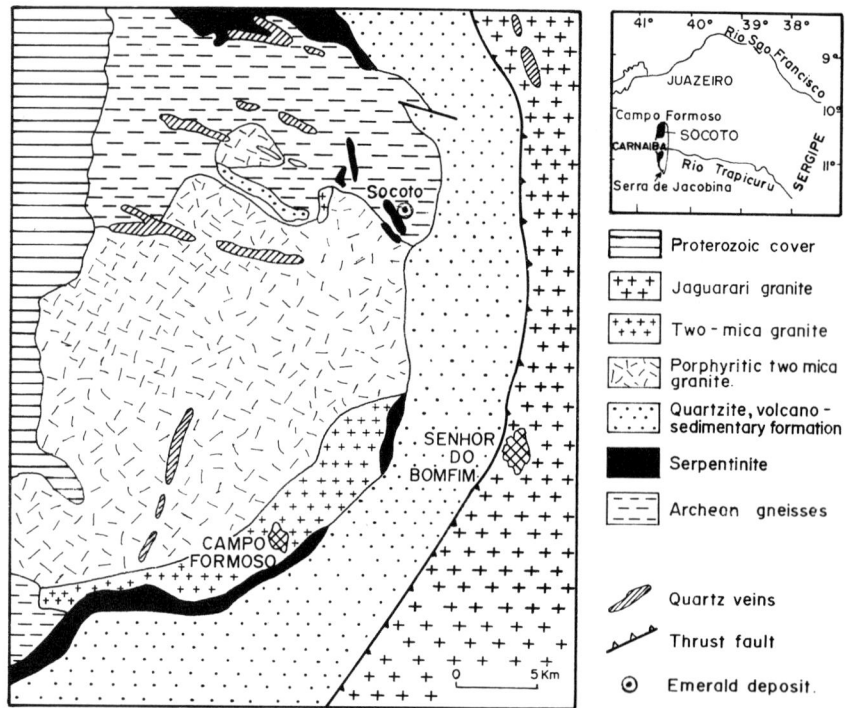

Figure 8.33. Sketch map showing location and geology of the Socoto emerald deposits (adapted from Couto and Almeida, 1982).

Extensive pegmatite dikes and quartz veins are associated with the Campo Formoso granite. The pegmatites largely comprise potassium feldspar, quartz, muscovite, garnet, and tourmaline. These pegmatites have formed similar metasomatic zoning as in the case of Carnaiba pegmatites; likewise at Socoto, the emeralds are largely found in the biotite-phlogopite schist at the contact of pegmatites and country rock, resulting from metasomatism of the ultramafic rocks.

Based on study of inclusions in the Socoto and Carnaiba emeralds, Schwartz (1984, 1986) categorized them with the Ural Mountains type. Rudowski and others (1987) expressed the view that Carnaiba and Socoto emeralds formed through the incorporation of Cr, Fe, and V in the beryl lattice during the percolation of hydrothermal fluids in the serpentinites. They are the product of the interaction between pegmatitic hydrothermal fluids and serpentinites, resulting in metasomatic zoning and formation of the emeralds.

Geology of the Salininha "emerald" deposit

Salininha "emeralds" were probably the very first ones to be commercially exploited. Mining of this deposit started in 1963 (Sauer, 1982). Also known as the Pilao Arcado deposit, it is located at longitude 42°32′ 40″ W and latitude 10°09′11″S, along the Sao Francisco River, near the town of Salininha, midway between Remanso and Xique-Xique (fig. 8.29).

The Salininha deposit occurs in Archean crystalline rocks which are blanketed by post-Triassic platform cover (fig. 8.29). The Archean rocks largely comprise highly folded gneisses, intruded by anatectic granites. Amphibolites, quartzites, mica schists, lenses of soapstone, marble, and mafic and ultramafic rocks are also present (Sauer, 1982).

The "emeralds" occur in a kaolinized pegmatite which intruded gneissic rocks. The wall rock was altered to chlorite-talc-carbonate schist. The "emeralds" are commonly found in the talc schist along the contact zone. Despite the fact that like other Brazilian emeralds, the Salininha emeralds are also associated with pegmatitic intrusions in mafic/ultramafic rocks, their green color is mainly due to vanadium (0.15%) because their chromium content is only in traces (0.0003%, Sinkankas, 1981). They are therefore not emeralds but green beryls.

Geology of the South Bahia emerald deposits

Three relatively small and economically insignificant emerald deposits occur in the southern part of the state of Bahia located near the town of Brumado and are commonly known as the Brumado, Acude Sossego, and Fazenda do Pombo emerald deposits. These deposits are associated with the Proterozoic metamorphic rocks south of the Sao Francisco craton (fig. 8.34).

Brumado emerald deposit:—This deposit is located 16 km west of Brumado town, in the Sierra das Eguas mountains in a region that is predominantly Precambrian gneisses and schists, which were intruded by granites and amphibolite dikes. The crystalline sequence is covered by dolomites and quartzites. The area is better known for its magnesite deposit.

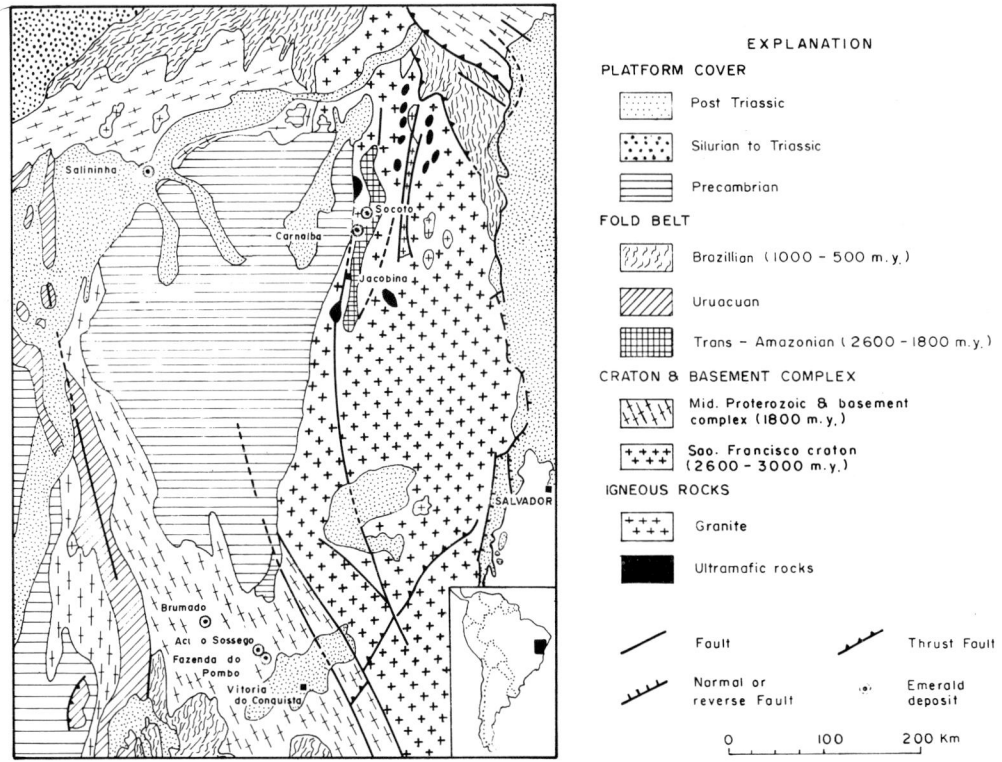

Figure 8.34. Map showing location and geology of the south Bahia emerald deposits, Brazil (geology from Almeida, 1978).

The Brumado emeralds occur in presumably Early Paleozoic altered dolomitic marble, which overlies the Precambrian gneisses and schists. The marble is intruded by amphibolite and rhyolite and contains extensive sheets of talc. The emeralds are primarily found in the geodes and cavities in the talcose rock and are associated with tourmaline and topaz. The talcose rocks and the cavities therein also contain kyanite, phlogopite, rutile, monazite, xenotime, spodumene, albite, lepidolite, magnesite, specular hematite, kyanite, beryl, and dolomite.

Acudo Sossego emerald deposit:—This deposit is located about 47 km northwest of the city of Vitoria da Conquista, in the Anage county (fig. 8,34). It is also known as Madureira, Nozinho or Segredo deposit. It occurs in the same geotectonic setting as the Brumado emerald. However, instead of being associated with dolomitic marbles, the Acudo Sossego emeralds are found in Precambrian mica schists intruded by small pegmatite veins. In this region, highly micaceous soft schistose rocks (metaultramafic?) have been intruded by granitic rocks. According to Sauer (1982) this deposit is also of the contact metamorphic type.

Fazenda do Pombo emerald deposit:—This deposit is situated about 3 km away from the Acude Sossego deposit and about 44 km northwest of Vitoria da Conquista (Sauer, 1982). The deposit is located on a dome-shaped outcrop of granite gneiss which is intercalated with layers of amphibolite and mica schist which were intruded by pegmatite veins that are upto 5 meters thick. The emeralds occur in the contact zone between the pegmatites and the decomposed greenschists which are the relics of altered ultramafic rocks.

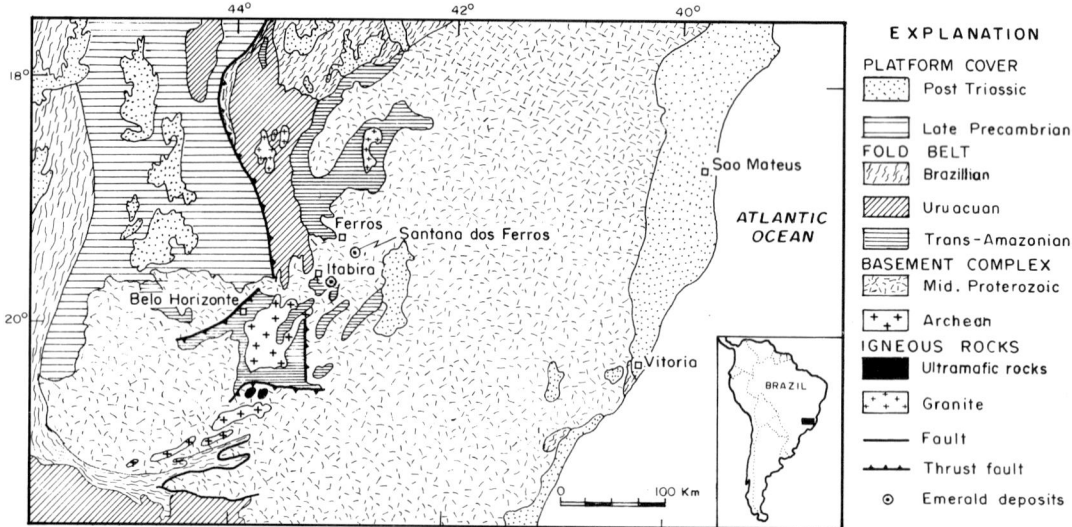

Figure 8.35. Map showing location and geology of Minas Gerais emerald deposits, Brazil (geology from Almeida, 1978).

Geology of the Minas Gerais emerald deposit

In the State of Minas Gerais emeralds occur at two principal localities namely Itabira and Santana dos Ferros, northeast of Belo Horizonte (fig. 8.35). The occurrence of emeralds at Santana dos Ferros was reported in 1922, while the Itabira deposits were discovered in 1978 (Bastos, 1981). These deposits are located in the lower to middle Precambrian metamorphic rocks of the east Brazilian fold belt.

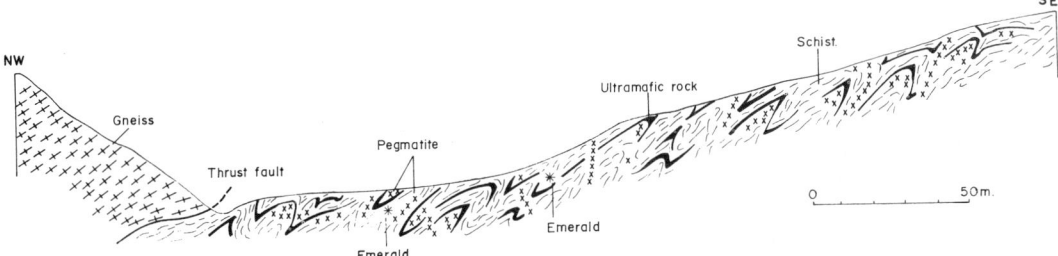

Figure 8.36. Geological cross-section of the Itabira emerald mine area, Brazil (after Sauer, 1982).

Itabira emerald deposit:—This deposit is located about 100 km northeast of Belo Horizonte and 18 km southeast of the town of Itabira (fig. 8.35). It is also referred to as the Oliveira Castro mine (Bastos, 1981) and was discovered in 1978.

The Itabira region largely comprises gneisses, schists, and quartzites. The gneiss crops out north and west of the mine and probably has been thrust southeastward over a tightly folded sequence of biotite-chlorite schists interlayered with altered and metamorphosed mafic and ultramafic rocks (fig. 8.36). The schists and the metamafic/ultramafic rocks have been invaded by concordant layers and veins of quartz and have been intruded by pegmatite veins which are more abundant near the contact with gneissic rocks (Sauer, 1982). The pegmatites are largely quartz and kaolin and apart from emeralds they also contain garnet and kyanite (Mumme, 1982). Emeralds are however largely found in pinkish, altered biotite schist or biotite-chlorite schist which was intruded or rather intensely permeated by small pegmatite veins.

Santana dos Ferros emerald deposit:—This deposit is located about 50 km northeast of Itabira, 20 km southeast of Ferros and 2 km from the village of Esmeraldas (fig. 8.35). The deposit has been known since 1920 and owing to the relatively low quality of the emeralds it has been mined very superficially and intermittently (Sauer, 1982).

The geological setting of the Santana dos Ferros deposit is similar to the Itabira deposit. The Precambrian gneissic country rock contains large bodies of a biotite-vermiculite schist which has been intruded by small pegmatite lenses (Da Cunha, 1961). The pegmatite is largely comprised of kaolin and quartz and it contains beryl, goshenite, and some emerald. According to Sauer (1982), the emeralds are always imbedded in quartz or a biotite-quartz mass.

Geology of the Emerald deposits of the State of Goias

The emerald deposits of the State of Goias are located in the Central Brazilian fold belt. This fold belt was multiply deformed during Precambrian Uruacuan (1000 m.y.) orogeny and Brazilian (600-500) folding and magmatism (fig. 8.37). Elements of the Lower and Middle Precambrian crystalline basement (largely gneisses) and Precambrian metasediments and ophiolites are present. In this belt emeralds have been found at three localities namely at Fazenda das Lages, Serra Dourada, and Santa Terezinha de Goias. These are briefly described below.

Fazenda das Lages emerald deposit:—This is the earliest known emerald deposit in the state of Goias and was discovered by Senator Antopio Ramos Gaiado in 1920. The deposit is located 14 km southeast from the Goias city on the left bank of Bugre stream. This area is covered with weathered intercalations of talc schist, chlorite schists, actinolite schist, and quartzite of Precambrian age. The emerald mineralization is found in "an eluvial concentration of clay and clastic sediments of angular altered schists, limonite after pyrite cubes, and limonite concentration" (Sauer, 1982). The washed material from this weathered rock contains abundant hematite, limonite, chlorite, and microcline. The paragneisses suggest the likelihood that the Fazenda das Lages is a typical "schist type" emerald deposit where beryllium-bearing pegmatites have intruded chromium-bearing metamafic/ultramafic rocks resulting in the formation of emeralds.

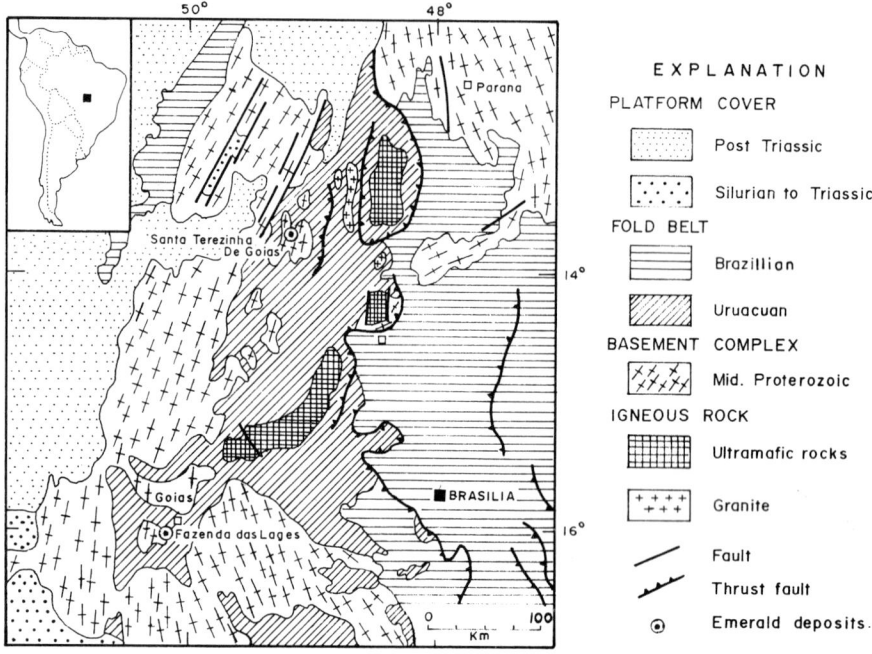

Figure 8.37. Map showing location and geology of the emerald deposits of Goias, Brazil (geology from Almeida, 1978).

Serra Dourada emerald deposit:—This deposit was discovered in 1974 in the Serra Dourada area but it turned out to be a small noncommercial deposit.

Santa Terezinha de Goias emerald deposit:—This deposit is located near the village of Santa Terezinha de Goias, about 25 km north of the Santa Terezinha village, about 310 km north of the city of Goiania. The deposit was discovered in 1981 and it has proved to be very productive (Hanni and Kerez, 1983). The Santa Terezinha area is blanketed by a lateritic cover, under which a sequence of highly altered sericite schist, talc schist, hematite schist, and quartzite is found. Sauer (1982) has observed that there is only one exposed pegmatite vein in the area which cuts the schistose sequence and that the emerald mineralization is confined to only one schistose layer. This mineralized schist is located at a relatively large distance from the pegmatite. Large amounts of emerald crystals are found randomly disseminated in the schist which is about 50 times richer in its emerald content than the mineralized schists at the Itabira deposit.

REFERENCES

Almeida, F.F.M.de, 1978, Tectonic map of South America (1:5,000,000): Published by Ministry of Mines and Energy, Brazil, in collaboration with UNESCO and Commission for the Geological Map of the World.

Anderson, S.M., 1976a, A Note on the occurrence of emerald at Mayfield farm, Fort Victoria, Rhodesia: Journal of Gemmology, v. 15, p. 80-82.

Anderson, S.M., 1976b, Notes on the occurrence and mineralogy of emeralds in Rhodesia: Rhodesia Southern, Geologic Survey, Annals, v. 2, p. 50-55.

Anderson, S.M., 1978, Notes on the occurrence and mineralogy of emeralds in Rhodesia: Journal of Gemmology, v. 16, p. 177-185.

Anhaeusser, C.R., 1976, Archean metallogeny in South Africa: Economic Geology, v. 71, p. 16-43.

Bank, H., 1974, The emerald occurrence of Miku, Zambia: Journal of Gemmology, v. 14, p. 8-15.

Barrie, J., and Kalix, Z., 1959, Gemstones: Summary Report of the Australian Bureau of Mineral Resources, Geology & Geophysics, no. 43 (Canberra), 42 p.

Barros, J.C., 1985, The chemistry of emeralds and associated beryls in Porangatu Deposit, Brazil: Memorias VI Congreso Latinoamericano de Geologia, Bagota, Colombia, v. 6, no. 3, p. 332-341.

Bastos, F.M., 1981, Emeralds from Itabira, Minas Gerais, Brazil: Lapidary Journal, v. 35, p. 1842-1848.

Behier, J., 1957, Minerals de Provincia de Mocambique: Provincia Mocambique, Services de Industria-e-Geologia Boletim, v. 22, p. 135.

Beus, A.A., 1965, *Geochemistry of Beryllium and Genetic Types of Beryllium Deposits:* W.H. Freeman and Company, San Francisco, California, p. 401.

Beus, A.A., 1979, Sodium: a geochemical indicator of emerald mineralization in the Cordillera Oriental, Colombia: Journal of Geochemical Exploration, v. 11, p. 195-208.

Broughton, P.L., 1974, Emerald deposits of western North Carolina: Earth Science, v. 27, p. 222-228.

Broughton, P.L., 1979, Economic geology of Australian gemstone deposits: Minerals Sciences and Engineering (Johannesburg), v. 11, p. 3-20.

Brown, G., 1984, Australia's first emeralds: Journal of Gemmology, v. 19, p. 320-335.

Burke, K., Dewey, J.F., and Kidd, W.S.F., 1977, World distribution of sutures—the sites of former oceans: Tectonophysics, v. 40, p. 69-99.

Cooper, D.G., 1964, The geology of the Bikita pegmatite: *in* Houghton, S.H., (editor), The geology of some ore deposits in Southern Africa: Geological Society of South Africa, v. 2, p. 441-461.

Couto, P., and Almeida, J.J., 1982, Geologica e mineralizacao na area do garimpo de Carnaiba (Bahia): Sociedade Brasileira de Geologia, Anais do Congresso 32, Salvador, Bahia, v. 3, p. 850-861.

Da Cunha, O.L., 1961, Emeraldas da Fazenda do Sossego, Santana de Ferros, Minas Gerais, Brasil: Gemologia, v. 25, p. 9-14.

Draper, T., 1963, A new source of emeralds in Brazil: Gems and Gemology, v. 11, p. 111-113.

Ferreira, E.O., 1972, Tectonic map of Brazil, explanatory note: v. 1, Departamento Nacional Da Producao Mineral, Boletim 1, p. 1-19.

Fersman, A.E., 1923, Izumrudnye Kopi na Urale Sbornik statei i Materialov 46 Materialy dlya izucheniya estestvennykh proizvoditelinykh sil Rossi: Komisseiy Pri Rossisskoi Akademi Nauk, Petrograd (St. Petersburg), 82 p.

Fersman, A.E., 1929, Geochemische Migration der Elemente, III, Smaragdgruben im Uralgebirge: Abhandl. Prakt. Geologie und Bergwirtschaftslehre, v. 18, p. 74-116.

Garstone, J.D., 1981, The geological setting and origin of emerald deposits at Menzies, Western Australia: Journal of Royal Society, Western Australia, v. 64, pt. 2, p. 53-64.

Gemuts, I., and Theron, A., 1973. The stratigraphy and mineralization of the Archaean between Norseman and Coolgardie, Western Australia: ANZAAS 45th Congress, Abstracts, v. 3, p. 88.

Ghera, A., and Lucchesi, S., 1987, An unusual vanadium-beryl from Kenya: Neues Jahrbuch für Mineralogie, Monatschefte, v. 6, p. 263-274.

Glikson, A.Y., and Lambert, I.B., 1976, Vertical zonation and the petrogenesis of the early Precambrian crust in Western Australia: Tectonophysics, v. 30, p. 55-89.

Graindorge, J.M., 1974, A gemmological study of emerald from Poona, W.A.: Astralian. Gemmologist, v. 12, p. 75-80.

Graziani, G., Gübelin, E., and Lucchesi, S., 1979, Einschlüsse und Genese cines Vanadium beryls von Salininha, Bahia, Brasilien: Deutschen Gemmologischen Gesellschaft, Zeitschrift, v. 28, p. 134-145.

Grundmann, G., 1985, Die Mineralien des Smaragdvorkommens im Habachtal: Lapis, Austria, v. 10, p. 13-33.

Grundmann, G., 1985, Emeralds from Sandawana: Journal of Gemmology, v. 6, p. 340-354, Also in Gems and Gemology, v. 9, p. 195-203.

Gübelin, E.J., 1960, Emerald from Sandawana: Gemologist, 29, p. 8-16.

Gübelin, E.J., 1974, The emerald deposit at Lake Manyara: Lapidary Journal, v. 28, p. 338-344, 346-347, 359-360.

Hanni, H.A., and Kerez, C.J., 1983, Neues vom Smaragdvorkommen von Santa Terezinha de Goias, Goias, Brasilien: Deutshen Gemmologischen Gesellschaft, Zeitschrift, v. 32, p. 50-80.

Hanni, H.A., and Klein, H.H., 1982, Ein Smaragdvorkommen in Madagaskar: Deutschen Gemmologischen Gesellschaft, Zeitschrift, v. 21, p. 71-77.

Hanni, H.A., and Klein, H.H., 1983, Un gisement d'emeraudes a Madagascar: Univ. Vale, inst. mineral, Basel 4056, CHE, Revue de Gemmologie, v., 74, p. 3-5.

Hickman, A.C.J., 1972, The Miku emerald deposit: Economic Report no. 27, Geological Survey of Zambia, Lusaka, p. 1-35.

Hickman, A.H., 1980, Archaean geology of the Pilbara Block: International Archaean Symposium, 2nd, Perth, Western Australia, Excursion Guide, p. 55.

Hidden, W.E., 1883, The discovery of emeralds in North Carolina, U.S. Geological Survey, Mineral Research (1882), p. 500-502.

Hintze, J., 1979, The emerald deposits of Colombia: Aufschluss, v. 30, p. 83-92.

Hoernes, S., and Friedrichsen, H., 1974, Sauerstoffisotopen-Untersuchungen an metamorphen Gesteinen der Westlichen Hohen Tauern (Osterreich): Schweizerische Mineralogische und Petrographische Mitteilungen, v. 54, p. 769-788.

Höll, R., 1975, Die Scheelitlagerstatte Felbertal und der Vergleich mit anderen Scheelitvorkommen in den Ostalpen: Akademie der Wissenschaften, Munich, Mathematisch—Naturwissenschaftliche Klasse, Abhandlungen, v. 154, p. 114.

Holms, A, 1951, The sequence of Pre-Cambrian orogenic belts in south and central Africa: International Geological Congress v. 14, p. 254-269.

Jäger, E., Karl, F., and Schmidegg, O., 1969, Rubidium-Strontium-Altersbestimmungen an Biotit— Muskovit-Granitgneisen (Typus Augen— und Flasergneise) aus dem nordlichen Grobvenedigerbereich (Hohe Tauern), Tschermaks Mineralogische und Petrographische Mitteilungen, v. 13, p. 251-272.

Jordan, H., 1972, Die Minas—Gruppe in Nordost Bahia, Geologische Rundschau v. 61, p. 441.

Kafarskiy, A.K., Ghmyriov, V.M., Dronov, V.I., Stazhilo-Alekseev, K.F., Abdullah, J., and Seikovskiy, V.S. 1976, Geological map of Afghanistan, Geological Survey of India, Miscellaneous Publication no. 41, Himalayan Geology Seminar, New Delhi, 1976.

Keller, P.C., 1981, Emeralds of Colombia, Gems & Gemology, p. 80-92.

King, P.B., and Beikman, H.M., 1974, Geologic map of the United States, (1:2,500,000), U.S. Geological Survey.

Konigsberger, A., 1913, Versuch einer Einteilung der Ostalpinen Minerallagerstatten: Zeitschrift für Kristallographie, v. 52, p. 151-174.

Kozlowski, A., Metz., P., and Jaramillo, H.A.E., 1988, Emeralds from Somondoco, Colombia: Chemical composition, fluid inclusions and origin: Neues Jahrbuch für Mineralogie, Abhandlungen, v. 159, p. 23-49.

Kreuzer, H., Harre, W., Muller, P., Rasse, P., and Raith, M., 1973, K/Ar mineral ages for the Venediger area and the eastern Zillerthal Alps (Hohe Tauern): Meeting Geotraverse 1A, Trento.

Kriewaldt, M., 1970, Explanatory notes to 1:250,000 Geological Series Maps: Sheets SH/51-5, Menzies.

Kunz, G.F., 1894, A new locality of true emerald (Bakersville, N.C.): American Journal of Science, v. 3, p. 429-430.

Kutina, J., 1970, Structural analysis of ore genesis in the Caribbean Islands Arc, the North Coast of South America and East Coast of Central America: United Nations, New York, (unpublished report).

Lacroix, A., 1922, Mineralogie de Madagascar: Tome III, Paris, Societe d'editions Geographiques, Maritimes et Coloniales.

Lavrinenko, L.F., Levenshteyn, M.L., Polunovskiy, R.M., Rozanov, K.I., and Rozenberg, D.Sh., 1971, Nakhoda izumruda via Ukraine: Mineralogiya Heskii Sbornik, no. 25, pt. 1, p. 85-87.

Le Grange, J.G., 1930, The Barbara beryls: a study of an occurrence of emeralds in the northeastern Transvaal with some observations on metallogenic zoning in the Murchison Range: Transactions of Geological Society, South Africa, v. 32, p. 1-25.

Leitmeier, H., 1937, Das Smaragdvorkommen im Habachtal in Salzburg und Seine Mineralien: Zeitschrift für Kristallographie, Mineralogie, und Petrographie, Abteilung A., v. 49, p. 245-368.

Leonov, Y.G., and Khain, V.E., 1982, Tectonic map of the World, (1:45,000,000), Commission for the Geologic Map of the World, Cartographic Group of the Commission for International Tectonic Maps of the USSR Academy of Sciences.

Martin, H.J., 1962, Some observations on Southern Rhodesian emeralds and chrysoberyl: Chamber of Mines Journal 4, no. 10, p. 34-38.

Morteani, G., 1974, Petrology of Tauern Window, Austrian Alps, Excursion B 9., Fortschritte der mineralogie, Beiheft, 52, v. 1, p. 195-220.

Morteani, G., and Grundmann, G., 1977, The emerald porphyroblasts in the penninic rocks of central Tauern Window: Neues Jahrbuch für Minerallogie, Monatschefte, v. 11, p. 509-516.

Morteani, G., and Rasse, P., 1974, Metamorphic plagioclase crystallisation and zones of equal anorthite content in epidote—bearing, amphibole free rocks of western Tauern Fenster, eastern Alps: Lithos, v. 7, p. 101-111.

Muhling, P.C. and de Laeter, J.R., 1971, Ages of granitic rocks in the Poona-Dalgaranga area of Yilgarn Block, Western Australia: Geological Society of Australia, Special Publication, no. 3, p. 25-31.

Mumme, I.A., 1982, *The Emerald, its occurrence, discrimination and valuation:* Mumme Publications, 46 Turriell Point Road, Port Hacking NSW, Australia, 135 p.

Neves, J.M.C. 1978, Estudo de ocorrencias de esmeralda na provincia da Zambezia—Mocambique: Sociedade Brasileira de Geologia, Anais do Congresso, 30, v. 3, p. 1141-1155.

Nisbet, E.G., and Chinner, G.A., 1981, Controls of the eruption of mafic and ultramafic lavas, Ruth Well Ni-Cu Prospect, West Pilbara: Economic Geology, v. 76, no. 6, p. 1729-1735.

Okrusch, M., Richter, P. and Gurkan, A., 1981, Geochemistry of blackwall sequences in the Habachtal emerald deposits, Hohe Tauern, Austria, part 1: Presentation of Geochemical Data: Tschermaks Mineralogische und Petrographische Mitteilungen, v. 29, p. 9-13.

Oppenheim, V., 1948, The Muzo emerald zone, Colombia, S.A: Economic Geology, v. 43, p. 31-38.

Ottaway, T.L., Wicks, F.J., Bryndzia, L.T., and Spooner, E.T.C., 1986, Characteristics and origin of Muzo emerald deposit; Colombia International Mineralogical Association, Abstracts with Programs, Stanford, California, p. 193.

Palache, C., Davidson, S.C., and Goranson, E.A., 1930, The Hiddenite deposit in Alexander County, North Carolina: American Mineralogist, v. 15, p. 280-302.

Palfreyman, W.D., Addario, G.W.D., Swoboda, R.A., Bultitude, J.M., and Lamberts, I.T., 1976, Geology of Australia: Bureau of Mineral Resources, Geology and Geophysics, Canberra, A.C.T., Australia (Geological Map 1:2,500,000).

Petrusenko, S. and Arnaudov, V., 1980, Emeralds from desilicified pegmatites in Bulgaria, Gem Minerals, Proceedings of the 11th General Meeting of International Mineralogical Association, Novosibirsk, 4-10 Sept, 1978, p. 74-79.

Pittman, E.F., 1901, The mineral resources of New South Wales: Geological Survey of New South Wales, Australia, 487. p.

Platt, J.P., 1980, Archaean greenstone belts: a structural test of tectonic hypotheses: Tectonophysics, v. 65, p. 127-150.

Pohl, W., Horkel, A., Neubauer, W., Niedermayer, G., Okelo, R.E., Wachira, J.K., and Werneek, W., 1980, Notes on the geology and mineral resources of the Mtito Andei-Taita Area (Southern Kenya): Oesterreichische Geologische Gesellschaft, Mitteilungen, v. 73, p. 135-152.

Pratt, S., and Texpet, Col., 1979, The Muzo emerald mine, Geological field trips Colombia, 1959-1978: Colombian Society for Petrology, Geology and Geophysics, p. 33-63.

Prochaska, W., and Pohl. W., 1983, Petrochemistry of some mafic and ultramafic rocks from the Mozambique Belt, North Tanzania: Journal of African Earth Sciences, v. 1, p. 183-191.

Read, P., 1984, Gem studies in Brazil: Australian Gem and Treasure Hunter, v. 92, p. 8-11.

Renders, P.J., and Anderson, G.M., 1987, Solubility of kaolinite and beryl to 573K: Applied Geochemistry, v. 2, p. 193-204.

Rolff, P.A.M., 1970, The geological environment for emerald in Brazil: Lapidary Journal, v. 23, p. 1488-1490, 1492-1502.

Rossovskiy, L.N., 1980. Gemstone deposits of Afghanistan (in Russian). Geologikila Rudnykh Mestorozhdenift, v. 3, p. 74-88.

Roy, B.C., 1955, Emerald deposits in Mewar and Ajmer-Merwara: Geological Survey of India, Records, v. 86, pt. 2, p. 377-401.

Rudowski, L., Giuliani, G., and Sabate, P., 1987, Metallogenie les phlogopitites a emeraude au voisinage des granites de Campo Formoso et Carnaiba (Bahia, Bresil): un exemple de mineralisation proterozoique Be, Mo et W dans des ultrabasites metasomatise es: Academic des Sciences Paris, Comptes Rendus, v. 304, sr. II, no. 18, p. 1129-1134.

Sauer, D.A., 1982, Emeralds from Brazil: Proceedings of the International Gemological Symposium: Eash, D.M., editor, Gemological Institute of America, Santa Monica, p. 357-377.

Schmetzer, K., 1982, Absorptionsspektren und Farben Vanadium-und Chromhaltiger Minerale: Deutschen Gemmologischen Gesellschaft, Zeitschrift, v. 31, p. 125-130.

Schmetzer, K., and Bank, H., 1981, An unusual pleochroism in Zambian emeralds, Journal of Gemmology, v. 17, p. 443-446.

Schmetzer, K., Berdesinski, W., and Bank, H., 1974, Uber die Mineralart Beryll, ihre Farben und Absorptionsspektren: Deutschen Gemmologischen Gesellschaft, Zeitschrift, v. 23, p. 5-39.

Schwartz, D., 1984, Inclusoes em esmeraldas: Revista Escola de Minas Ouro Preto, v. 37, p. 25-34.

Schwartz, D., 1986, Classificao genetica das ocorrecias de esmeralda: Boletim no. 1, Sociedade Brasileira de Geologia, Congresso 34, Resumo, p. 184.

Shackleton, R.M., Ries, A.C., Graham, R.H. and Fitenes, W.R., 1980, Late Precambrian ophiolite melange in the eastern desert of Egypt. Nature, v. 285, p. 472-474.

Simpson, E.S., 1948, Minerals of Western Australia: Perth Government Printer, v. 1, p. 195-207.

Sinkankas, J., 1981, *Emerald and other beryls,* Chilton Book Co., Radnor, Pennsylvania, USA, 665 p.

Sinkankas, J. 1982, Alpine type beryl-emerald deposits near Hiddenite, North Carolina: American Mineralogist, v. 67, p. 181.

Soliman, M.M., 1986, Ancient emerald mines and beryllium mineralization associated with Precambrian stanniferous granites in the Nugrus—Zabara area, southeastern desert, Egypt: Arab Gulf Journal of Scientific Research, v. 4, p. 529-548.

Sterrett, D.B., 1912, An occurrence of emerald in North Carolina: Washington Academy of Science, proceedings, Geological Society, p. 360-361.

Stocklin, J., 1977, Structural correlation of the alpine ranges between Iran and Central Asia: Societe Geologique de France, Memoires, no. 8, p. 333-353.

Tether, J., Patney, R.K., and Malik, J.I., (undated ?), The distribution and provenance of Zambia's coloured gemstones, Occasional Paper 110, Geological Survey of Zambia, Lusaka.

Thurm, R., 1972, Emaragde vom Lake Manyara in Tanzania: Deutschen Gemmologischen Gesellschaft, Zeitschrift, v. 21, p. 9-12.

Vail, J. R., 1983, Pan-African crustal accretion in northeast Africa: Journal of African Earth Sciences, v. 1, p. 285-294.

Van Eeden, O.R., Partridge, F.C., Kent, L.E., and Brandt, J.W., 1939. The mineral deposits of the Murchison Range, east of Leysdorp: Memoir of Geological Survey of South Africa, no. 36, p. 163.

Vlasov, K.A., and Kutukova, E.I., 1960, Izumrudnye Kopi: Moscow Akademiya Nauk SSGR, 252 p.

Weinschenk, E., 1896, Die Minerallagerstatten des Grobvenedigerstockes in den Hohen Tauern: Zeitschrift für Kristallographie, v. 26, p. 337-508.

Whitfield, G.B., 1975, Emerald occurrence near Menzies, Western Australia: Australian Gemmologist, v. 12, p. 150-152.

Williams, I.R., 1973, Structural subdivision for the Eastern Goldfields Province, Yilgarn block, Western Australia: Geological Survey of Western Australia, Annual Report, p. 53-58.

Wood, D.L., and Nassau, K., 1968, The characterization of beryl and emerald by visible and infrared absorption spectroscopy, American Mineralogist, v. 53, p. 777-800.

Zeck, H.P., and Malling, S., 1976, A major global suture in the Precambrian of Sweden: Tectonophysics, v. 31, p. 35-40.

Origin and Classification of Pakistani and World Emerald Deposits

Lawrence W. Snee and Ali H. Kazmi

INTRODUCTION

Emerald, the result of the substitution of a small amount of chromium for aluminum in the beryl crystal structure (Deer and others, 1986), is the product of special geological conditions. Even though beryl is the most common mineral that has beryllium as a major constituent, it is itself rare and is generally found in significant abundance only in late-stage igneous rocks, such as pegmatites. Beryl is rare because beryllium (with an ionic charge of +2) has both a small atomic radius (0.3 Å; Shannon and Prewitt, 1969; hence it is excluded from the crystal structure of most minerals) and a small crustal abundance (less than 5 ppm as estimated by Wedepohl, 1978, or less than 3.5 ppm according to Beus, 1965). Emerald is even scarcer because chromium, the coloring agent of emerald, and beryllium are geochemically incompatible. Chromium is found only in significant amounts in "primitive" rocks such as ultramafic igneous rocks; in these, beryllium is absent. Thus, special circumstances are necessary to bring chromium and beryllium together and therefore, emerald is one of the rarest and most precious gemstones.

In Pakistan, a series of special geological events resulted in the formation of emerald. In the first three chapters of this book, we summarize the large-scale events. In chapters four through seven, we discuss additional clues on the origin of the emeralds; these clues are revealed in physical and chemical characteristics of the emeralds as well as in the morphology and physicochemical conditions of the formation of tiny inclusions within the crystals. In chapter 8, we review the geology of other known world emerald deposits. In all, an intriguing picture about the general geologic setting and the origin of not only Pakistani, but world emeralds is unfolding. In this final chapter, the characteristics of Pakistani emeralds are summarized and are used to formulate a theory on their origin. Next, a classification scheme based on the general geologic setting of world emeralds is proposed and discussed and is developed into an hypothesis for the origin of all emeralds. Finally, we propose some additional studies that would shed new light on the origin of emeralds.

CHARACTERISTICS AND ORIGIN OF PAKISTANI EMERALDS

As summarized in the first two chapters, the special geological situation that allowed the formation of emeralds in Pakistan was the collision of the Indian and the Asian continental plates in Cretaceous and Tertiary time. Trapped in between these two enormous crustal masses were an island arc and pieces of ultramafic rocks stripped from a once-intervening ocean floor. The collision and suturing of India and Asia formed the Himalaya and juxtaposed chromium- and beryllium-bearing rocks along a broad belt now exposed in northern Pakistan. Although the two chemically different groups of rocks that contain the critical elements for formation of emerald were brought together by this collision, the formation of emerald could not occur until beryllium was chemically transported to the chromium-bearing host rocks.

Kazmi and others show in chapter 3 that Pakistani emeralds are found in two geological settings and the detailed geology of the emerald deposits in both settings provides us with insight into the origin of emerald. The economically most important emerald deposits are located in shear zones that cut across talc-carbonate schists. The schistose chromium-rich host rocks were derived from "primitive" oceanic rocks now included in melange zones that were formed during the collision of the Indian and Asian continental plates during the Cretaceous and Tertiary time periods. These melange zones are near Indian plate continental rocks that have been intruded after collision by chemically evolved granites. The presently economically less important emerald deposits in Pakistan are in pegmatites that intruded pods of chromium-rich mafic schist, gneiss, and amphibolite. The pegmatites were derived from the partial melting of Indian plate continental rocks and were emplaced along the fault zone between the Indian plate and the oceanic rocks. In both occurrences it is clear that the emeralds formed after most other geologic activity associated with continental collision had ceased. In other words, the emeralds in the melange zones formed after metamorphism and after post-metamorphic faulting had formed pathways for the infiltration of beryllium-bearing solutions; in comparison, the emeralds in the pegmatites were late magmatic precipitates that formed because beryllium-rich evolved magma, which was derived from the melting of collision-thickened continental crust, interacted with pods of chromium-bearing country rock.

The physical characteristics of Pakistani emeralds are also important indicators of the geologic processes that formed them. As Gübelin points out in chapter 4, emeralds from the numerous Pakistani deposits have color and clarity ranging from excellent to poor. This is expected if growth of emeralds was dependent on an influx of either (or both) chromium and beryllium from different sources. The presence of color-zoning in the emeralds is proof that their growth was not instantaneous and therefore the quality and color of the gems was controlled by the rate of chemical transport and the compositional changes of the mineralizing fluids. Gübelin also describes an interesting array of inclusions in emeralds from all Pakistani localities. The presence of coexisting liquid and vapor in these inclusions indicates that the emeralds formed from a hydrothermal fluid. The various solid-phase inclusions were derived both from saline hydrothermal fluids and from the host rocks. The observations and conclusions of Gübelin are supported by Seal's work in chapter 7. In addition, Seal's heating and cooling studies on inclusions in both emerald and quartz provide evidence that the hydrothermal fluids from which the emeralds precipitated were hot (greater than 250°C but less than 450°C) but the fluid did not boil. Seal's work also shows the presence of at least two hydrothermal fluids: an earlier, higher temperature, higher salinity fluid that was the source of the emeralds and a later lower temperature, lower salinity fluid associated with late quartz formation. The introduction of this later fluid marked the end of emerald formation.

The chemical study by Snee and others, chapter 5, and the microprobe work by Hammarstrom, chapter 6, quantify the dependence of emerald color on chromium content and prove that Pakistani emeralds have elevated chromium contents. Furthermore, the chemistry of Pakistani emeralds shows differences from other emeralds of the world; in fact, chemical differences are documented from place to place even within Pakistan. These differences in emerald chemistry reflect differences in geologic settings and processes of formation. Snee and others chemically document two hydrothermal fluids and Hammarstrom shows that color zonation in emeralds directly reflects the chromium content. In addition, the chemical studies provide direct evidence that the source of the beryllium-bearing hydrothermal fluids was within evolved continental rocks and therefore was external to the host chromium-rich oceanic rocks.

Thus, in all Pakistani emerald deposits except Khaltaro, a hot, high salinity, beryllium-bearing hydrothermal fluid, that was probably derived from post-collisional, chemically evolved granites, travelled through chromium-bearing host rocks of oceanic origin. Beryllium in the hydrothermal fluids was probably transported in the form of fluoride-, chloride-, or hydroxy-complexes. When the hydrothermal fluid reached the chromium-bearing oceanic host rocks, its composition was appropriate to incorporate chromium allowing formation of emerald which probably precipitated when the pH of the hydrothermal fluids increased due to wallrock alteration. Even though a later, lower temperature, less saline hydrothermal fluid also incorporated some chromium before it solidified to form quartz, it produced no emerald because it contained no beryllium. At Khaltaro, in contrast, the beryllium-bearing fluid, that intruded chromium-rich rocks, was pegmatitic magma. Before the magma cooled, enough chromium was incorporated from the mafic schists to form emerald. In both cases however, the juxtaposition of chromium-rich oceanic rocks and beryllium-bearing continental rocks was a direct result of the plate tectonic processes associated with the collision of the Indian and Asian continents. Furthermore, the formation of evolved hydrothermal fluids and pegmatites was likely related to emplacement of granites after collision; these granites may have formed from melting of thickened continental crust.

CLASSIFICATION OF WORLD EMERALDS

From the preceding discussion it is apparent that it is uncommon to find chromium and beryllium, the two critical constituents of emerald along with aluminum, silicon, and oxygen, together in normal crustal environments. It is also apparent that in Pakistan, geologic processes brought beryllium-and chromium-bearing rocks close enough together so that beryllium could be transported by later remobilization processes to the chromium-rich host rocks. In chapter 8, from summary of the geologic setting of world emerald deposits, it is also apparent that special geologic processes brought chromium and beryllium together. Thus in our classification scheme of world emerald deposits, we think it is necessary to focus on two geologically controlled constraints; the source of beryllium and the source of chromium.

A broad generalization, which is obvious but nonetheless important, is that emeralds are found on the continents. In continental crust, chromium is a rare elemental constituent. Of the entire Earth's crust, chromium averages only 50 ppm (Wedepohl, 1978) and by far the majority of this occurs in oceanic crust. Chromium is relatively common in ultramafic rocks in which several thousand ppm chromium may occur. Thus we feel that some geologic process must bring chromium-bearing rocks, generally (but not necessarily) ultramafic, in contact with continental crustal rocks. So the first part of our classification scheme is based on this geologic process that brings chromium-bearing rocks in contact with continental crust. However, just bringing these two types of rocks together is not adequate for the formation of emerald. It is also commonly recognized that emerald is found in veins, vugs, dikes, or pegmatite that cross-cut, or are very near chromium-rich rocks. Therefore we think that a second, not necessarily related geologic process, is necessary to mobilize beryllium and transport it to the chromium-bearing host rocks where the transporting medium can pick up chromium before the precipitation of beryl occurs. The second part of our classification scheme is based on the transport mechanism that carries the beryllium to the chromium-rich host.

Sources of chromium

Let us focus first on the source of chromium. Because chromium is uncommon in continental crust, special geologic processes must juxtapose chromium-bearing rocks with continental crust. There appear to be three geologic environments that exhibit this juxtaposition and that may contain anomalously high concentrations of chromium. These environments are *suture zones, granite-greenstone terrains,* and sedimentary sequences that contain *shales or metamorphosed shales* derived in part from chromium-rich source rocks. Emeralds have been reported from all three.

Suture zones, such as that exposed in northern Pakistan, are elongate or belt-like zones marking a region of deformation that formed during the collision of two or more of the crustal plates of the Earth's surface. A map of the distribution of world suture zones is available in Burke and others (1977) and is reproduced in figure 9.1 on which some of the important emerald deposits near suture zones are identified. Because ocean-floor rocks may be trapped between continental plates during the process of collision, it is common to find ultramafic rocks exposed within major suture zones.

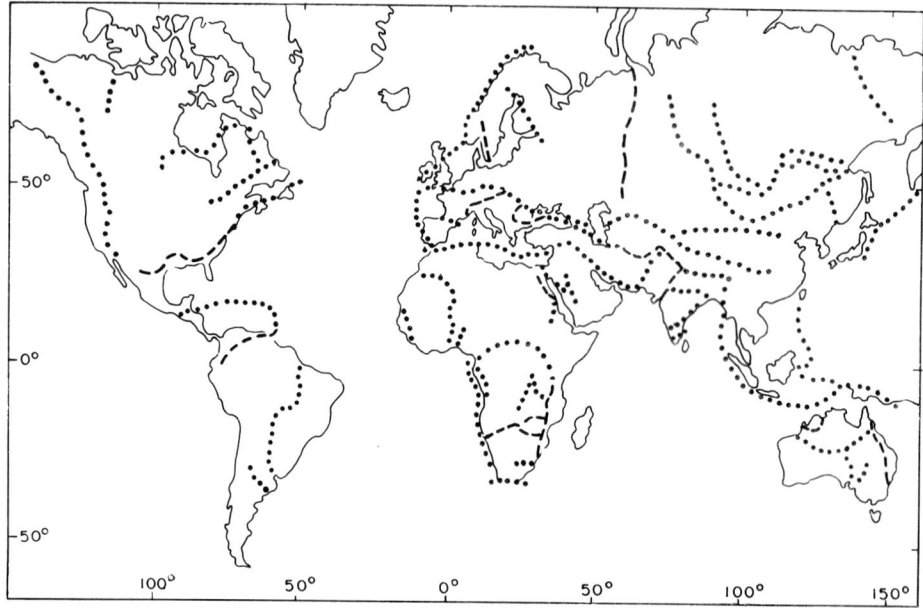

Figure 9.1. Distribution of major Phanerozoic and Proterozoic suture zones (dotted) modified from Burke and others (1977). Suture Zones that are found near emerald deposits are highlighted. (dashed lines).

Greenstone portions of *granite-greenstone terrains* are typically elongate belts of volcanic and sedimentary rocks generally at low metamorphic grade. The name greenstone is derived from the common occurrence of a large amount of mafic volcanic rocks that were metamorphosed to low–grade epidote-chlorite-actinolite rock; these rocks are named greenstone because of their green massive appearance. Greenstone belts commonly wrap around granitic batholiths and are generally found in Archean-aged geologic regions. Granite-greenstone belts have been described from every continent except Antarctica. The origin of granite-greenstone terrains is the subject of much debate. Nisbet (1987) synthesized geologic data and arrived at the conclusion that the geologic style of granite-greenstone terrains was formed during a three-stage tectonic process resulting from major horizontal motions that took place within the Archean crust because of crustal plate interaction. The first stage of formation was the deposition of greenstones on gneissic basement with subsequent intrusion by tonalite

Table 9.1. Emerald Classification

Geologic Setting		Source of Chromium		
		Suture Zone	Granite/Greenstone terrain	Shale/Metashale
Source of Beryllium	Pegmatite	India Khaltaro, Pak United Arab Rep. Northern Carolina Ural Mountains Austria Tanzania Afghanistan Bulgaria Ukraine (?)	Western Australia Brazil South Africa Zimbabwe Madagascar Zambia Mozambique Kenya	Colombia—Muzo
	Hydrothermal or Connate/ Meteoric Fluids	Pakistan Afghanistan	(Brazil—St. Terezinha de Goias) (Mozambique)	Colombia—Chivor (Muzo)
	Metamorphic/ Tectonic	(Austria)		(Colombia)

Note: Emerald deposits of Nigeria and Ghana are not included in this table because their geologic occurrences have not been published.

diapirs. During the second stage of development, greenstones were infolded with the basement and with syntectonic plutons during regional compression. A later regional compressional event resulted in cross-folding which produced the complex regional style of the terrains. Furthermore, it is likely that they were formed from plate tectonic processes similar to those in operation today, i.e., continental accretion and associated suturing or rifting (Nisbet, 1987).

Shales and their metamorphic equivalents were formed from sediments derived from the erosion of a provenance, or source area. Most provenances commonly include several types of rocks. If the source area of a shale contained a suture zone or a granite-greenstone terrain, it would be expected that chromium would be found in the shale. Some shales contain several hundred to several thousand ppm chromium.

Sources of beryllium

As stated above, once one of the above three processes puts chromium-rich rocks into the continental crust, it is necessary for beryllium to come in contact with the chromium in order to form emerald. There are three possible mechanisms that could transport beryllium to chromium-bearing host rocks. The two likeliest are emplacement of beryllium-rich pegmatitic magma and the infiltration of hydrothermal or heated meteoric/connate fluids. Most of the emerald deposits of the world received their beryllium by one of these two processes. A third possible transport medium of beryllium is fluid that was mobilized during metamorphism; this type of fluid has been proposed as the primary source of beryllium for Austrian and Colombian emeralds (Kozlowski and others, 1988; Okrusch and others, 1981; Morteani and Grundman, 1977).

Emerald classification and origin

Table 9.1 shows our classification for world emerald deposits. We feel this is a preliminary scheme that needs to be tested by additional study of the various deposits, but available data support our contention that all known emerald deposits fit within this classification. The only exceptions to this generalization are a few deposits of green beryls that receive their color from vanadium, e.g., the Emmaville "emerald" of Australia. In addition, within the classification, itself, some deposits listed under one category alternatively may fit in a different category. For equivocal cases, the deposits are listed under two categories; the less likely alternative, in our view, is enclosed in parentheses.

From table 9.1 and chapter 8, it is clear that the majority of known emerald deposits are found in pegmatites that intruded suture zones or granite-greenstone terrains. This is true for all known deposits except those of Colombia, Afghanistan, Pakistan (excluding Khaltaro), and possibly Austria, Mozambique, and one Brazilian occurrence (St. Terezinha de Goias). Of these, all except the Colombian deposits occur in or near suture zones or granite-greenstone belts; thus, their source of chromium is the same as the majority of emerald deposits. These deposits are different, however, because of their conspicuous lack of pegmatites. Transport of beryllium for these deposits seems to have occurred in hydrothermal fluids, which formed emerald-bearing veins and vugs.

Okrusch and others (1981) and Morteani and Grundmann (1977) proposed an additional mechanism for the formation of emerald. Based on their studies, emeralds of Austria formed during metamorphism (see chapter 8) from beryllium-rich metamorphic fluids in the presence of chromium. Although it is possible that metamorphic fluids mobilize beryllium and that emerald may form under appropriate metamorphic conditions, we think it is equally possible that Austrian emeralds were premetamorphic crystals formed by hydrothermal processes that affected the host rocks before metamorphism. (We base this suggestion on the fact that because the emeralds are deformed crystals within the schists, critical premetamorphic textural relationships may have been completely obliterated during metamorphism and deformation that affected the host rocks and enclosed emeralds.) Because Austrian emeralds were apparently formed under unusual conditions, more study is necessary.

The sole possible exception to our classification is, of course as nature would have it, the most important emerald deposits in the world—those of Colombia. Despite the importance and numerous studies of these deposits, their origin remains unclear. Colombian emeralds occur within deformed shales. Although few analyses of Colombian shales are available, it is reasonable to conclude that chromium contents of the shales were high enough to color beryl. It is also reasonable to suggest that some chromium-bearing rocks exist within the provenance of the shales because this portion of South America is (and was) a tectonically active region near the intersection of several crustal plates (see figure 9.1). In our view, the most problematic aspect of Colombian emeralds is the source of beryllium. Beus and Mineev (1974) concluded that beryllium was drived from an "endogeneous deep-seated source". Recently, however, Kozlowski and others (1988) proposed that Somondoco emerald, Columbia, were formed entirely from beryllium and chromium both mobilized from the shale host by heated "connate" fluids. In their study, fluid inclusions within the emeralds recorded homogenization temperatures of about 450°C without pressure corrections. (After homogenization temperatures are corrected for pressure, the resultant formation temperatures of the emeralds will be 450°C or higher.) The fluid inclusions studied were rich in chlorine. Kozlowski and others (1988) concluded that chlorine in the hot fluids, which were derived from an unknown source, leached beryllium from the shales to ultimately form emerald when brought in contact with sufficient chromium. Although the hypothesis that beryllium was leached from the shale host may be correct, we suggest that fluid temperatures of greater than 450°C strongly indicate a hydrothermal origin for the fluids and possibly the beryllium. Renders and Anderson (1987) have shown that beryllium is transported in hydrothermal fluids as hydroxy-, chloride-, or fluoride-complexes. In any case, the origin of the world's most important emeralds is not yet clear.

ADDITIONAL STUDIES

Literature on world emerald deposits is extensive. (A comprehensive list of that literature follows this chapter.) However, despite the numerous studies conducted on emeralds and emerald deposits, much work remains to be done, particularly of a geological and geochemical nature, before we can understand emerald genesis. We feel that our study of Pakistani emeralds and our evaluation of the geology of other emerald deposits have led to a reasonable hypothesis for the origin of emeralds. Some additional studies, which could provide valuable data for testing our hypothesis or for defining new and better theories, are listed below.

More detailed geologic mapping of emerald deposits and petrography of host rocks and emerald-bearing ore are needed to provide a better understanding of the geologic framework of the deposits. Of critical importance are identification of the sources of chromium and beryllium and characterization of textural relationships between emerald and host. Beyond this, regional evaluations of the geology of the emerald deposits within the framework of modern tectonic theories are needed.

More comprehensive chemical studies should be conducted. The chemistry of the emeralds, their host rocks, and fluid inclusions within emeralds can provide information to both characterize emeralds with respect to region as well as shed light on emerald genesis.

Careful microprobe studies should be conducted to determine the nature of chemical substitutions within the emerald crystal lattice.

Stable isotope geochemical studies of emeralds and host rocks could provide important information on the source of fluids that formed the emeralds. Perhaps our understanding of the origin of Colombian emerald deposits would benefit most from a better characterization of the isotopic character of the emerald-bearing fluids.

Finally, we think it is time to evaluate the internal structures within emeralds in light of geological and geochemical data. Most existing studies were done within a gemological context. Important information on the genesis of emeralds may be revealed when a geological/geochemical evaluation of the morphology of internal emerald structures is conducted.

CONCLUSION

Because emerald is naturally occurring, its magnificence stems not only from its beauty and gemological characteristics but also from the secrets of Earth history locked within each crystal. Beyond this, it is fitting that emerald is of great value both because of its beauty and because it is scarce. This scarcity, in turn, is geologically and geochemically controlled. Beryllium and chromium are generally partitioned into geologically different environments. Thus, the mere presence of emerald in a rock indicates that extraordinary geologic circumstances existed. We feel that all known emerald deposits, except perhaps those of Colombia, exist as a result of crustal plate motions that juxtaposed chromium-bearing ultramafic oceanic rocks and beryllium-bearing felsic continental rocks. Emerald deposits in relatively young rocks are found within sutures that mark the zone of collision between continental plates. Emerald deposits in ancient rocks are generally within or near the greenstones of granite-greenstone terrains; these, too, may mark such collision zones. Thus, we marvel at each emerald crystal because of its natural beauty *and* of its untold tale of continental collision. We wonder how many undiscovered emerald deposits may be waiting within the few slivers of ancient ocean floor locked within some unexplored suture zone or greenstone belt.

REFERENCES

Beus, A.A., 1965, *Geochemistry of Beryllium and Genetic Types of Beryllium:* W.H. Freeman and Company, San Francisco, California, U.S.A., 401 p.

Beus, A.A., and Mineev, D.A., 1974, Geology and geochemisty of the emerald-bearing zone Muzo-Coscues, Cordillera Oriental, Colombia: Geologiya Rudnykh mestorozhdenii, Moscow, v. 16, no. 4, p. 18-30 (in Russian).

Burke, K., Dewey, J.F., and Kidd, W.S.F., 1977, World distribution of sutures—the sites of former oceans: Tectonophysics, v. 40, p. 69-99.

Deer, W.A., Howie, R.A., and Zussman, J., 1986, *Rock-forming Minerals,* Volume 1B, Disilicates and ring silicates: Longman Scientific and Technical Limited, London, England, p. 372-409.

Kozlowski, A., Metz, P., Jaramillo, H.A.E., 1988, Emeralds from Somondoco, Colombia: chemical composition, fluid inclusions and origin: Neues Jahrbuch für Mineralogie, Abhandlungen, v. 159, p. 23-49.

Morteani, G., and Grundmann, G., 1977, The emerald porphyroblasts in the penninic rocks of the central Tauern window: Neues Jahrbuch für Mineralogie, Monatschefte, v. 11, p. 509-516.

Nisbet, E.G., 1987, *The Young Earth—An Introduction to Archaean Geology:* Allen and Unwin, Inc., Winchester, Massacusetts, U.S.A., 402p.

Okrusch, M., Richter, P., and Gurkan, A., 1981, Geochemistry of Blackwall Sequences in the Habachtal emerald deposit: Hohe Tauern, Austria, part 1: Presentation of Geochemical Data: Tschermaks Mineralogische und Petrographische Mitteilungen, v. 29, p. 9-31.

Renders, P.J., and Anderson, G.M., 1987, Solubility of kaolinite and beryl to 573K: Applied Geochemistry, v. 2, p. 193-203.

Shannon, R., and Prewitt, C., 1969, Effective ionic radii in oxides and fluorides: Acta Crystallographica, B25, p. 925-946.

Wedepohl, K.H., 1978, *Handbook of Geochemistry:* Springer-Verlag, Berlin, Germany.

Selected Bibliography on World Emeralds

Ali H. Kazmi and Sadaqat A. Jafary

Acknowledgement: The compilers are grateful to Elizabeth Yeates, Chief Librarian, U.S. Geological Survey, and Barbara A. Chappel and Gertrude M. Sinnott of U.S. Geological Survey for their help and assistance, without which compilation of this bibliography would not have been possible.

Amstutz, G.C., and Bank, H., 1977, Geologic, petrographic and mineralogic observations in some mines of emerald, tanzanite, tsavorite and ruby in Tanzania and Kenya: Deutschen Gemmologischen Gesellschaft, Zeitschrift, v. 26, p. 118-127.

Anderson, B.W., 1950, The absorption spectra of emerald and alexandrite: Gems and Gemology, v. 6, p. 263-266, 291.

Anderson, B.W., 1966, Chromium as a criterion for emerald: Journal of Gemmology, v. 10, p. 41-45.

Anderson, S.M., 1975, A note on the occurrence of emerald at Mayfield Farm, Fort Victoria: Rhodesia, Southern, Geological Survey, Annals 1, p. 60-61, Journal of Gemmology, v. 15, p. 80-82.

Anderson, S.M., 1976, Notes on the occurrence and mineralogy of emeralds in Rhodesia: Rhodesia, Southern, Geological Survey, Annals, v. 2, p. 50-55.

Anderson, S.M., 1978, Notes on the occurrence and mineralogy of emeralds in Rhodesia: Journal of Gemmology, v. 16, p. 177-185.

Anderson, S.M., 1980, Emeralds: South African Lapidary Magazine, v. 14, p. 84-86.

Anderton, R.W., 1950-1951, Report on Chivor Emerald Mines: Gems and Gemology: v. 12, p. 376-377, 379.

Andres-Gayon, J.I., 1978, A new synthesis: Instituto Espanola de Gemologia, v. 17, p. 37-38.

Andreyenko, E.D., and Solodova, Y.P., 1986, Inclusions in gemstones as typomorphic and diagnostic indicators, typomorphism of minerals and mineral associations: Izdatal stvo Nauka, p. 104-111.

Anonymous, 1909, Emerald Mines of Muzo (Colombia): Mining Journal., v. 85, p. 763-764.

Anonymous, 1937, Gemstone Names, Recommendations by International Conference: Gemmologist, v. 6, p. 188-191.

Anonymous, 1959, First annual field conference, Barco Concession: Santander Norte Colombian Society of Petroleum Geologists and Geophysicists, p. 35.

Anonymous, 1974, Mining: An important activity for Minas Gerais in Brazil: Mineracao Metalurgia, v. 38, p. 8-35.

Anonymous, 1978, Other gemstones: Western Australia Department of Mines, Annual Report, p. 128.

Anonymous, 1978, Precious and semi-precious stones in Western Australia: Geological Survey of Western Australia, Bulletin, v. 11, p. 20.

Anonymous, 1981, Gemstones: Report of the Sixth annual commodity meeting, Institution of Mining and Metallurgy, Transactions, Section A, Mining Industry, v. 91, p. A110-A140.

Anonymous, 1981, The Aga Khan; an Australian emerald mine: Mining Magazine, v. 145, p. 77.

Anonymous, 1982, Crabtree emerald mine in production again: Jewelry Making, Gems and Minerals, v. 538, p. 5-6.

Anonymous, 1982, Emerald production revives in Australia: Australian Gem and Treasure Hunter, v. 73, p. 37.

Anonymous, 1983, Emerald fever strikes Socoto (Brazil): Latin American Mining Letters, (San Francisco).

Anonymous, 1983, Gem industry has made recent major progress: World Mining, v. 36, p. 99, 101.

Arbab, M. S. H., 1972, Note on the emerald of Mohmand Agency (N.W.F.P.): Pakistan Geological Survey, Geonews, v. 2, p. 56-58.

Asmus, B., 1976, Buying emeralds in Colombia: Lapidary Journal, v. 29, p. 2288-2289.

Babu, S.K., 1972, Occurrence of chlorapatite at emerald mine, Rajgarh village, Ajmer: Mineralogical Magazine, v. 38, p. 972-974.

Bagchi, T.C., 1958, The Geology of the Bubani emerald mine near Ajmer, Rajasthan, with a note on the origin of emerald: Proceedings 45th Indian Scientific Congress, pt. 3, p. 243-244, Indian Mining Journal, v. 6, no. 3, p. 1-4.

Bancroft, P., 1986, Exotic crystal mines: Lapidary Journal, v. 40, p. 21-37.

Bank, H., 1973, A new emerald find in Zambia (Miku-Deposit): Deutschen Gemmologischen Gesellschaft, Zeitschrift, v. 22, p. 60-61.

Bank, H., 1973, Ein neues Smaragdvorkommen in Zambia (Miku-Deposit): Deutschen Gemmologischen Gesellschaft, Zeitschrift, v. 22, p. 60-61.

Bank, H., 1974, Smaragd, Alexandrit und Rubin als Komponenten einer Paragenese vom Lake Manyara in Tansania: Deutschen Gemmologischen Gesellschaft, Zeitschrift, v. 23, p. 62-63.

Bank, H., 1974, The emerald occurrence of Miku, Zambia: Journal of Gemmology, v. 14, p. 8-15.

Bank, H., 1974, The refraction index of Brazilian emeralds: Deutschen Gemmologischen Gesellschaft, Zeitschrift, v. 23, p. 297-299.

Bank, H., 1976, Colorless topaz overgrown by synthetic emerald from Lechleitner: Deutschen Gemmologischen Gesellschaft, Zeitschrift, v. 25, p. 109-110.

Bank, H., 1976, Natural colorless beryl overgrown by synthetic emerald: Deutschen Gemmologischen Gesellschaft, Zeitschrift, v. 25, p. 107-108.

Bank, H., 1976, Synthetic emerald with a beryl seed crystal mistaken for a doublet: Deutschen Gemmologischen Gesellschaft, Zeitschrift, v. 25, p. 111.

Bank, H., 1980, Emerald with very high refracting index from Zambia: Deutschen Gemmologischen Gesellschaft, Zeitschrift, v. 29, p. 101-103.

Bank, H., 1980, Low refracting beryl covered with synthetic emerald according to Lechleitner: Deutschen Gemmologischen Gesellschaft, Zeitschrift, v. 29, p. 197.

Bank, H., 1980, Synthetic emerald made using the raw form of natural emerald from Zambia: Deutschen Gemmologischen Gesellschaft, Zeitschrift, v. 29, p. 90-91.

Bank, H., and Bank, M., 1980, Special effects of refractometer reading of synthetic emerald (according to Lechleitner) with covered beryl core: Deutschen Gemmologischen Gesellschaft, Zeitschrift, v. 29, p. 198.

Bank, H., and Bank, M., 1980, Very high refracting beryl core covered with synthetic emerald according to Lechleitner: Deutschen Gemmologischen Gesellschaft, Zeitschrift, v. 29, 199 p.

Bank, H., and Gübelin, E.J., 1976, The emerald and alexandrite occurrence of Lake Manyara, Tanzania: Deutschen Gemmologischen Gesellschaft, Zeitschrift, v. 25, p. 130-147.

Barbosa, J. E., 1973, The analytic profile of emeralds, (abstr.): Congresso Brasileira Geologia, v. 2, p. 128-129.

Barbosa, J.E., and de Almeida, C., 1973, Analytical profile of emeralds: Brazil, Departamento Nacional Da Producao Mineral, Boletim, v. 12, p. 29.

Barclay, G. C., 1939, Gem sources of the South Atlantic States: Gems and Gemology, v. 3, p. 9-10 and p. 27-28.

Barrie, J., and Kalix, Z., 1959, Gemstones: Summary Report of Australian Bureau of Mineral Resources, Geology and Geophysics, (Canberra) no. 43, 42 p.

Barros, J. C., 1985, The Chemistry of emeralds and associated beryls in Porangatu Deposits, no. 3, Brazil: Memorias VI Congreso Latinoamericano de Geologia, v. 6, p. 332-341.

Barros, J. C., 1985, The chemistry of emeralds and associated beryls in Porangatu Deposits, Brazil: Memorias VI Congreso Latinoamericano de Geologia, v. 6, p. 415-424.

Barton, M.D., 1986, Phase equilibria and thermodynamic properties of minerals in the Be O—Al_2O_3—Si O_2—H_2O (BASH) system, with petrologic applications: American Mineralogist, v. 71, p. 277-300.

Bassett, A.M., 1979, Hunting for gemstones in the Himalayas of Nepal: Lapidary Journal, v. 33, p. 1492-1494, 1496, 1500, 1502, 1506, 1508, 1510, 1512, 1514-1520.

Bastin, E.S., 1911, Geology of the pegmatites and associated rocks of Maine, including feldspar, quartz, mica and gem deposits: U.S. Geological Survey, Bulletin 445, p. 152.

Bastos, F.M., 1981, Emeralds from Itabira, Minas Gerais, Brazil: Lapidary Journal, v. 35, p. 1842-1848.

Bazarov, L. Sh., Klyakhin, V.A., and Senina, V.A., 1974, Primary inclusions of melts solutions of Ural emeralds: *in* Daldov, Yu, A., Sobolev, V.S., and Chepurov, A. I., editors, Mineralogiya Endogennykh Obrazovanii, p. 96-108.

Behier, J., 1957, Minerals de Provincia de Mocambique: Provincia Mocambique, Services de Industria-e-Geologia Boletim, v. 22, p. 135.

Benavides, K., 1985, Forms of natural dissolution of beryl crystals (emeralds) in Colombia: Zapiski Vsesoyuznogo Mineralogicheskogo Obschestva, v. 114, p. 591-593.

Bernauer, F., 1926, Die Sog, Smaragddrillinge von Muzo und ihre Optischen Anomalien: Neues Jahrbuch für Mineralogie, v. 54 A, p. 205-242.

Beus, A. A., 1965, *Geochemistry of Beryllium and Genetic Types of Beryllium Deposits:* W.H. Freeman and Company (Publishers), San Francisco, p. 401.

Beus, A.A., 1979, Sodium: a geochemical indicator of emerald mineralization in the Cordillera Oriental, Colombia: Journal of Geochemical Exploration, v. 11, p. 195-208.

Beus, A.A., and Mineev, D.A., 1974, Geology and geochemistry of the Muzo-Cosquez, emerald bearing zone, eastern cordillera, Colombia: Geologiya Rudnykh Mestorozhdenii, Moscow, v. 26, p. 18-30.

Bhola, K.L., 1934, Short note on the beryl deposits of Ajmer-Merwara: Mining Geology and Metallurgical Institute of India, Transactions, v. 29, p. 127-139.

Black, P., 1985, Emerald: Australian Gem and Treasure Hunter, v. 104, p. 24-26.

Bolick, K., 1982, We mined for emeralds: Lapidary Journal, v. 36, p. 410-412.

Bose, S.K., 1959, A note on the geology and origin of the Rajgarh emerald deposit, Ajmer, Rajasthan: Indian Scientific Congress, 46th Proceedings, pt. 3, p. 229-230.

Bostos, F.M., 1981, Emerald from Itabira, Minas Gerais, Brazil: Lapidary Journal, v. 35, p. 1842-1848.

Bowersox, G. W., 1985, A status report on gemstones from Afghanistan: Gems and Gemology, v. 21, p. 192-204.

Bowersox, G. W., and Anwar, J., 1989, The Gujar Killi emerald deposit, Northwest Frontier Province, Pakistan: Gems and Gemology, v. 25, p. 16-24.

Bridges, C.R., 1974, Green grossularite garnets ("Tsavorites") in East Africa: Gems and Gemology, v. 14 p. 290-295.

Brough, B.H., 1904, Cantor Lectures on the Mining of Non Metallic Minerals: Journal of the Society of Arts, v. 52, p. 113-122, 139-148, 152-163, 167-179, also Quarry, v. 9, p. 201-205, 270-273.

Broughton, P.L., 1974, Emerald deposits of Western North Carolina: Earth Science, 27, p. 222-228.

Broughton, P.L., 1979, Economic geology of Australian gemstone deposits: Minerals Sciences and Engineering, Johannesburg, v. 11, p. 3-20.

Brown, G., 1982, Leichleitner coated beryl inclusions: The Australian Gemmologist, v. 14, p. 274-275.

Brown, G., 1984, Australia's first emeralds: Journal of Gemmology and Proceedings of the Gemmological Association of Great Britain, v. 19, p. 320-335.

Brown, G., and Snow, J., 1984, Inclusions in Biron synthetic emeralds: The Australian Gemmologist, v. 15, p. 167-171.

Bruce, 1814, Emerald: American Mineralogist, v. 1, p. 263-265.

Buis, O., 1966, Guidebook to the geology of the Chivor emerald mine: Colombian Society of Petrology, Geology, and Geophysics, p. 31.

Bukin, G.V., Godovikov, A.A., Klyakhin, V.A., and Sobolev, V.S., 1977, Growing emerald crystals: Vsesoiuznoe Soveshchanie Po Rostu Kristallokhimii, Tezisy Dokladov, v. 2, p. 15-16.

Bukin, G.V., Godovikov, A.A., Klyakhin, V.A. and Sobolev, V.S., and Samotsvety, 1980, Synthetic emerald: *in* Sidorenko, A.V., editor, Gem mineral, XI Sézda Mezhdunarodnaya Mineralogicheskaya Assotsiatsiya, Novosibirsk, USSR Sept. 4-10, 1978, p. 36-44.

Bukin, G.V., Godovikov, A.A., and Sobolev, V.S., 1980, Methods for growing emerald crystals: Rost Krist, v. 13, p. 215-223.

Bukin, G.V., and Maslov, V.A., 1974, Coloration of beryls: Eksp. Issled Mineral., Godovikov, A.A., Sobolev, V.S., and Fursenko, B.A. (editors)., Academia nauk SSSR Sibirskoe otdelenie, Institut geologii i geofiziki, p. 109-114.

Butcher, J., and White, E.A.D., 1964, A study of the hydrothermal growth of ruby: Mineralogical Magazine, v. 33, p. 974-985.

Cambell, I.C.C., 1974, Where is the dividing line between emerald and green beryl: Journal of Gemmology, v. 14, p. 177-180.

Campbell, I.C.C., 1973, Emeralds reputed to be of Zambian origin: Journal of Gemmology, v. 13, p. 169-179.

Campbell, I.C.C., 1978, A study of emeralds from an unsubstantiated African source of origin: Journal of Gemmology, v. 16, p. 93-108.

Campbell, I.C.C., 1980, Personal notes on gems, minerals and related aspects, Zimbabwe (Rhodesia): Lapidary Journal, v. 33, p. 2626-2635.

Carbonnel, J.P., 1976, A visit to the Mingora Emerald Mine, Swat, Pakistan: Lapidary Journal, v. 30, p. 1236-1238.

Carbonnel, J.P., Selo, M., and Poupeau, G., 1973, Fission track age of the gem deposit of Pailin (Cambodia) and recent tectonics in the Indochianan Province: Modern Geology, v. 4, p. 61-64.

Carrel, R.P., 1974, The precious stones of Pakistan: Association Francaise Gemmologie, Bulletin, v. 41 7 p.

Cassedanne, J., 1985, Emerald country: Le Monde les Mineraux, v. 66, p. 16-20.

Cassedanne, J.P., and Cassedanne, J.O., 1974, The emerald mine at Carnaiba: Association Francaise Gemmologie, Bulletin, v. 40, p. 4-8.

Cassedanne, J.P., and Cassedanne, J.O., 1977, Axinite, hydromagnesite, amethyst and other minerals from near Victoria da Conquista, Brazil: Mineralogical Record, v. 8, p. 382-387.

Cassedanne, J.P., and Sauer, D.A., 1981, Emeralds of Santa Terezinha: Revue de Gemmologie, v. 71, p. 4-8.

Cassedanne, J.P., and Sauer, D.A., 1984, The Santa Terezinha de Goias emerald deposit: Gems and Gemology, v. 20, p. 4-13.

Cassedanne, J.P., Cassedanne, J.O., and De Mello, Z.F., 1976, Occurrence of emerald at Acude Sossego (Anage Municipio-Bahia), Brazil: Mineracao Metalurgia, v. 40, p. 36-42.

Castor, G.R., 1966, Gems, synthetic: Kirk-Othmer Encyclopedia of Chemical Technology, 2nd ed., v. 10, p. 509-519.

Cemic, L., Langer, K., and Franz, G., 1986, Experimental determination of melting relationships of beryl in the system $BeO—Al_2O_3—SiO_2—H_2O$ between 10 and 25 K bar: Mineralogical Magazine, v. 50, p. 55-61.

Chambers, B., 1976, Emerald, three for $ 150: Australian Gemmologist, v. 12, p. 365-370.

Chaudhari, M.W., 1969, Unusual emerald, (Composite crystals with poor quality hexagonal core and gem quality overgrowths, cell parameters, refractive indices, material reportedly from Colombia): Schweizerische Mineralogische und Petrographische Mitteilungen, v. 49, p. 569-575.

Chepurnova, G.A., and Il'in, A.G., 1984, Study of microhardness of synthetic and natural beryls: *in* Kolonin, G.R., editor, *Fizika-Khimicheskie Issedovaniya Sul'fidnykh Silik Sistem*, Academia Nauk SSSR, Sibirskoe, otdelenie, Institut geologii i geofiziki, p. 150-154.

Chikayama, A., 1974, Natural and synthetic gemstones newly appeared after the 2nd World War: Gemmological Society of Japan, v. 1, p. 65-70.

Chizhik, O.E., and Lekukh, Z.V., 1980, Genesis of emeralds in glimmerite-type ore deposits: *in* Petrov, V.P., editor, *Dragotsennye Tsvetnye Kamni*, Izdatel'stvo, Nauka, Moscow, p. 158-174.

Claremont, L., 1913, Prehistoric emerald mines: Knowledge, v. 36, p. 124-127.

Clements, T., 1941, The Emerald Mines of Muzo, Columbia, South America: Gems and Gemology, p. 130-134.

Codazzi, R.L., 1925, Catalogo Descriptivo de los Minerales de Muzo: v. 8, p. 1-25 (Bogota).

Codazzi, R.L., 1929, Minas de Esmeraldes: Boletin de Minas y Petroleo, Bogota, v. 1, p. 114-143.

Comen, P., 1982, Go right to the source for emeralds, Colombia: Lapidary Journal, v. 36, p. 1604-1605.

Cotton, W.L., 1970, A Trip to the Carnaiba Emerald Mines of Brazil: Lapidary Journal, v. 23, p. 1360-1363.

Cox, D.P., 1986, Descriptive model of emerald veins: *in* Cox, D.P., and Singer, D.A. (editors), U.S. Geological Survey, *Mineral deposit models,* Bulletin 1693, p. 219.

Crookshank, H., 1947, Emeralds in Mewar: Indian Minerals, v. 1, p. 28-30.

Crookshank, H., 1948, Minerals of the Rajputana pegmatites: Mining Geology and Metallurgical Institute of India, Transactions, v. 42, p. 105-189.

Crowningshield. R., 1975, Developments and highlights of GIA's lab in New York: Gems and Gemology, v. 15, p. 89-94.

Da Cunha, O.L., 1961, Esmeraldas da Fazenda do Sossego, Santana de Ferros, Minas Gerais, Brasil: Gemologia, v. 25, p. 9-14.

Dallow, L., 1983, Pakistan: Australian Gem and Treasure Hunter, v. 82, p. 26-31.

Darragh, P.J., and Hill, R.E.T., 1986, The Warda Warra emerald deposit: The Australian Gemmologist, v. 16, p. 82-83.

Datta, A.K., 1966, Geological milieu of emeralds in Rajgarh area, Ajmer District, Rajasthan: Geological Society of India, Bulletin, v. 3, p. 29-33.

Dave, Y.N., 1970, Some considerations on the origin of emeralds in Rajasthan: Indian Scientific Congress Association, 57th Session, Proceedings part 3, p. 216-217.

David, E.S., 1881, On the emerald-green spodumene from Alexander County, North Carolina: American Journal of Science, v. 22, p. 179-182.

Davies, R.G., 1962, A green beryl (emerald) near Mingora, Swat state: Panjab University Geological Bulletin, v. 2, p. 51-52.

de Bruin, J.d.R., 1964, Gemstone varieties of Republic of South Africa: Lapidary Journal, v. 18, p. 226-228.

Deer, W. A., Howie, R.A., and Zussman, J., 1986, *Rock-forming Minerals,* Volume 1B, Disilicates and ring silicates: Longman Scientific and Technical Limited, London, England, p. 372-409.

Dele-Dubois, M., Dhamelincourt, P., and Schubnel, H., 1980, Study of inclusions in diamonds, sapphires and emeralds by Raman spectroscopy: Revue de Gemmologie, A.F.G., 64, p. 13-16.

Dele-Dubois, M.L., Dhamelincourt, P., and Schubnel, H.J., 1980, Raman spectroscopy of inclusions in diamonds, sapphires and emeralds: Revue de Gemmologie, A.F.G., 63, p. 11-14.

Deshpande, M.L., 1978, Gemstones and semi-precious stones: Indian Minerals, v. 32, p. 1-17.

Draper, D., 1928, The Western Transvaal diamond and emerald areas: South African Mining Journal, v. 38, p. 643, 651-652.

Draper, T., 1963, A new source of emeralds in Brazil: Gems and Gemology, v. 11, p. 11i-113.

Duyk, F., 1971, A few observations related to synthetic emerald: Journal of Gemmology, v. 12, p. 253-255.

Edgar, A., and Hutton, D.R., 1978, Exchange-coupled pairs of chromium (3+) ions in emerald: Physics, v. 11, p. 5051-5063.

Elizarov, L.I., 1978, Characteristics of magnetic properties of emerald containing micas: Razvedka i Okhrana Nedr, no. 6, p. 57-60.

Eppler, W.F., 1935, Mining Emeralds in Colombia: Gemmologist, v. 4, p. 201-207.

Eppler, W.F., 1962, Three-phase inclusions in emerald, aquamarine and topaz: Journal of Gemmology, v. 8, p. 245-250.

Eppler, W.F., 1963, Emerald from Burbar, Colombia: Journal of Gemmology, v. 9, p. 123-126.

Eppler, W.F., 1968, Another Lechleitner made synthetic emerald: Journal of Gemmology, v. 12, p. 37-41.

Escobar, R. 1978, Geology and geochemical expression of Gachala emerald district, Colombia: Geological Society of America, Abstracts with Programs, v. 10, p. 397.

Escobar, R., 1979, Geology and geochemistry of the Gachala emerald mines, Cundinamarca: Bol. Geol. Inst. Nac. Invest. Geol. (INGEOMINAS), v. 22, p. 117-153.

Escobar, R., and Mariano, A.N., 1976, On the origin of Colombian emeralds: Second biennial M.S.A.-F.M. symposium; Crystal chemistry and paragenesis of the gem minerals, Tucson, Ariz., United States, Feb. 15-17, 1976, Mineralogical Society of America and Friends of Mineralogy, Symposium 4, p. 2.

Fallick, A.E., and Barros, J.G., 1987, A stable-isotope investigation into the origin of beryl and emerald from the Porangatu deposits, Goias State, Brazil: Chemical Geology, Isotope Geoscience Section, v. 66, p. 293-300.

Farm, A.E., 1975, Emeralds and beryls: Journal of Gemmology, v. 14, p. 322-323.

Feininger, T., 1970, Emerald mining in Colombia; history: Journal of Gemmology, v. 12, p. 253-255.

Feininger, T., 1970, Emerald mining in Colombia; history and geology: Mineralogical Record, v. 1, p. 142-149.

Fersman, A.E., 1923, Izumrudnye Kopi na Urale Sbornik statei Materialov 46 Materialy dlya izucheniya estestvennykh proizvoditelinykh sil Rossi: Komisseiy Pri Rossisskoi Akademi Nauk, Petrograd (St. Petersburg), 82 p.

Fersman, A.E., 1929, Geochemische Migration der Elemente, III, Smaragdgruben im Uralgebirge: Abhandlungen praktische, Geologische und Bergwirtschaftslehre, v. 18, p. 74-116.

Fillman, K., and Banerjee, A., 1987, Smaragde; Chemismus, Einschluesse, Gefuege: Fortschritte der Mineralogie, Beiheft 65, 48 p.

Flanigen, E.M., Breck, D.W., Mumbach, N.R., and Taylor, A.M., 1965, New hydrothermal emerald: Gems and Gemology, v. 11, p. 259-264, 286.

Flanigen, E.M., Breck, D.W., Mumbach, N.R., and Taylor, A.M., 1967, Characteristics of synthetic emeralds: American Mineralogist, v. 52, p. 744-772.

Fonger, W.H., and Struck, C.W., 1975, Temperature dependences of Chromium (3+) ion radiative and non radiative transitions in ruby and emerald: Physical Review B., v. 11, p. 3251-3260.

Frere, A., 1975, Aventurine quartz as 'emerald' and 'jade': Deutschen Gemmologischen Gesellschaft, Zeitschrift, v. 24, p. 163-164.

Fryer, C.W., 1969-70, New nonfluorescent high property synthetic emeralds: Gems and Gemology, v. 13, p. 106-111.

Fujisaki, Y., 1976, Inclusions in emerald from Muzo, Colombia: Hoseki Gakkaishi, v. 3, p. 157-164.

Fumey, P., 1982, Muzo, capital of emeralds: Revue de Gemmologie, 72, p. 16-17.

Furbish, W.J., 1972, Unusual quartz inclusions in North Carolina emerald: Gems and Gemology, v. 14, p. 34-37.

Gaines, R.V., 1951, The sapphire mines of Kashmir: Rocks and Minerals, v. 26, p. 464-472.

Galia, W., 1972, Diagnostic criterion for the Linde synthetic emerald: Deutschen Gemmologischen Gesellschaft, Zeitschrift, v. 21, p. 112-117.

Garcia-Gimenez, R., and Leguey-Jimenez, S., 1980, Application of the Cauchy equation to the identification of gems: Revue de Gemmologie, A.F.G., 65, p. 8-10.

Garstone, J.D., 1981, The geological setting and origin of emerald deposits at Menzies, Western Australia: Journal of Royal Society, Western Australia, v. 64, p. 53-64.

Gentile, A.L., Cripe, D.M., and Andres, F.H., 1963, The flame fusion synthesis of emerald: American Mineralogist, v. 48, p. 940-944.

Gerbaux, X., and Hadni, A., 1970, Far infrared absorption spectra of three emeralds: Journal de Chimie Physique et de Physico-Chimie Biologique, v. 67, p. 1674-1675.

Ghera, A., and Lucchesi, S., 1987, An unusual vanadium beryl from Kenya: Neues Jahrbuch für Mineralogie, Monatschefte, v. 6, p. 263-274.

Ghiordanescu, V., and Cerchez, M.V., 1985, The effects of iron on ultraviolet protection and color characteristics of emerald green glasses: Glass Technology, v. 26, p. 26-62.

Gibbs, G.V., Breck, D.W., and Meagher, E.P., 1968, Structural refinement of hydrous and anhydrous synthetic beryl Al_2 $(Be_3 Si_6)$ O_{18} and emerald, $Al_{1.9}$ $Cr_{0.1}$ $(Be_3 Si_6)$ O_{18}: Lithos, v. 1, p. 275-285.

Gibbs, H.L., 1945, Beryl: Utah Mineralogical Society News Bulletin, v. 6, p. 29-33.

Gilles, V.A., 1979, The geology of Chivor emerald mines: Geological field trips, Colombia, 1959-1978: Colombian Society of Petrology, Geology, and Geophysics, p. 161-169.

Ginzburg, A.I., 1955, On the question of the composition of beryl: Trudy Mineralogischeskogo Muzeya, Akad. Nauk SSSR, Leningrad, v. 7, p. 56-69.

Gourlay, A.J., 1981, Gemstones: Australian Mineral Industry Annual Review, 1979, p. 123-127.

Graindorge, J.M., 1974, A gemmological study of emerald from Poona, W.A: Australian Gemmologist, v. 12, p.75-80.

Granadchikova, B.G., Andreenko, E.D., Solodova, Yu. P., Bukin, G.V., and Klyakhin, V.A., 1983, Diagnostics of natural and synthetic emeralds, Izvestiia Vyssikh Uchebnykh Zavedenii, Geologiya i Razvedka, v. 26, p..87-93.

Graziani, G., Gübelin, E.J., and Lucchesi, S., 1979, Einschluse und Genese eines Vanadium beryls von Salininha, Bahia, Brasilien: Deutschen Gemmologischen Gesellschaft, Zeitschrift, v. 28, p. 134-145.

Graziani, G., Gübelin, E.J., and Lucchesi, S., 1982, Genesis of emeralds in Kitwe, Zambia: Proceedings of the Congress of the Italian Society of Mineralogy and Petrology, Societa Italiana di Mineralogia, e Petrologia, Congresso, Abano, Padua, Italy, Rendiconti della Societa Italiana di Mineralogia-e-Petrologia, v. 38, p. 906-907.

Graziani, G., Gübelin, E.J., and Lucchesi, S., 1983, The genesis of an emerald from the Kitwe district, Zambia: Neues Jahrbuch für Mineralogie, Monatscheft, v. 4, p. 175-186.

Graziani, G., Gübelin, E.J., and Lucchesi, S., 1984, Emerald from the Kitwe district, Zambia: The Australian Gemmologist, v. 15, p. 227-234.

Griffon, J.C., Kremer, M.R., and Misi, A., 1967, Structure and genesis of the Carnaiba emerald deposit, Bahia: Brasillian Academy of Science, Annals, Brazil, v. 39, p. 153-161.

Grigor'ev, N.A., Aizikovich, A.N., and Sherstobitova, L.A., 1980, Kaolin formations of the emerald containing weathering crust: Doklady Akademii Nauk SSSR, v. 253, p. 957-959.

Grundmann, G., 1981, Inclusions of beryls and phenakite in emerald deposits of Habach Valley, Salzburg: Technische Universität Berlin, Fachgebiet Petrologie, v. 311, p. 227-237.

Grundmann, G., 1981, Inclusions of beryl and phenakites, in emerald deposits of Habach Valley (Salzburg Province, Austria): Karinthin, v. 84, p. 227-237.

Grundmann, G., 1985, Die Mineralien des Smaragdvorkommens im Habachtal: Lapis, Austria, v. 10, p. 13-33.

Grundmann, G., 1985 Emeralds from Sandawana: Journal of Gemmology, v. 6, p. 340-354; also in Gems and Gemology, v. 9, p. 195-203.

Grundmann, G., and Koller, F., 1979, The aeschynites and their zonal structure from beryllium-mineral parageneses of the emerald deposit near Leckbachscharte, Habachtal, Salzburg, Austria: Neues Jahrbuch für Mineralogie, Abhandlungen, v. 135, p. 36-47.

Grundmann, G., and Morteani, G., 1982, Geology of the Habach Valley emerald deposits, Salzburg: Institut für Angewandte, Geophysik und Petrologie, p. 71-107.

Gübelin, E.J., 1940, Differentiation between Russian and Colombian Emeralds: Gems and Gemology, v. 3, p. 89-92.

Gübelin, E.J., 1951, Some additional notes on Indian emeralds: Gems and Gemology, v. 7, p. 13-22.

Gübelin, E.J., 1956, The emerald from Habachtal: Gems and Gemology, v. 8, p. 295-309.

Gübelin, E.J., 1958, Notes on the new emeralds from Sandawana: Gems and Gemology, v. 9, p. 195-203.

Gübelin, E.J., 1960, The emeralds from Sandawana: Gemmologist, v. 29, p. 8-16.

Gübelin, E.J., 1964, Two new synthetic emeralds: Gems and Gemology, v. 11, p. 139-148.

Gübelin, E.J., 1969, On the nature of mineral inclusions in gemstone: Journal of Gemmology, v. 11, p. 146-192.

Gübelin, E.J., 1972, Inclusions in gemstones: Australian Gemmologist, v. 11, p. 3-14.

Gübelin, E.J., 1974, *Internal World of Gemstones:* ABC Zurich, Switzerland, 234 p.

Gübelin, E.J., 1974, The emerald deposit at Lake Manyara, Tanzania: Lapidary Journal, v. 28, p. 338-344, 346-347, 359-360.

Gübelin, E.J., 1981, The emerald and the ruby/spinel resources of Pakistan: Hoseki Gakkaishi, v. 8, p. 61-66.

Gübelin, E.J., 1982, The precious stones of Pakistan; II, The emerald deposits of Swat Valley: Lapis (Meunchen), v. 7, p. 19-26.

Gübelin, E.J., 1982, Gemstone of Pakistan: emerald, ruby, and spinel: Gems and Gemology, v. 18, p. 123-139.

Gübelin, E.J., 1989, Gemological characteristics of Pakistani emeralds: *in* A. H. Kazmi and L.W. Snee, editors, *Emeralds of Pakistan: Geology, Gemology and Genesis,* Van Nostrand Reinhold, New York, p. 75-92.

Gübelin, E.J., and Koivula, J.I., 1986, *Photoatlas of Inclusions in Gemstones:* ABC Zurich, Switzerland, 532 p.

Gübelin, E.J., and Schiffmann, C.A., 1978, Report on gemmologic investigation methods: Deutschen Gemmologischen Gesellschaft, Zeitschrift, v. 27, p. 33-40.

Guiliani, G., Couto, P., and Orstom, N., 1988, O metassomatismo de infiltracao e sua importancia nos depositos de emeralda do Brasil: Anais do VII Congresso Latino-Americano de geologia, v. 7, p. 457-459.

Guiliani, G., and Weisbrod, A., 1988, Scanning electron Microscopy (SEM) and its applications; determinations of solid and daughter minerals in fluid inclusions from some Brazilian emerald deposits. Annals of the VII Latinamerican Geologic Conference, v. 7, p. 445-458.

Halford-Watkins, J.F., 1935, Kashmir sapphires: Gemmologist, v. 4, p. 167-172; v. 5, p. 7-14.

Hamlin, A. C., 1972, The emerald: Mineralogical Digest, v. 3, p. 17-32.

Hammarstrom, J.M., Snee, L.W., and Kazmi, A.H., 1988, Mineral chemistry of emeralds from NW Pakistan; Geologic setting, chemical signature, and zoning: Geological Society of America, Abstracts with Programs, v. 20, p. A102.

Hammarstrom, J.M., 1989, Mineral chemistry of emeralds and some associated minerals from Pakistan and Afghanistan; an electron microprobe study: *in* A.H. Kazmi, and L.W. Snee, editors, *Emeralds of Pakistan: Geology, Gemology and Genesis,* Van Nostrand Reinhold, New York, p. 125-150.

Hanni, H.A., 1982, A contribution to the separability of natural and synthetic emeralds: Journal of Gemmology and Proceedings of the Gemmological Association of Great Britain, v. 18, no. 2, p. 138-144.

Hanni, H.A., 1983, Chemical comparision of natural and synthetic emeralds: Revue de Gemmologie, v. 76, p. 6-8.

Hanni, H.A., and Klein, H.H., 1982, Ein Smaragdvorkommen in Madagaskar: Deutschen Gemmologischen Gesellschaft, Zeitschrift, v. 21, p. 71-77.

Hanni, H.A., and Klein, H.H., 1983, Emerald deposit in Madagascar: Revue de Gemmologie, v. 74, p. 3-5.

Hanni, H.A., and Kerez, C.J., 1983, Neues vom Smaragdvorkommen von Santa Terezinha de Goia's, Goia's Brasilien: Deutschen Gemmologischen Gesellschaft, Zeitschrift, v. 32, p. 50-58.

Hasan, Z., Keany, S.T., and Manson, N.B., 1986, Spectrul energy transfer and fluorescence line narrowing in emerald: Journal of Physics C: Solid State Physics, v. 19, p. 6381-6387.

Henderson, E.P., 1945, A cat's eye emerald: Gems and Gemology, v. 5, p. 222.

Hermann, F, and Wussow, D., 1935, The emerald deposits of the World: Mining Magazine, v. 53, p. 20-25.

Heron, A.M., 1930, The gemstones of the Himalaya: Himalayan Journal, v. 2, p. 21-28.

Hickman, A.C.J., 1972, The Miku emerald deposit, Zambia: Zambian Geological Survey, Economic Report, no. 27, p. 1-35.

Hicks, T., 1988, Learning to collect in North Carolina: Lapidary Journal, v. 41, p. 53-60.

Hidden, W.E., 1882, The discovery of emeralds in North Carolina: New York Academy of Science, Treatise 1, p. 101-105.

Hidden, W.E., 1882, The discovery of emeralds in North Carolina: U.S. Geolgical Survey, p. 500-502.

Hidden, W.E., 1886, Recent discovery of emeralds and hiddenite in North Carolina: American Journal of Science, v. 3, p. 483-484.

Hintze, J., 1979, The emerald deposits of Colombia: Aufschluss, v. 30, p. 83-92.

Holms, A., 1951, The sequence of Pre-Cambrian orogenic belts in south and central Africa: International Geological Congress, v. 14, p. 254-269.

Hore, M.K., and Jadia, S.K., 1981, Geological, petrological and geochemical controls of the emerald mineralization in Tikhi-Kaliguman area, Udaipur District, Rajasthan: Symposium on three decades of developments in petrology, mineralogy and petrochemistry in India, Geol. Survey of India, 133 p.

Howie, R.A., 1984, Gemstones: Transactions of the Leicester Literary and Philosophical Society, v. 78, p. 22-27.

Hudson, S., 1986, Gemstones in Dixie: Rock and Gems, v. 16, p. 48-51.

Hutton, D.R., and Barrington, E.N., 1977, Electron spin resonance of emeralds: Australian Gemmologist, v. 13, p.107-108, 117-118.

Ichev, M., 1986, A new emerald deposit in Bulgaria: Geological Institute of Sofia, Bulgaria, Doklady Bolgarskoy Akademiya Nauk, v. 39, p. 81-83.

Ingeominas, 1981, Colombian emeralds: Bibliografia (Bogota), v. 1, p. 1.

Ives, R.L., 1976, Colombian emerald sources, Rocks and Minerals, v. 51, p. 507-510.

Jenkins, W.J., 1977, A Soude emerald from Colombia, a new presentation of an old pretender: Lapidary Journal, v. 31, p. 1630-1632.

Jobbins, A., 1978, Gemstones: *in* A. Woolley, editor, *the Illustrated Encyclopedia of the Mineral Kingdom,* Larousse Co., New York, p. 183-195.

Johnson, P.W., 1961, All about emeralds, natural or synthetic: Lapidary Journal, v. 15, p. 118-220.

Johnson, P.W., 1961, The Chivor emerald mine: Journal of Gemmology, v. 8, p. 126-152.

Just, E., 1926, Emeralds at Bom Jesus dos Meiras, Bahia, Brazil: Economic Geology, v. 21, p. 808-810.

Kanis, J., 1980, Gemstone news from Southern Africa: Deutschen Gemmologischen Gesellschaft, Zeitschrift, v. 29, p. 55-57.

Kashayev, N.I., 1973, Geochemical indicators of emerald occurrence in micaceous rocks: Razvedka i Okhrana Nedr, no. 3, p. 24-29.

Katz, E.F., 1981, A tour of African countries: Lapidary Journal, v. 34, p. 2320.

Kazmi, A.H., 1985, Development of gemstone resources of Pakistan: Muslim (Special Supplement), Feb. 24, Islamabad.

Kazmi, A.H., 1989, A brief overview of the geology and metallogenic provinces of Pakistan: *in* A.H. Kazmi, and L.W. Snee, editors, *Emeralds of Pakistan: Geology, Gemology and Genesis,* Van Nostrand Reinhold, New York, p. 1-11.

Kazmi, A.H., Peters, J.J., and Obodda, H.P., 1985, Gem pegmatites of Shingus-Dusso area, Gilgit, Pakistan: Mineralogical Record, v. 16, p. 393-411.

Kazmi, A.H., Lawrence, R.D., Anwar, J., Snee, L.W., and Hussain, S., 1986, Mingora emerald deposits (Pakistan), suture-associated gem mineralization: Economic Geology. v. 81, p. 2022-2028.

Kazmi, A.H., and Snee, L.W., (editors), 1989, *Emeralds of Pakistan: Geology, Gemology and Genesis,* Van Nostrand Reinhold, New York, 269 p.

Kazmi, A.H., Anwar, J., Hussain, S., Khan, T., and Dawood, H., 1989, Emerald deposits of Pakistan, *in* A.H. Kazmi, and L.W. Snee, editors, *Emeralds of Pakistan: Geology, Gemology and Genesis,* Van Nostrand Reinhold, New York, p. 39-74.

Kazmi, A.H., and Snee, L.W., 1989, Geology of the world emerald deposits, a brief review: *in* A.H. Kazmi, and L.W. Snee, editors, *Emeralds of Pakistan: Geology, Gemology and Genesis,* Van Nostrand Reinhold, New York, p. 165-228.

Kazmi, A.H., and Jafary, S.A., 1989, Selected bibliography on world emeralds: *in* A.H. Kazmi, and L.W. Snee, editors, *Emerald of Pakistan Geology,: Gemology and Genesis,* Van Nostrand Reinhold, New York, p. 237-254.

Keith, M., 1981, Japanese synthetic emeralds: Journal of Gemmology, v. 17, p. 290-291.

Keller, P.C., 1981, Emeralds of Colombia: Gems and Gemology, p. 80-92.

Keller, P.C., and Kampf, A.R., 1984, Gemstones and their origins: Terra, v. 23, p. 3-12.

Kellner, G.J., 1927, Reitrage zum Smaragdbergbau in der Republik Kolumbien, Sudamerika: Zeitschrift für praktische geologie, v. 35, p. 70-74.

Kiyevlenko, Y.Y., 1980, Precious stones; raw material resources and some results of geological studies: *in* Petrov, V.P., editor, *Gems and precious stones,* Nauka, Izdatel stvo, p. 5-10.

Koivula, J.I., 1980, The three-phase inclusion-a product of environment: Gems and Gemology, v. 16, p. 338-342.

Koivula, J.I., 1980-1981, Carbon Dioxide as a fluid inclusion: Gems and Gemology, v. 16, p. 386-390.

Koivula, J.I., 1982, Tourmaline as an inclusion in Zambian emeralds: Gems and Gemology, v. 18, p. 225-227.

Koivula, J.I., 1984, Mineral inclusions in Zambian emeralds: The Australian Gemmologist, v. 15, p. 235-239.

Koller, F., 1978, Not only emeralds; mineral occurrences in the Habach Valley: Lapis, v. 3, p. 39-46.

Koller F., and Niedermayr, G., 1977, Habach Valley, emerald deposit in the Lech River Channel: Oesterreichische Mineralogische Gesellschaft; Exkursion, Wieden P. (editor), Oesterreiche Mineralogische Gesellschaft Geotechnisches Institute, p. 15-24.

Korpershoek, H.R., 1968, Sapphire in the Taua emerald deposit, Ceara: Mineracao Metalurgia, v. 48, p. 36-42.

Kourimsky, J., 1977, Synthetic emeralds: Geologicky Pruzkum, v. 19, p. 219.

Kovac, C., 1984, Emeralds of Pakistan: Australian Gem and Treasure Hunter, v. 96, p. 8-11.

Kovac, C., 1984, Pakistan; dazzling gems and minerals from a rugged land: Australian Gem and Treasure Hunter, v. 94, p. 19-21.

Kovac, C., 1986, Minerals and gemstones of Pakistan: The Australian Gemmologist, v. 16, p. 57-59.

Kozlowski, A., Metz, P., and Jaramillo, H.A.E. 1988, Emeralds from Somondoco, Colombia: chemical composition fluid inclusions, and origin: Neues Jahrbuch für Mineralogie, Abhandlungen, v. 159, p. 23-49.

Kraus, P.D., 1976, Emerald; birthstone for May: Lapidary Journal, v. 30, p. 480-486.

Kunz, G.F., 1894, A new locality of true emerald: American Journal of Science, v. 3, p. 429-430.

Kunz, G.F., 1896, Mineral resources of the United States, Precious Stones: U.S. Geological Survey, Annual Report, v. 18 (1986-1897), pt. 5., p. 1183-1217.

Kwast, K., Keilhoff, U., Engelmann, F., Petrova, M.G., and Belakovskii, D.I., 1986, Emerald from calciphyres: Izvestiia Vyssikh Uchebnykh Zavedenii, Geologiya i Razvedka, no. 12, p. 114-115.

Laflamme, J. C. K., 1885, Note sur un gisement d'emeraude au Saguenay: Royal Society of Canada, Proceedings and Transactions, 2, iv, p. 231-232.

Landero, C.F., 1923, Las Esmeraldas y otras Gemas Glucinicas: Memorias de la Sociedad Cientifico "Antonio Alzate", v. 42, p. 473-488.

Latham, E.B., 1911, The Newly Discovered Emerald Mines of "Somondoco": School of Mines Quarterly, v. 32, p. 210-214.

Lavrinenko, L.F., Levenshtein, M.L., Polunovskii, R.M., Rozanov, K.I., and Rozenberg, D. Sh., 1971, Emerald find in the Ukraine: L'vovskoe geologicheskoe, Mineralogicheskii Sbornik, no. 25, p. 85-87.

Lawrence, R.D., Kazmi, A.H., and Snee, L.W., 1989, Geological Setting of the emerald deposits: *in* A.H. Kazmi and L.W. Snee, editors, *Emeralds of Pakistan: Geology, Gemology and Genesis,* Van Nostrand Reinhold, New York, p. 13-38.

Lebeau, P., 1895, Sur la Traitement de I' Emeraude et la Preparation de la Glucine pure: Academic des Sciences, Paris, Comptes Rendus, v. 121, p. 641.

Lefever, R.A., Chase, A.B., and Sobon, L.E., 1962, Synthetic emeralds: American Mineralogist, v. 47, p. 1450-1453.

Le Grange, J.G., 1930, The Barbara beryls; a study of an occurrence of emeralds in the northeastern Transvaal, with some observations on metallogenic zoning in the Murchison Range: Transactions of the Geological Society of South Africa, v. 32, p. 1-25.

Leitmeier, H., 1937, Das Smaragdvorkommen im Habachtal in Salzburg und seine Mineralien: Zeitschrift für Kristallographie, Mineralogie, und Petrographie, Abteilung A., v. 49., p. 245-368.

Leitmeier, H.C., Doelter, 1930, Das Samagdvorkommen im Habachtal: Mineralogische und Petrographische Mitteilungen, v. 49, p. 11-15.

Liddicoat, R.T. Jr., 1964, Developments in the synthetic-emerald field: Gems and Gemology, v. 11, p. 131-138.

Liddicoat, R.T., Jr., 1973, Development and highlights at GIA's lab in Los Angeles: Gems and Gemology, v. 14, p. 200-207.

Lind, T., Schmetzer, K., and Bank, H., 1984, Gem-quality blue and green beryls (Aquamarine and emeralds) from Nigeria: Deutschen Gemmologischen Gesellschaft, Zeitschrift, v. 33, p. 128-i38.

Lind, T., Schmetzer, K., and Bank, H., 1986, Blue and green beryls (aquamarines and emeralds) of gem quality from Nigeria: Journal of Gemmology and Proceedings of the Gemmological Association of Great Britain, v. 20, p. 40-48.

Lindstein, D.C., 1975, The emerald: Lapidary Journal, v. 28, p. 1694-1700.

Lindsten, D.C., 1985, Emerald; the rarest beryl: Lapidary Journal v. 39, p. 25-32.

Little, H.P., 1917, An ancient reference to the emeralds: Science, v. 45, p. 291-292.

Lokhova, G.G., Ripinen, O.I., Bukin, G.V., Veis, M.E., and Solntsev, V.P., 1977, The quantitative estimation of natural emerald color characteristics: Zapiski Vsesoyuznogo Mineralogicheskogo Obschestva, v. 6, p. 704-707.

Loeffler, B.M., and Burns, R.G., 1977, Shedding further light on the color of minerals and gems: Geological Society America, Abstracts with Programs, v. 9, p. 1072-1073.

Lowery, R., 1972, Emerald rush on the Adams Farm: Rock and Gem, v. 2, p. 12-17.

Mac Nevin, A.A., and Holmes, G.G., 1980, Gemstones: Mineral Industry of New South Wales, v. 18, p. 119.

Mariano, A.N., 1987, Geochemical characterization of mineral deposits—Pakistan: United Nations Department of Technical Cooperation for Development, unpublished report, 43 p.

Martin, H.J., 1962, Some observations on southern Rhodesian emeralds and chrysoberyls: Chamber of Mines Journal, 4, no. 10, p. 34-38.

Martin, H.J., 1963, Observations on southern Rhodesian emeralds and chrysoberyl: Symposium on Pegmatites, Rhodesia, Salisbury, S. Rhodesia, p. 35-39.

McKague, H. L., 1964, Trapiche emeralds from Colombia: Gems and Gemology, v. 11, p. 210-213, 223.

Mendes, J.C., and Schwarz, D., 1985, Emerald deposits of Brazil, geology and mineralization: University of Federal Ouro Preto, Sexto Congreso Latinoamericano de geologia; Bogota, Colombia, p. 291-299.

Metson, N.A., and Taylor, A.M., 1977, Observations on some Rhodesian emerald occurrences: Journal of Gemmology, v. 15, p. 422-234.

Mian, I., 1970, Chromium-bearing minerals of the Northwest Frontier Province: University of Peshawar, Geological Bulletin, v. 5, p. 13.

Michel, H., 1927. Der Brasilianische Smaragd: Centre für Mineralogie, p. 218-220.

Miers, H.A., 1893, Quartz from the emerald and Hiddenite mine, N.C.: American Journal of Science, v. 46, p. 420-424.

Mil'grom, G.B., and Musafronov, V.M., 1980, Some features of geologic-economic evaluation of deposits of precious stones: *in* Petrov, V.P., (editor), Gems and Precious Stones, Nauka Izdatel stvo, p. 40-44.

Mitchell, J.R., 1984, Field trip; Hiddenite, North Carolina: Rock and Gem, v. 14, p. 48-51.

Miyata, T., Hosaka, M., and Chikayama, A., 1987, On the inclusions in emeralds from Santa Terezinha de Goias; Brazil: Journal of Gemmology, v. 20, p. 377-379.

Moeller, R., 1983, Colorimetric study on green varities of beryl emeralds: Gemologia (Barcelona), v. 18, p. 5-53.

Moir, J., 1912, Notes on the spectra of the precious emerald and other gemstones: Royal Society of South Africa, Transactions, 2, p. 273.

Moir, J., 1918, Spectrum phenomena in the chromium compounds: Part iv of the spectrum of the ruby and emerald, Royal Society of South Africa, Transactions, 7., p. 129-130.

Moroz, I.I., 1978, Main features of the geochemistry of emerald—containing glimmerites of the Urals: Izvestiia Vyssikh Uchebnykh Zavedenii, Geologiya i Razvedka, v. 21, p. 157.

Moroz, I.I., 1981, Forms of indicator elements in primary aureoles of smaragdite mica schists in Ural deposits: Izvestiia Vyssikh Uchebnykh, Zavedenii, Geologiya i Razvedka, v. 6, p. 61-67.

Moroz, I.I., 1983, Forms of occurrence of element-indicators in the primary haloes of emerald-bearing biotites in an Uralian deposit: International Geology Review, v. 25, p. 1021-1026.

Moroz, I.I., and Lobanov, V.K., 1980, Geochemical zoning of endogene aureoles in emerald-bearing micas from the Urals: *in* Petrov, V.P., editor, *Gem and Precious Stones*, Nauka, Izdatel' stvo, p. 175-181.

Morteani, G., and Grundmann, G., 1977, The emerald porphyroblasts in the penninic rocks of the central Tauern Window: Neues Jahrbuch für Mineralogie, Monatschefte, 11, p. 509-516.

Muktinath, N.C., 1967, Some recent mineral development work in the Aravalli belt, Rajasthan: Symposium of the Upper Mantle Project, Proceedings, p. 363-391.

Muktinath, N.C., Challopadhyay, N., and Banerjee, S.N., 1969, Emerald mineralization in Rajgarh-Chat-Bithur area, Ajmer district, Rajasthan: Indian Scientific Congress Association, 56th Session, Proceedings, part 3, p. 237.

Mumme, I.A., 1982. *The emerald; its occurrence, discrimination and valuation:* Mumme Publications, 46, Australia., 135 p.

Mumme, I.A., 1985, Emeralds: facts and legends: Australian Gem and Treasure Hunter, v. 98, p. 13-16.

Mumme, I.A., 1985, Modes of occurrence of emeralds in Australia: The proceedings of the 20th International Gemmological Conference; part 2, Hicks, B., (editor), The Australian Gemmologist, v. 16, p. 106-108.

Munoz, G.O., and Villalba, A.M.B., 1948, Esmeraldas de Colombia: Banco Republic, Bogota, p. 135.

Namiki, M., 1979, Two new synthetic emerald: Hoseki Gakkaishi, v. 6, p. 36-46.

Nassau, K., 1976, Synthetic emerald: the confusing history and the current technologies: Journal of Crystal Growth, v. 35, p. 211-222.

Nassau, K., 1977, The history of emerald synthesis, crystal growth and habit: Third Mineralogical Society of America-Friends of Mineralogy Symposium, p. 13-14.

Nassau, K., and Jackson, K.A., 1970, Trapiche emeralds from Chivor and Muzo, Colombia: American Mineralogist, v. 55, p. 416-427.

Nassau, K., and Jackson, K.A., 1970, Trapiche emeralds from Colombia, Correction: American Mineralogist, v. 55, p. 1808-1809.

Neves, J.M.C., 1978, Estudo de ocorrencias de esmeralda na provincia da Zambezia—Mocambique: Sociedade Brasileira de Geologia, Anais do Congresso, 30, v. 3, p. 1141-1155.

Niedermayr, G., 1978, Beryllium minerals from the Pinzgau: Lapis, v. 3, p. 60-62.

Niedermayr, G., 1988, Mineralien und Smaragd bergau im Habachtal: Emser Hefte, v. 9, p. 1-48.

Nottes, G., 1981, A voyage to the emerald mines of Colombia: Der Aufschluss, v. 32, p. 325-334.

O'Donoghue, M.J., 1971, Trapiche Emerald: Journal of Gemmology, v. 12, p. 329-332.

Ohkura, H., Hashimoto, H., Mori, Y., Chiba, Y., and Isotani, S., 1987, The luminescence and ESR of a synthetic emerald and the natural ones mined from Santa Terezinha in Brazil: Japanese Journal of Applied Physics, part 1, v. 26, p. 1422-1428.

Okrusch, M., Richter, P., and Gurkan, A., 1981, Geochemistry of blackwall sequences in the Habachtal emerald deposits: Hohe Tauern, Austria, part 1: Presentation of Geochemical Data: Tschermaks Mineralogische und Petrographische Mitteilungen, v. 29, p. 9-31.

Olden, C., 1912, Emeralds; their mode of occurrence and methods of mining and extraction in Colombia: Institution of Mining and metallurgy, Transactions, v. 21, p. 193-203.

Oppenheim, V., 1948, The Muzo emerald zone, Colombia. S.A.: Economic Geology, v. 43, p. 31-38.

Ottaway, T.L., Wicks, F.J., Bryndzia, L.T., and Spooner, E.T.C., 1986, Characteristics and origin of the Muzo emerald deposit, Colombia: International Mineralogical Association, Abstracts with Programs, Stanford, California, p. 193.

Oughton, J.H., 1975, New synthetic gems set a problem: Australian Gemmologist, v. 12, p. 222-226.

Palache, C., Davidson, S.C., and Goranson, E.A., 1930, The Hiddenite deposit in Alexander County, North Carolina: American Mineralogist, v. 15, p. 280-302.

Ponahlo, J., and Koroschetz, T., 1986, Quantitative cathodoluminescence of gemstones: The Australian Gemmologist, v. 16, p. 64-71.

Pearl, R.M., 1978, Gems from the rocks: Earth Science, v. 31, p. 36-38.

Peisley, D., and Driver, S.P., 1979, Painter Province; Land of promise, part I: Australian Gems and Crafts Magazine, v. 41, p. 6-11.

Petrusenko, S., and Arnaudov, V., 1978, Emeralds from desilicified pegmatites in Bulgaria: *in* Sidorenko, A.V., editor, *Gem Minerals,* Nauka, Izdatel'stvo, Leningrad otd., USSR, p. 74-79.

Petrusenko, S., Arnaudov, V., and Kostov, I., 1966, Emerald Pegmatite from the Urdini Lakes, Rila Mountain: Godishnik na Sofiiskiya Universitet, Geologo-Geografski Fakultet, v. 59, p. 247-268.

Petrusenko, S., Arnaudov, V., and Kostov, I., 1971, Comparative study of beryls in Bulgaria: Bulgarska Akademiia na Naukite, Sofia, Geologicheski Institut, Izvestiia, Seriia Geokhimiia, Mineralogiia, i Petrologiia, v. 20, p. 45-68.

Pittman, E.F., 1901, The mineral resources of New South Wales, Sydney: Geological Survey of New South Wales, 487 p.

Platonov, A.N., Dorfman, M.D., Taran, M.N., and Tarashchan, A.N., 1978, Spectroscopic research on emeralds from different deposits: Konstitutsiia i Svoistva Mineralov, v. 12, p. 115-121.

Pogue, J.E., 1961, The emerald deposits of Muzo, Colombia: American Institution of Mining and Engineering, Transactions, 4, p. 910-933.

Pohl, W., Horkel, A., Neubauer, W., Niedermayer, G., Okelo, R.E., Wachira, J.K., and Werneek, W., 1980, Notes on the geology and mineral resources of the Mtito Andei-Taita Area (Southern Kenya): Metteilungen der Osterreichischen Geologische Gesellschaft, v. 73, p. 135-152.

Poirot, J. P., 1971, Some notes on growth-disturbances found in Colombian emeralds: Journal of Gemmology, v. 12, p. 271-274.

Pratt, S., and Texpet, Col., 1979, The Muzo emerald mine, Geological field trips Colombia, 1959-1978: Colombian Society for Petrology, Geology, and Geophysics, p. 33-63.

Pyatnitski, P., 1929, Geological explorations in the emerald region of the Urals , genetical interrelations of the rocks of the emerald schist series: Bulletin of the Commission of Geology of Leningrad, v. 48, p. 283-306.

Pyatnitski, P.P., 1932, Geological exploration in the emerald district in the Urals, answer to the question of the genesis of emeralds: Geological Survey of U.S.S.R., Transactions, no. 75, p. 1-77.

Pyatnitski, P.P., 1932, Geological explorations in the emerald region of the Urals, III, Geological conditions of the emerald occurrence out of U.S.S.R.: Geological Survey of U.S.S.R.: Transactions, no. 189, p. 1-46.

Pyatnitsky, P., 1934, Emeralds, their occurrences and genesis: Ukraine Geological, Hydrological and Geodetical Trust, Transactions, p. 1-47.

Rafiq, M., and Jan M.Q., 1983, A discovery of emerald near Bucha, Mohmand Agency: University of Peshawar, Geological Bulletin, v. 16, p. 188.

Rafiq, M., and Jan, M.Q., 1985, Emerald and green beryl from Bucha, Mohmand Agency, NW Pakistan: Journal of Gemmology and Proceedings of the Gemmological Association of Great Britain, v. 19, p. 404-411.

Rainier, P.W., 1929, The Chivor-Somondoco emerald mines of Colombia: American Institution of Mining and Metallurgical Engineering, Technical Publication, v. 258, p. 21.

Ralls, B.W., 1977, The Chivor emerald mine: Gems and Minerals, v. 482, p. 10-12, 72-73.

Ralls, B.W., 1978, Micro news: Gems and Minerals, v. 489, p. 60-61.

Randolph, G. C., 1934, Emeralds, large flawless specimens, symbols of perfection: Oregon Mineralogist, v. 2, p. 7-8.

Read, P., 1980, Introduction to gemmology, part 9, Detection of artificial coloration luminescence: Australian Gem and Treasure Hunter, v. 52, p. 10-12.

Read, P., 1984, Gem studies in Brazil: Australian Gem and Treasure Hunter, v. 92, p. 8-11.

Read, P.G., 1980, Visual colorimetry and comparison grading: Journal of Gemmology, v. 17, p. 29-42.

Remaut, G., and Vochten, R., 1985, Blue-green apatite from Gravelotte, South Africa: The Proceedings of the 20th International Gemmological Conference; part 2, Hicks, B., (editor), The Australian Gemmologist, v. 16, p. 115.

Renders, P.J., and Anderson, G. M., 1987, Solubility of kaolinite and beryl to 573 K: Applied Geochemistry, v. 2, p. 193-203.

Rife, D., 1965, Regarding emerald and beryl: Rocks and Minerals, no. 40, p. 569-570.

Ringsrud, R., 1988, Muzo emerald: Lapidary Journal, v. 41, p. 27-84.

Rippinen, O.I., Solntsev, V.P., and Lokhova, G.G., 1977, Study of the color of grown emeralds: Akademiia nauk Gzuzinskoi SSR, Tiflis; Tezisy Dokladov Vsesoiuznogo Soveshchaniia po Rostu Kristallografiia, v, 5, p. 237-278.

Roberts, C., 1981, Beryl; color and variety: Australian Gem and Treasure Hunter, v. 64, p. 35.

Robertson, A.D., MacNevin, A.A., Tan, S.H., Johns, R.K., Carter, J.D., Threader, V.M., Brooks, J.H., Hiern, M.N., and Connolly, R.R., 1976, Precious stones: *in* Knight, C.L., editor, Economic geology of Australia and Papua New Guinea: Industrial Minerals and Rocks, v. 4, p. 305-324.

Rogers, A.F., and Sperisen, F.J., 1942, American synthetic emerald: American Mineralogist, v. 27, p. 762-768.

Rolff, P.A.M., 1970, The geological environment for emerald in Brazil: Lapidary Journal, v. 23, p. 1488-1490, 1492-1502.

Root, E., 1986, Gems and minerals of the U.S.S.R.: Lapidary Journal, v. 40, p. 42-47.

Rossman, J., 1936, The emerald deposits of Colombia: Mineralogist, v. 4, p. 5-6, 25-30.

Rossovskiy, L.N., 1980, Gemstone deposits of Afghanistan: Geologikila Rudnykh Mestorozhdenift, v. 22, p. 77-88.

Rothstein, J., 1982, Just what is an emerald: Lapidary Journal, v. 36, p. 734-743.

Roy, B.C., 1955, Emerald deposits in Mewar and Ajmer-Marwara: Geological Survey of India, Records, v. 86, p. 377-401.

Rudowski, L., 1988, Petrologie et geochimie des series evolutives de granites peralumineux des massifs de Campo Formoso et Carnaiba et leurs relations avec les mineralisations a emeraudes (Serra de Jacobina, Bahia, Bresil): Anais do VII Congresso Latino—Americano de Geologia, v. 7, p. 567-575.

Rudowski, L., Giuliani, G., and Sabate, P., 1987, Metallogenie les phlogopitites a'émeraude au voisinage des granites de Campo Formoso et Carnaiba (Bahia, Bresil): un exemple de mineralisation prote rozoique a'Be, Mo et W dans des ultrabasites metasomatiseés: Academie des Sciences, Paris, Comptes Rendus, v. 304, Sr. II, no. 18, p. 1129-1134.

Sakamoto, C., Fujii, S., Sugie, Y., and Hanamoto, T., 1985, Crystal growth of emerald and chromium (3+) ion doping conditions in the flux process, Yogyo Kyokaishi, v. 93, p. 732-738.

Sakikawa, N., 1973, Formation of natural gem stones: Kagaku Kogyo, v. 24, p. 483-488.

Samostsvety, 1978, *Gem minerals:* Sidorenko, A.V., editor, xi S'ezda Mezhdunarodnaya Mineralogicheskaya Assotsiatsiya Novosikiisk, Union of Socialist Republic, Sept. 4-10, p. 150.

Sauer, D.A., 1982, Emeralds from Brazil: Proceedings of the International Gemological Symposium. Eash, D.M., (editor), Gemological Institute of America, p. 357-377.

Schedenhelm, W.R.C., 1986, United States gemstones: Rocks and Gems, v. 16, p. 18-21.

Scheibe, R., 1926, Beitrage zur Geologie und Mineralogie von Kolombien (Von O. Stutzer), III, Das Smaragdvorkommen Von Nemocon: Neues Jahrbuch für Mineralogie; B-B., v. 53, p. 321-324.

Scheibe, R., 1926, Beitrage zur Geologie und Mineralogie von Kolombien (Von O. Stutzer), IV, Die Smaragdlagerstatte von Muzo (Kolombien): Und ihre nahere Umgebung, Neues Jahrbuch für Mineralogie, B-B., v. 50, p. 419-447.

Schiffmann, C., 1968, Unusual Emeralds: Journal of Gemmology, v. 11, p. 105-114.

Schiffmann, C., 1969, Unusual Emeralds, Lapidary Journal, v. 23, p. 828-831.

Schlenker, J.L., Gibbs, G.V., Hill, E.G., Crews, S.S., and Myers, R.H., 1977, Thermal expansion coefficients for indialite, emerald and beryls: Physics and Chemistry of Minerals, v. 1, p. 243-255.

Schlussel, R., 1984, Structural defects in natural and synthetic emeralds: Revue de Gemmologie, v. 81, p. 13-18.

Schmetzer, K., 1982, Absorptionsspektren und Farben Vanadium-und Chromhaltiger Minerale: Deutschen Gemmologischen Gesellschaft, Zeitschrift, v. 31, p. 125-130.

Schmetzer, K., and Bank, H., 1980, Emeralds from Zambia with unusual pleochroism: Deutschen Gemmologischen Gesellschaft, Zeitschrift, v. 29, p. 149-151.

Schmetzer, K., and Bank, H., 1981, An unusual pleochroism in Zambian emerald: Journal of Gemmology and Proceedings of Gemmological Association of Great Britain, v. 17, p. 443-446.

Schmetzer, K., Berdesinski, W., and Bank, H., 1974, Uber die Mineralart Beryll, ihre Farben und

Absorptionsspektren: Deutschen Gemmologischen Gesellschaft, Zeitschrift, v. 23, p. 5-39.

Schmetzer, K., and Brezina, J., 1975, Emeralds from Ghana: Deutschen Gemmologischen Gesellschaft, Zeitschrift, v. 24, p. 94.

Schmetzer, K., and Krupp, H., 1979, New discoveries of precious stones in East Africa: Deutschen Gemmologischen Gesellschaft, Zeitschrift, v. 28, p. 35-38.

Schrader, H.W., 1983, Contributions to the study of the distinction of natural and synthetic emeralds: Journal of Gemmology, v. 18, no. 6, p. 530-543.

Schrader, H., 1985, A "three-phase inclusion" in an emerald from South Africa: Journal of Gemmology and Proceedings of the Gemmological Association of Great Britain, v. 19, p. 484-485.

Schrader, H.W., 1987, Farbzentren durch Erstaz von silicium in Beryl, $Be_3 Al_2 Si_6 O_{18}$: Fortschritte der Mineralogie, Beiheft 65, p. 171.

Schrader, H., and Henn, U., 1986, On the problems of using the gallium content as a means of distinction between natural and synthetic gemstones: Journal of Gemmology and Proceedings of the Gemmological Association of Great Britain, v. 20, p. 108-113.

Schwarz, D., 1984, Inclusions in emeralds, Revista Escola de Minas, v. 37, p. 25-34.

Schwartz, D., 1986, Classificao genetica das ocorrecias de emeralda: Sociedade Brasileira de Geologia, Congresso 35, Resumo, Boletim No. 1, p. 184.

Schwartz, D., Hanni, H.A., Martin, F.L. Jr., Fischer, M., 1988, The emeralds of Fazenda Boa Esperanca, Taua, Ceara, Brazil; occurrence and properties: Journal of Gemmology, v. 21, p. 168-178.

Schwartz, D., and Mendes, J.C., 1985, Classification of the inclusions in Brazilian emeralds from the mines of Socoto/BA, and Itabira, MG: Congreso Latinoamericano de Geologia, Memorias, v. 6, p. 400-414.

Seal, R.R. II., 1989, A reconnaissance study of the fluid inclusion geochemistry of the emerald deposits of Pakistan and Afghanistan: *in* A.H. Kazmi and L.W. Snee, editors, *Emeralds of Pakistan: Geology, Gemology and Genesis,* Van Nostrand Reinhold, New York, p. 151-164.

Selig, B.L., 1965, Carnaiba emerald mine, Gems and Minerals, no. 331, p. 22-24.

Selset, R., 1963, Emerald Locality of Mjosa Lake, Norway: Rocks and Minerals, v. 38, p. 608-609.

Shaefer, W., 1984, The Colombian emerald deposits at Muzo and Chivor: Lapis, v. 9, p. 9-23.

Shatskii, V.S., Lebedev, A.S., and Klyakhin, V.A., 1980, Micromorphology of synthetic hydrothermal emerald crystals: Neodnorodnost, Mineralov Materialy, II-go S'ezda M.M.A. Novosibirsk, 1978, M., p. 255-261.

Shelton, W., 1986, The mines and minerals of Russia; IV: The Earth Science News (Earth Science Club of Northern Illinois), v. 37, p. 14-15.

Sherstyuk, A.I., 1975, Ore-controlling significance of metasomatites, correlation of auto-and allometamorphism processes to the emerald content and the use of chromium titanium and other indicator ratios for evaluating the future outlook of deposits: *in* Kazitsyn, Yu. V., and Landa, E.A., editors, Metasomatizm Rudoobraz, "Nedra", Moscow, USSR, p. 219-225, 260-274.

Shire, M., 1982, Emeralds: Proceedings of the International Gemological Symposium, Eash, D.M., (editors), Gemological Institute of America, p. 349-354.

Sinkankas, J., 1977, Historical notes on South American gemstones, Gems and Gemology, v. 15, p. 334-344.

Sinkankas, J., 1981, Alpine type beryl-emerald deposits near Hiddenite, North Carolina: *in* Brown, G.E., editor, *The mineralogy of pegmatite,* American Mineralogist, v. 67, p. 9-10.

Sinkankas, J., 1981, *Emerald and other beryls:* Chilton Book Co., Radnor, Pennsylvania, USA., 665 p.

Sinkankas, J., 1982, Alpine type beryl-emerald deposits near Hiddenite, North Carolina: *in* Brown, G.E., Jr. editors, The Mineralogy of Pegmatites, American Mineralogist, v. 67, p. 181.

Sliwa, A.S., and Nguluwe, C.A., 1984, Geological setting of Zambian emerald deposits: Precambrian Research, v. 25, p. 213-228.

Smith, F.G., 1951, Emerald from Colombia: Progress Report: Temperature-Pressure Research of Hydrothermal Mineral Deposits, v. 3, p. 104.

Snee, L.W., Foord, E.E., Hill, B., and Carter, S.J., 1989, Regional chemical differences among emeralds:

and host rocks of Pakistan and Afghanistan: implications for the origin of emerald: *in* A.H. Kazmi, and L.W. Snee, editors, *Emeralds of Pakistan: Geology, Gemology and Genesis,* Van Nostrand Reinhold, New York, p. 93-124.

Snee, L.W., and Kazmi A.H., 1989, Origin and classification of Pakistani and world emeralds deposits: *in* A.H. Kazmi, and L.W. Snee, editors, *Emeralds of Pakistan: Geology, Gemology and Genesis,* Van Nostrand Reinhold, New York, p. 229-236.

Soliman, M.M., 1986, Ancient emerald mines and beryllium mineralization associated with Precambrian stanniferous granites in the Nugrus-Zabara area, southeastern desert, Egypt: Arab Gulf Journal of Scientific Research, v. 4, p. 529-548.

Sposito, E., 1934, Las Esmeraldas de Venezuela: Venezuela University Center, Carcas Annals; v. 22, p. 152-166.

Staatz, M.H., Griffiths, W.R., and Barnett, P.R., 1965, Differences in the minor element composition of beryl in various environments: American Mineralogist, v. 50, p. 1783-1795.

Sterrett, D., 1958, "Old Plantation" emerald mine: Rocks and Minerals, v. 33, p. 302-307.

Sterrett, D.B., 1912, An occurrence of emeralds in North Carolina: Washington Academy of Science, v. 2, p. 360-361.

Sterrett, D.B., 1976, July-August, 1958, "Old plantation" emerald mine: Rocks and Minerals, v. 51, Retrospective 50th anniversary, p. 521-524.

Stockton, C.M., 1984, The chemical distinction of natural from synthetic emeralds: Gems and Gemology, v. 20, p. 141-145.

Stockton, C.M., 1987, The separation of natural from synthetic emeralds by infrared spectroscopy: Gems and Gemology, v. 23, p. 96-99.

Stroh, R., 1982, New finds in the Alps: Lapis, v. 7, p. 34-35.

Sunagaw, H., 1982, Gem materials, natural and artificial: Current Topics in Materials Science, v. 10, p. 353-497.

Sunagawa, I., 1964, A distinction between natural and synthetic emeralds: American Mineralogist, v. 49, p. 785-792.

Sunagawa, I., 1975, Surface microtopograph as a tool of distinguishing natural and synthetic emeralds: Fortschritte der Mineralogie v. 52, Special Issue: International Mineralogical Association, Nineth general meeting, p. 515-520.

Sunagawa, I., 1981, Crystal growth; physics of minerals; electronic microscopy; 12th general assembly of the International Mineralogical Association, Bulletin de Mineralogie, v. 104, p. 128-132.

Sunagawa, I., 1981, Natural and synthetic gem materials, a comparison: Bulletin of Mineralogy, v. 104, p. 128-132.

Sunagawa, I., and Hamada, M., 1978, Lattice images of synthetic emerald: Gemmological Society of Japan, Journal, v. 5, p. 11-14.

Superchi, M., and Roland, V., 1980, A proposal for delimiting ruby (from rose and violet corundum) and emerald (from light green and dark green beryl): Deutschen Gemmologischen Gesellschaft, Zeitschrift, v. 29, p. 68-70.

Sviridov, D.T., Sviridova, R.K., and Grum-Grzhimailo, S.V., 1970, Chromium (III) in a trigonal field: Grum-Grzhimailo, S.V. (editor), Spektroskopiya Kristallov, Materialy Simpoziuma, 2nd, p. 266-270.

Sviridova, R.K., Shchetkov, A.A., and Cherepanov, V.I., 1975, Polarization dependences of the intensities of the R-lines in ruby and emerald: Vsesoiuznyi Sbornik Spektroskopii Kristallografiia, p. 246-54.

Swarz, D., 1986, Synthetic emeralds manufactured by hydrothermal processes: Revista Escola de Minas, v. 39, Federal University of Ouro Preto., p. 35.

Takubo, H., 1979, Characteristics of synthetic hydrothermal emeralds from the USA and the USSR, Hoseki Gakkiashi, v. 6, p. 113-128.

Takubo, H., Kitamura, Y., Nakazumi, Y., and Koizumi, M., 1979, Internal textures and growth conditions of flux-grown emeralds from the USSR: Hoseki Gakkaishi, v. 6, p. 22-28.

Takubo, H., and Koizumi, M., 1977, Specific gravity, refractive indices and depth of formation of natural emeralds: Gemmological Society of Japan, v. 4, p. 24-33; Hoseki Gakkaishi, v. 4, p. 168-177.

Takubo, H., Muguruma, A., and Koizumi, M., 1977, Relation between internal textures and growth conditions of flux-grown emerald: Hoseki Gakkaishi, v. 4, p. 3-12.

Taylor, A.M., 1976, African clues to the great emerald mystery: Lapidary Journal, v. 30, p. 692-694, 696, 718, 720, 722.

Taylor, A.M., 1977, Emeralds and emeralds: Journal of Gemmology v. 15, p. 372-376.

Tenhagen, J. W., 1972, Muzo emerald mine; a visit: Gems and Gemology, v. 14, p. 77-81.

Tether, J., Patney, R.K., and Malik, J.I., (undated). The distribution and provenance of Zambia's coloured gemstones: Occasional Paper 110, Geological Survey of Zambia. Lusaka.

Theisen, V., 1969, Structural phenomena of gems: Goldschmiede Zeitung, v. 67, p. 16-18.

Themelis, T., 1987, Inclusion of the month; trapiche emerald: Lapidary Journal, v. 41, p. 19.

Themelis, T., 1989, Gemology, new East African deposits: Lapidary Journal, v. 42, p. 34-36, 38-39.

Thurm, R.E. 1972, Emerald from Lake Manyara, Tanzania: Deutschen Gemmologischen Gesellschaft, Zeitschrift, v. 21, p. 9-12.

Thurm, R.E., 1972, The Lake Manyara emeralds of Tanzania: Journal of Gemmology, v. 13, p. 98-99.

Tkachenko, A.A., 1980, Structure of emerald crystals grown by the flux method: Vyrashchivanie Kristallov Berillievykh Mineralov i Issled, ikh Svoistv., p. 41-44.

Touray, J.C., and Poirot, J.P., 1968, Observations on primary fluid inclusions in emeralds and their relationship with solid inclusions: Academie des Science, Paris, Comptes Rendus, Ser. D., v. 266, p. 305-308.

Trapp, F., 1968, The Land of the sky and emeralds: Lapidary Journal, 22, p. 586, 588-592, 594-596.

Ulloa, M., and Carlos, E., 1980, Geology of the emerald deposits of Colombia: Boletin de la Sociedad Geologica del Peru, v. 65, p. 157-170.

Ushio, M., 1976, Study on the synthesis of industrial large single crystals, VII, Vanadium pentoxide flux-growth of emerald single crystals: Nippon Kagaku Kaishi, no. 5, p. 748-751.

Ushio, M., 1976, Study on the synthesis of industrial large single crystals, IX. Studies on temperature distribution and thermal convection in a molten vanadium pentoxide flux for emerald single crystal synthesis: Nippon Kagaku Kaishi, no. 5, p. 752-256.

Ushio, M., 1977, Appearance and disappearance of crystal faces of emerald during crystal growth by vanadium (V) oxide flux method: Hoseki Gakkaishi, v. 4, p. 51-59.

Ushio, M., 1977, Study on the synthesis of industrial large crystals, X. Appearance and disappearance of crystal faces of emerald during crystal growth by divanadium pentoxide flux method, Nippon Kagaku Kaishi, no. 2, p. 194-199.

Ushio, M., and Sumiyoshi, Y., 1972, Synthesis of industrial large single crystals, II. Flux-growth of emerald single crystals: no. 9, p. 1648-1655.

Ushio, M., and Sumiyoshi, Y., 1973, Synthesis of industrial large single crystals, III. Growth rate of each plane of synthetic emerald single crystal by the vanadium (V) oxide flux method: Nippon Kagaku Kaishi, no. 3, p. 506-513.

Ushio, M., and Sumiyoshi, Y., 1973, Synthesis of large single industrial crystals, IV. Observation of synthetic emerald single crystal surfaces grown by the vanadium flux method: Nippon Kagaku Kaishi, no. 5, p. 941-947.

Ushio, M., and Sumiyoshi, Y., 1975, Synthesis of large single crystals for industrial use, VII, Thermal expansion of synthetic emerald single crystals: Nippon Kagaku Kaishi, no. 10, p. 1730-1733.

Vakanjac, B., 1978, Gems and ornamental raw materials in Yugoslavia: Proceedings of the 9th Geologic Congress Yugoslavia, Cicic, S., (editor), p. 603-610.

Valdiri, W. J., and Carmona, R.A., 1978, Emeraldas: *in* Carmona, R.A., editor, recursos minerales de Colombia: Publicaciones Geologicas Especiales de INGEOMINAS, 1, p. 179-192.

Vasilenko, M.V., Bukin, G.V., Gavrilov, F.F., Kruzhalov, A.V., Mizgulin, V.N., Neshov, F.G., and Ogneva, O.K., 1979, Distribution of a chromium impurity in surface layers of artificial emerald single crystals: Zhurnal Prikladnoi Spektroskopii, v. 30, p. 527-530.

Vasilenko, M.V., Kruzhalov, A.V., Glazyrin, M.P., 1977, Growth of beryl and emerald single crystals by a solution-melt method: Khimiya Tverdogo, Tela, no. 1, p. 52-53.

254

Veremeichik, T.F., 1984, Similarity of ground and excited state of absorption spectra of chromium (3+) ions in a strong crystal field: Physica Status Solidi B, v. 124, p. 719-729.

Vertushkov, G.N., Zhernakov, V.I., Laskovenko, A.F., and Konyukhova, N.P., 1978, Glimmerites from the Ural-type emerald deposits: Mineralogiia Petrografiia, Urala, v. 1, p. 3-10.

Viand, J., Tuccillo, R., and Rolleri, E.O., 1978, Gemmology; Study of gemmologic minerals: Geological congress of Argentina, Actas 7, Tomo 1, p. 107-117.

Victor, G., 1965, The gemstones of Russia: Lapidary Journal, v. 18, p. 1296-1301.

Viticoli, S., Gastaldi, L., Flamini, A., and Grubessi, O., 1984, Unusual EPR properties of Miku emeralds: Journal of Gemmology and Proceedings of the Gemmological Association of Great Britain, v. 19, p. 160-163.

Vlasov, K.A., and Kutukova, E.I., 1960, Izumrudnye Kopi: Moscow Akademiya Nauk SSGR, 250 p.

Walton, J., 1950, Unusual Emerald: Gemmologist, v. 19, p. 123-125.

Webster, R., 1955, The Emerald: Journal of Gemmology, v. 5, p. 185-221.

Weinschenk, E., 1986, Die Minerallagerstatten des Grobvenedigerstockes in den Hohen Tauern: Zeitschrift für Kristallographie, v. 26, p. 337-508.

Weisbach, K., 1971, Emeralds; hunt them yourself: verlag Das Bergland Buch, p. 192.

Whitfield, G.B., 1975, Emerald occurrence near Menzies, Western Australia: Australian Gemmologist, v. 12, p. 150-152.

Wild, G.O., and Klemm, R., 1926, Mitteilungen uber Spektroskopische Untersuchungen an Mineralien, VI, Smaragd: Centre für Mineralogie, p. 21-22.

Wood, D.L., and Nassau, K., 1968, The characterization of beryl and emerald by visible and infrared absorption spectroscopy: American Mineralogist, v. 53, p. 777-800.

Yagi, I., and Adachi, N., 1979, Syntheses of emerald and other color-varieties of beryl minerals: Hoseki Gakkaishi, v. 6, p. 67-72.

Yelizarov, L.I., 1978, Characteristic magnetic features of emerald bearing mica: Razvedka i Okhrana Nedr v. 6, p. 57-60.

Yu, R.M., 1974, Growth features in South African emerald crystals: Journal of Gemmology, v. 14, p. 120-131.

Zaveri, C., 1961, Gemstones of India: Gemmologist, v. 30, p. 46-52, 143-146.

Zeb, Aurang, 1985, Geological structure of emerald deposits of the Swat region (Pakistan) and characteristics of their genesis: Voprosy Orudeneniya Ul'tramafitakh, editor, Romanovich, I.F., Nauka, Moscow, USSR, p. 152-155.

Zeitner, J.C., 1979, Gems of Brazil: Lapidary Journal, v. 33, p. 142-155.

Zeitner, J.C., 1982, The quest for green bolts: part I; Lapidary Journal, v. 36, p. 378, 380, 382-383, 402-410.

Zeitner, J.C., 1982, The quest for green bolts: part II; Lapidary Journal, v. 36, p. 558-566.

Zeitner, J.C., 1989, Green as in emeralds: Lapidary Journal, v. 42, p. 22-40.

Zhernakov, V.I., 1975, Morphology of chromium-containing beryl crystals from micaceous complexes; Trudy Sverdlovskogo Gornogo Instituta, v. 106, p. 107-110.

Zhernakov, V.I., 1976, X-ray diffraction study of phlogopites from emerald-containing micaceous veins: Ural skii Nauchno Tsentral nyi Akademiia Nauk SSSR, Trudy instituta Geologii i Geokhimiia, v. 118, p. 126-129.

Zhernakov, V.I., 1984, Genesis of emerald in the Urals, in Chesnokov, B.V., and Popov, V.A., editors, the mineralogy of ore deposits in the Urals: Ural skii, Naucho Tsentral nyi Adademiia Nauk SSSR, p. 49-58.

Zirkl, E. J., 1966, Der Smaragdbergbau im Habachtal: Universum, v. 21, p. 252-255.

Authors' Biographies

ALI HAMZA KAZMI

Kazmi has served as a geologist with the Geological Survey of Pakistan for over 40 years. He received his B.Sc. degree from Lucknow, India (1948) followed by D.I.C. from the Imperial College of Science and Technology, London (1954) where he researched on the stratigraphy of Ziarat area (northeastern Baluchistan). He was a Nuffield Fellow at the Cambridge University, England (1964-1965) where he did research on the Quaternary geology of the Indus Plain. As a Senior Fulbright Fellow and a Courtesy Professor at the Oregon State University (1984) he carried out research on the emerald deposits of Swat including the tectonics and stratigraphy of that area.

Kazmi joined the Geological Survey of Pakistan in 1949 and until 1955 carried out geological mapping, stratigraphic research, and mineral and groundwater exploration in different regions of Pakistan. From 1956 to 1959 he worked with the Groundwater Development Organisation (now part of WAPDA) on the geohydrology and Quaternary geology of the Punjab. He was also part time lecturer in geology at the University of the Punjab. During 1959-60 he was awarded a USAID Fellowship and worked with the Groundwater Branch of the USGS at Lawrence (Kansas), Denver, Phoenix and other places.

Back in Pakistan in 1960, Kazmi founded and headed the Water Resources & Engineering Geology Branch of the Geological Survey. Until 1973, besides geohydrology and engineering geology, he researched on the Quaternary of the Indus plain, active fault system of Pakistan and on tectonics. In 1974 he went on deputation to the newly formed Pakistan Mineral Development Corporation, as Chief of Planning and later as Chief Geologist. Until 1977 he formulated, guided and supervised the mineral exploration and development programs of the Corporation, including exploration for chromite, sulphur, kaolin, bauxite, placer gold, coal, fluorite and gemstones.

In 1978 the Government of Pakistan assigned to him the task of establishing the Gemstone Corporation of Pakistan and he worked as the Technical Director of this Corporation until 1985. During this period under his guidance and supervision a number of gemstone mines were developed and many new gemstone deposits were discovered.

He has travelled widely in the Middle East, Europe and North America and has published more than forty papers in national and international journals (including *Economic Geology*, *Mineralogical Records* and *Gems and Gemology*), and written more than fifty technical reports in the various geoscientific fields with which he has been associated. His more significant contributions are on the Quaternary geology of the Indus plain, the geohydrology of the Punjab, Seismotectonic Map of Pakistan, Tectonic Map of Pakistan, tectonics and stratigraphy of the Swat and Ziarat areas and what he likes to call Gem Geology. He was a member of the Geological Society of London. Presently he is the President of the Pakistan Institute of Geoscientists and a member of the Board of the International Geological Correlation Program (IUGS).

He returned to the Geological Survey of Pakistan in 1985 and at present he is the Director General of this Department.

LAWRENCE W. SNEE

Larry Snee's is a geologist-geochemist with the U.S. Geological Survey, Branch of Central Mineral Resources, Denver, Colorado. He received his B.S. degree from Florida State University in 1974 in geology, his M.S. and Ph. D. degrees in geology from the Ohio State University(1970 and 1982). Two years of post-doctoral work were done at the U.S. Geological Survey, Reston. He was an assistant professor of geology (igneous petrology) at Oregon State University from 1983 to 1986 before moving to the U.S.G.S. in Denver. Larry Snee's Ph.D. thesis was on the geology, geochemistry, and age of the Pioneer batholith in southwestern Montana, U.S.A. His postdoctoral work was on the strontium isotope geochemistry and paleomagnetism of plutonic rocks in southwestern Montana. His research topics have included shock metamorphism of lunar and terrestrial rocks; tectonics of the northwestern U.S.; age and duration of mineralization of the Panas queira, Portugal tin-tungsten deposit; age of volcanism and sedimentation in the eastern Cascade Mountains of Oregon; age of mineralization, plutonism, and metamorphism in Idaho and Montana; and others. For the past 5 years he has been working on the geology, age, and geochemistry of rocks in northern Pakistan and he has recently begun research on mineral deposits and plutons in southern China and on volcanic rocks of coastal Peru. Recently he has helped to develop and set-up an ^{40}Ar/^{30}Ar geochronology laboratory at the U.S.G.S. in Denver.

EDWARD J. GÜBELIN

Edward J. Gübelin is one of the world's foremost authorities on inclusions in gemstones and their imitations and is one of the leading gemologists of the world. He is a Certified Gemmologist (C.G.), Fellow of the Gemmological Association of Great Britain (F.G.A.), First Research Member of the Gemological Institute of America, Expert of the German Gemmological Association, Gem Expert of the Swiss Gemmological Association: Research Diploma holder of the Gemmological Association of Great Britain and a Member of the Swiss Chamber of Legal Experts. Born in Lucerne in 1913 he became interested in gemology in 1932 when his father established a gemological laboratory in Lucerne. He soon developed a fascination for inclusions in gemstones and developed into a master photographer of their "internal paragenesis". In fact he pioneered the application of inclusions to the identification of gemstones. He has published over 150 papers and 6 gemological books each of which is a classic by itself. His latest book "Photoatlas of Inclusions in Gemstone" is a monumental work crowning over 40 years of intensive research work. It contains a rich treasure trove of experience accumulated in a single volume and contains hundreds of exquisite and instructive colored photomicrographs of inclusions in gemstone, selected out of Gübelin's personal collection of over 10,000 transparencies. According to Dr. W.F. Eppler of Munich, these "photographs display such beauty and aesthetic quality of the inclusions as to invalidate their former designation as flaws."

Gübelin has made several discoveries and first observations, e.g. chrome pyroxene in diamonds, apatite and calcite in hessonite, and has been the first to introduce a systematic classification of the inclusions in gemstones. He has also designed several gemological instruments and accessories to improve the reliability of microscopy.

Edward Gübelin has the distinction of repeatedly visiting almost all the important gem deposits on five continents of the world and testing the inclusions of gemstones from all these sites and publishing his results in various gemological journals.

ROBERT D. LAWRENCE

Bob Lawrence is a structural geology Professor at Oregon State University in Corvallis, Oregon. He received his B.A. degree from Earlham College (1965) and his Ph.D. degree from Stanford University (1968). His Ph.D. thesis dealt with a petrofabric analysis of a shear zone along the Chewack-Pasayten fault, north-central Washington State, where what are now called S—C structures were studied to determine the sense of slip on the fault. After attending Stanford he taught for two years at Earlham College as part of a curriculum improvement program. In 1970 he joined the Department of Geology, Oregon State University where he is now an associate Professor. He teaches beginning geology, structural geology, tectonics, and geomorphology. His research has been concerned with the structure and tectonics of major faults. He has studied and published about the Chewack-Pasayten fault, Washington state; the Brothers Fault Zone and basin-and-range structure, Oregon; the Chaman fault, Pakistan; the Main Mantle thrust, Pakistan and the Raikot fault, Pakistan. He first became interested in Pakistan in 1978 through work on the Chaman fault with the Geological Survey of Pakistan and has been Fulbright Lecturer (1981-82) and a Fulbright Research Scholar (1986-87) with the University of Peshawar. He became involved in this study of emeralds from collaboration with Gemstone Corporation of Pakistan while he was at the University at Peshawar in 1981.

JAVED ANWAR

Javed Anwar got his B.Sc. Hons. (1968) and M.Sc. (1969) degrees in geology from the University of Peshawar. He has a wide and varied experience having worked as an engineering geologist at the Tarbela Dam Project (1969-71), as Mineral Development Officer with NWFP Government (1971), as an economic geologist with Pakistan Industrial Development Corporation and the Pakistan Mineral Development Corporation (1971-1981). Until recently he was the Chief Geologist of the Gemstone Corporation of Pakistan. He has now reverted to his parent organization—Pakistan Mineral Development Corporation.

He has written 12 papers on economic mineral deposits e.g. fluorite, talc, manganese, rock phosphate, chromite, graphite, rock salt and gemstones.

STEPHEN J. CARTER

Stephen J. Carter is a Master's candidate in geology at Oregon State University. He received his B.S. in geology from California State University, Chico, California in 1983. His Master's thesis research is on the geology of the Mingora area, Swat, Pakistan. He has worked for mining exploration companies; currently he works with a soils research firm in Newark, California.

HAMID DAWOOD

Hamid Dawood obtained his B.Sc. (1977) and M.Sc. (1979) degrees in geology from the Punjab University. For five years he worked as a gem exploration geologist with the Gemstone

Corporation of Pakistan (Peshawar), and also worked as Project Manager of the Gujar Kili Emerald Mines, Swat and the Katlang Topaz Mines, Mardan. He has published three papers on the geology of the Indus suture zone in Swat area and associated gem mineralization.

Presently he is working as a geologist with the Pakistan Natural History Museum, Islamabad.

EUGENE E. FOORD

Dr. Foord is a geologist-mineralogist with the U.S. Geological Survey, Central Mineral Resources Branch, Denver, Colorado. He received his A.B. degree from Franklin and Marshall College (1968) in geology, his M.S. degree in geology from Rensselaer Polytechnic Institute (1969), and his Ph.D. degree in geology-mineralogy from Stanford University (1976). One year of post-doctoral work was done at Stanford before coming to the U.S. Geological Survey. Dr. Foord's Ph.D. thesis was on the mineralogy and paragenesis of the Himalaya pegmatite-aplite dike system, Mesa Grande District, San Diego Co., California. His research has involved continuation of his southern California pegmatite studies; the mineralogy of niobium and tantalum; geology of quartz-huebnerite veins near Round Mountain, Nevada; chemistry, structure, and origin of the coloration of amazonite; mineralogy of Pb-Bi-Ag sulfosalts, mineralogy of the Black Range Tin District, Sierra and Catron Counties, New Mexico; mineral resource appraisal of wilderness areas in Nevada and Idaho; studies of various new or rare mineral species—rynersonite, cordoroite, hashemite, minasgeraisite, volkonskoite, planerite, aheylite, zimbabweite, scrutinyite, squawcreekite, chestermanite, durangite, iimoriite, jeremejevite, and others; petrology of alkalic rocks in Otero Co., New Mexico; and other topics. He also served as a co-investigator with Bruce Bohor and Pete Modreski on a Gilbert Fellowship investigating the Cretaceous-Tertiary (K/T) boundary event. He has written more than 50 articles for mineralogical journals including the American Mineralogist, Canadian Mineralogist, Mineralogical Magazine, Bulletin de Mineralogie, Gems and Gemology, the Mineralogical Record, Clays and Clay Minerals and others. He is currently working on a project to revise and update Dana's *Textbook of Mineralogy*.

JANE M. HAMMARSTROM

Jane Marie Hammerstrom is a geologist in the Branch of Resource Analysis at the U.S. Geological Survey in Reston, Virginia, U.S.A. She received a B.S. in geology from George Washington University, Washington, D.C., U.S.A. in 1972 and an M.S. degree from Virginia Polytechnic Institute and State University, Blacksburg, Virginia, U.S.A. in 1981. Her M.S. research was on the mineral chemistry of the Pioneer batholith, southwestern Montana, U.S.A. and she is currently studying the mineralogic variation in skarn deposits from north-central Nevada. She has recently worked on an empirical geobarometer for hornblende-bearing calcalkaline plutonic rocks and studies of the mineral chemistry of magmatic epidote-bearing plutons.

BRITTAIN HILL

Brittain Hill is a Ph.D. candidate at Oregon State University. He received his M.S. in geology from Oregon State University. His M.S. research was in geology and geochemistry of volcanic rocks in the Cascade Mountains. He has had extensive training in instrumental neutron activation analysis and he has worked with several mining and geothermal resource companies in the U.S. His Ph.D. research is on geology and geochemistry of volcanic rocks in the Cascade Mountains, Oregon.

SHAHID HUSSAIN

Shahid Hussain got his Masters degree in geology from the University of the Punjab. He attended a course in Mineral Exploration at Imperial College of Science and Technology London. For the past fourteen years as a senior geologist he has been mainly associated with mineral exploration, having worked with Messers Engineers Combine Ltd. and the Gemstone Corporation of Pakistan.

Presently he is working as an Associate Curator in the Earth Sciences Division of Pakistan Museum of Natural History, Islamabad. He has published fourteen research papers and written twelve geological reports (unpublished).

SADAQAT ALI JAFARY

Sadaqat Ali Jafary holds B.Sc. (Hons.) 1967, M.Sc. Geology (1968) from the University of the Punjab, Lahore. His specialization is in mineralogy and petrology. He spent a year at the Tarbela Dam Project as a geologist and later joined the Geological Survey of Pakistan in 1970 He is currently working as a Deputy Director. During his stay in the Geological Survey of Pakistan he was placed on deputation (1973-76) with Oil and Gas Development Corporation as a Analytical Laboratories at Karachi. Mr. Jafary worked in various branches of the Geological Survey of Pakistan at various places. He has many publications to his credit. Recently he visited Japan in connection with a collaborative project with the Geological Survey of Japan on "Research on Geological and Mineral Resources of the Collision Zone in Pakistan".

TAHSEENULLAH KHAN

Tahseenullah Khan obtained his B.Sc. Hons (1978) and M.Sc. (1979) degrees in geology from the University of Peshawar. He also took special courses and training in optical mineralogy, gemology and gem valuation, X-ray diffraction identification and analysis of clay minerals, carbonate petrology and structural geology.

He worked as a gem exploration geologist with the Gemstone Corporation of Pakistan (1981-1984) and is presently working as a geologist with the Geological Survey of Pakistan. He has published six papers on the geology of the Kot-Pran Ghar Melange Complex in the North West Frontier Provinces, gem pegmatites of Chitral, pink zeolite from Gilgit, turbidites at Sikandarabad (Gilgit), and computer applications to geologic problems.

ROBERT R. SEAL, II

Robert Seal is a Ph.D. candidate at the University of Michigan. He received his B.Sc. in geology from Virginia Polytechnic Institute and State University, Blacksburg, Virginia, U.S.A. in 1981 and his M.Sc. degree in Geology from Queen's University, Kingston, Ontario, Canada in 1984. He is currently involved in a regional study of a Ag-Pb-F district in south-central Idaho, U.S.A. in conjunction with the U.S. Geological Survey. His M.Sc. research was concerned with the genesis of a W-Mo-Sb deposit in New Brunswick. During summer field seasons, he has worked for companies, exploring for base, precious and strategic metals.

INDEX OF AUTHORS

Y

Z

INDEX OF SUBJECTS

A

B

C

D

E

F

G

H

I